Biodiversity of Armenia

George Fayvush
Editors

Biodiversity of Armenia

Editor
George Fayvush
Institute of Botany
National Academy of Sciences of Armenia
Yerevan, Armenia

ISBN 978-3-031-34331-5 ISBN 978-3-031-34332-2 (eBook)
https://doi.org/10.1007/978-3-031-34332-2

This Springer imprint is published by the registered company Springer Nature Switzerland AG
The registered company address is: Gewerbestrasse 11, 6330 Cham, Switzerland

Introduction

Armenia is a south Caucasian Republic that borders Georgia, Azerbaijan, Turkey, and Iran. It is a landlocked country with a total area of 29,740 km^2, located approximately 145 km from the Black Sea and 175 km from the Caspian Sea. It lies between 38°50′ and 41°18′ northern latitude and between 43°27′ and 46°37′ eastern longitude and measures 400 km along its main axis (northwest to southeast). Additionally, Armenia is generally a mountainous country, with 44% of Armenia being characterized by high-elevation mountains; specifically, the lowest point in Armenia is 375 m above sea level, while its highest point is 4095 m above sea level, and the average elevation of Armenia is 1850 m.

It is known that mountain areas in general are much richer in their biodiversity than lowland areas (Körner 2002, 2003). Thus, occupying about 12% of the land mass, mountains support an estimated one third of the terrestrial biological diversity (Spehn et al. 2011). Most biodiversity hotspots are mountainous (Hoorn et al. 2018). In this regard, Armenia is an amazing country. It occupies very small territory (less than 30,000 km^2) in the South Caucasus, but it has extremely rich landscape and biological diversity. All the main Caucasian ecosystems (besides humid subtropics) are represented in Armenia – deserts and semi-deserts, steppes, meadow-steppes, forests and open woodlands, sub-alpine and alpine vegetation, as well as intrazonal ecosystems. And they represent the habitats of amazing diversity of plant and animal species. About 3800 species of vascular plants grow here (about half of the flora of the whole Caucasus), and 428 species of algae, 399 species of mosses, 4207 species of fungi, 464 species of lichens, and 549 species of vertebrate animals are registered in Armenia. The estimated number of species of invertebrate animals is 17200 species (5th National report… 2014). Flora and fauna of Armenia include 146 plant and 479 invertebrate animal local endemic species!

Such wealth could not but attract the attention of researchers. Separate interesting data about plants and animals of Armenia are found in numerous books of ancient and medieval Armenian scientists (Barsegh Kesaraci, Eznik Koghbaci, Agatangehos, Lazar Parpeci, Movses Khorenaci, Anania Shirakaci, Amirdovlat Amasiaci, etc.). However, usually as a starting point for systematic investigations of biodiversity in Armenia, scientists indicate the French botanist P. Tournefort's trip to Ararat Mountain in 1700–1702, where he firstly noticed vertical zonation of vegetation. However, in the eighteenth century, studies of the nature of Armenia were irregular, random. More intensive research continued in the nineteenth and early twentieth centuries, when many prominent natural researchers from Russia and Europe worked here. However, a systematic targeted study of the biodiversity of Armenia began after the establishment of Soviet power and the formation of the Academy of Sciences with relevant institutions. In fact, throughout the second half of the twentieth century, serious studies of the flora, vegetation, fauna, and mycobiota of Armenia were going on.

By signing and ratifying the UN Convention on Biological Diversity in 1993, Armenia assumed a number of obligations. First of all, we are talking about the preservation of all species of plants and animals and their habitats, represented on the territory of the republic. However, as is completely clear, in order to save something, you need to know what you have. Yes, a large number of works have been published in Armenia on some components of biodiversity. For example, 11 volumes of Armenian Flora (1954–2010), Armenian Mycoflora, and Fauna of Armenian SSR series. But a coherent, comprehensive reference scientific work on all components of biodiversity has not been available so far. To some extent, data on the biodiversity of the republic are presented in six National Reports on Biodiversity, but in most cases, these data are limited only to statistical summaries without any serious analysis. In this work, we have tried to summarize all available information on the flora, ecosystems, fauna, and mycobiota of Armenia, analyze these data, identify gaps in our knowledge, identify the main threats to biodiversity, and outline the main directions for further research.

Department of Geobotany
and Ecological Physiology of the Institute of Botany
National Academy of Sciences of Armenia
Yerevan, Armenia

George Fayvush

References

5th National report to the Convention on Biological diversity of the Republic of Armenia (2014). Yerevan

Flora of Armenia (1954–2010) Ed. A. Takhtajan, vol 1–11

Hoorn C, Perrigo A, Antonelli A (eds) (2018) Mountains, climate and biodiversity. Wiley-Blackwell, Hoboken

Körner C (2002) Mountain biodiversity, its causes and function: an overview. In: Körner C, Spehn EM (eds) Mountains biodiversity a global assessment. Parthenon Publishing, London, pp 3–20

Körner C (2003) Alpine plant life – functional plant ecology of high mountain ecosystems. Springer, Berlin

Spehn EM, Rudmann-Maurer K, Körner C (2011) Mountain biodiversity. Plant Ecol Divers 4(4):301–302. https://doi.org/10.1080/17550874.2012.698660

Acknowledgements

I would like to thank the series editor for revising all manuscripts. I am thankful to all the authors of chapters who have contributed to this work, i.e., Karen Aghababyan, Alla Aleksanyan, Marine Arakelyan, Vardan Asatryan, Bardukh Gabrielyan, Arsen Gasparyan, Astghik Ghazaryan, Anahit Ghukasyan, Lusine Hambaryan, Hripsime Hovhannisyan, Mark Kalashian, Jakob Koopman, Lusine Margaryan, Siranush Nanagulyan, Astghik Poghosyan, Iren Shahazizyan, Helena Więcław, Noushig Zarikian.

The work was supported by the Science Committee of RA, in the frames of the research project nos. 21AG-1F004, 21SCZE-I-1F, 20RF-089, 21AG-1F004, 21T-1F281, 21T-1F334, 20TTSG-1F001.

Contents

Chapter 1
Natural Conditions of Armenia

George Fayvush

1.1 Location

Armenia is a South Caucasian republic, bordering with Georgia, Azerbaijan, Turkey, and Iran. It is a landlocked country with a total area of 29,740 km^2, at a distance of about 145 km from the Black Sea and 175 km from the Caspian Sea. It lies between 38°50′ and 41°18′ of northern latitude and between 43°27′ and 46°37′ eastern longitude and measures 400 km along its main axis (northwest to southeast). Armenia is generally a mountainous country, having its lowest point of 375 m above sea level and culminating at 4095 m with an average altitude of 1850 m (Fig. 1.1). Forty-four percent of the territory of Armenia are high mountainous areas, not suitable for inhabitation. The degree of land use is strongly unproportional. The zones under intensive development make 18.2% of the territory of Armenia with a concentration of 87.7% of total population. On these areas, the population density exceeds several times the ecological threshold index (200 person/km^2) reaching here up to 480–558 person/km^2. The poorly developed zones make 38.0% of the territory, where only 12.3% of total population resides with a very low density of 11–20 person/km^2. The zones under intensive development are provided with engineering-transportation infrastructures. In this zone, there are the most available public services, more human resources, and financial opportunities. Simultaneously, the poorly developed areas have rich natural resources with preserved unique natural ecosystems, beautiful landscapes, clean water and air, and biological resources (Gabrielyan 1986; Fayvush and Aleksanyan 2016).

The study area lies at the intersection of two phytogeographical regions (the Euro-Siberian and Irano-Turanian regions; Takhtajan 1986) and two biodiversity hotspots (Caucasian and Irano-Anatolian; Mittermeier et al. 2011). This position, in

G. Fayvush (✉)
Institute of Botany, National Academy of Sciences of Armenia, Yerevan, Armenia
e-mail: g.fayvush@botany.am

G. Fayvush (ed.), *Biodiversity of Armenia*,
https://doi.org/10.1007/978-3-031-34332-2_1

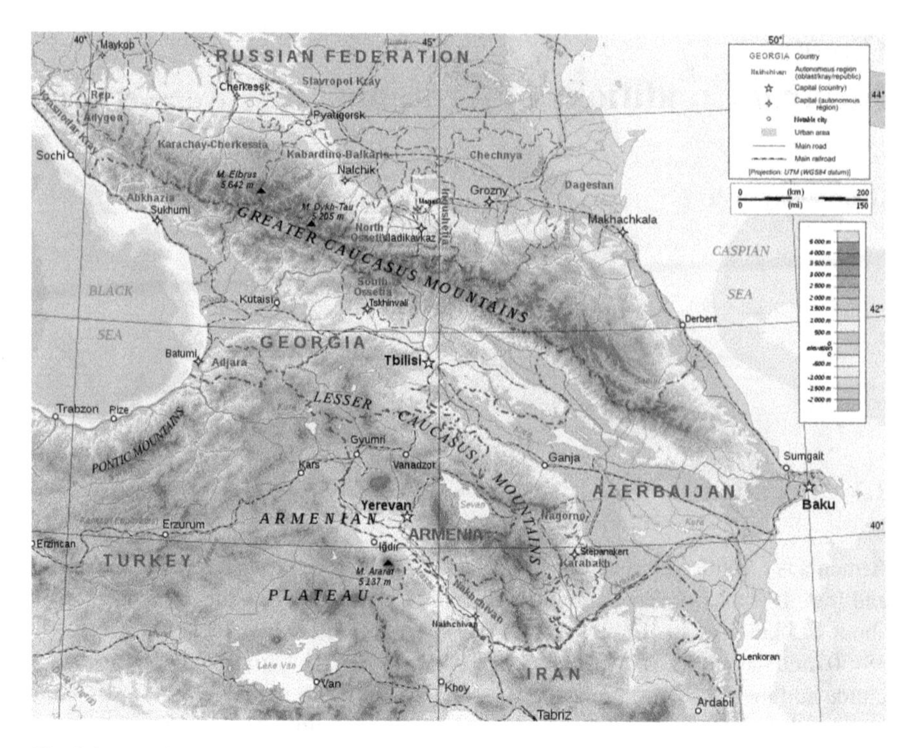

Fig. 1.1 Location of Armenia in the Caucasian Ecoregion

combination with its pronounced vertical zonation, is the cause of the great flora, vegetation, and fauna diversity. Thus, the location of the Transcaucasian Highlands at the intersection of these phytogeographical regions together with the diversity in climatic conditions and the active geological processes have resulted in the formation of a great diversity in ecosystems and plant diversity with a high level of endemism (Fayvush et al. 2013). As a result, on the small territory of the country (about 30,000 km^2), there are about 3800 species of vascular plants, 428 species of soil and water algae, 399 species of mosses, 4207 species of fungi, 464 species of lichens, 549 species of vertebrates, and about 17,200 species of invertebrate, many of which are considered endemics (The fifth… 2014).

1.2 Geology and Geomorphology

From orographic and physicogeographical points of view, Armenia form the northern edge of the system of folded-block mountains of the Armenian Highland. Unlike the Greater Caucasus, Armenia and the Lesser Caucasus are not a single, distinct watershed ridge. It is a system of coulisse-spaced ridges that merge with the mountain formations of the inner parts of the Armenian Highland and adjacent high areas. Since early geological epochs, the land surface of Armenia, and the surrounding

Armenian plateau, has been mountainous, with further mountain building occurring during the Cenozoic era (particularly after the Miocene). These complex tectonic shifts have resulted in a country dominated by a series of mountain massifs and valleys as well as in extensive volcanic activity. Climatic changes over the last million years also have left their mark on the country, with evidence of two glacial periods (Riss and Wurm) preserved on almost all mountains over 3000 m a.s.l. (Aslanyan 1958, 1985).

Four main geomorphological regions can be recognized within Armenia:

1. Mountain ridges and valleys in the northeast of the country which bear witness of extensive erosion.
2. Areas covered by lava of relatively recent (upper Pliocene) origin within Asia Minor characterized by gentle slopes with little evidence of erosion but in which larger rivers have carved out deep gorges and canyons.
3. A series of ridged mountains in the south of Armenia, which constitute the Lesser Caucasus system and show intense erosion.
4. The Ararat Valley representing the lowest part of the Ararat depression which is covered with alluvial and proluvial sediments (Aslanyan 1958; Gabrielyan 1962; Dumitrashko 1979).

1.3 Climate

A wide range of climatic zones are distinguished within Armenia (Fig. 1.2). This territory shows a pronounced vertical succession of six basic climate types – from dry subtropical up to severe alpine. The average annual temperature ranges from −8 °C in high-altitude mountainous regions (2500 m a.s.l. and higher) to 12–14 °C in low-traced valleys. In the lowlands, the average air temperature in July and August reaches 24–26 °C, but in the alpine belt, the temperature does not exceed 10 °C. January is the coldest month with an average temperature of −6.7 °C. The absolute minimum temperature is −42 °C. The overall climate is best characterized as dry continental, in some areas with an annual rhythm more or less similar to the Mediterranean climate regime. The average annual precipitation in Armenia is 592 mm. The most arid regions are the Ararat valley and the Meghri region with annual precipitations of 200–250 mm. The highest annual precipitation, 800–1000 mm, is observed in high altitude mountain regions. Major part of the precipitation falls in the spring. Long-lasting snow cover exists in the mountains above 1300 m, where the annual snowfall could attain 2 m. In the northern part of Armenia, humidity comes from the Black Sea in the west, in the southern part from the Caspian Sea in the east, while the central part lies in the rain shadow of mountain ridges and is the driest area (Bagdasaryan 1958; Third… 2015).

It is very important to note that global climate change on Earth seriously affects the climate of Armenia. Annual ambient temperature and precipitation in Armenia have been assessed for considerable periods of time. The results show that in recent decades, there has been a significant temperature increase. In the period of

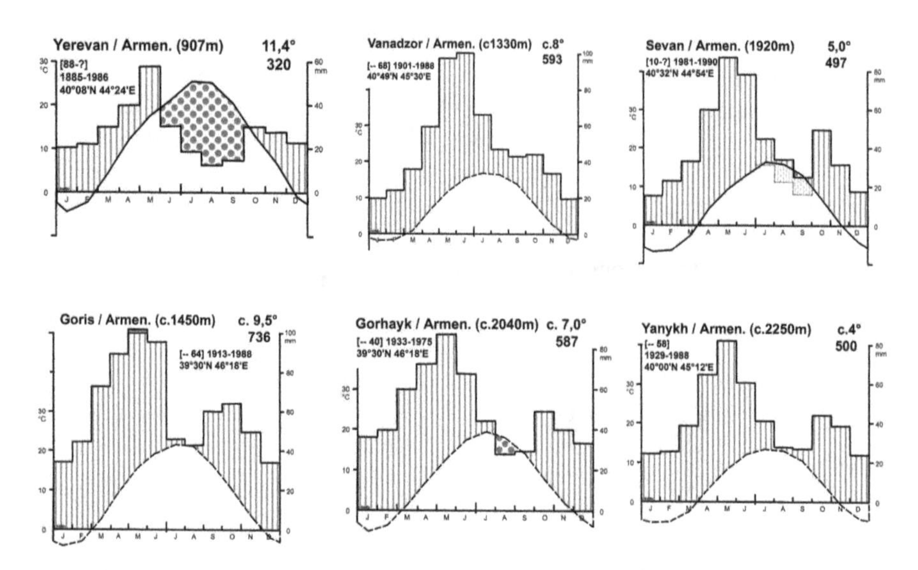

Fig. 1.2 Climatic diagrams of some meteostations of Armenia

1929–1996, the annual mean temperature increased by 0.4 °C; in 1929–2007, by 0.85 °C; in 1929–2012, by 1.03 °C; and in 1929–2016, by 1.23 °C (Fourth… 2020).

Changes in air temperature at different seasons have different trends. During 1966–2016, the average summer temperature has risen by about 1.3 °C, with extreme hot summers observed in Armenia during the last 20 years (2000, 2010, 2015). Changes in winter temperature demonstrate a completely different trend with very slight upward movement –0.4 °C. Extreme warm winters were observed in 1966 and in 2010, and extreme cold winter, in 1972.

Starting from 1935, comparison of estimated changes in the amount of precipitation over different periods shows that the decreasing trend in precipitation is maintained. Over the period of 1935–1996, the average annual precipitation decreased by 6%, and during the period of 1935–2016, by about 9%. During the period of 1935–2016, the average annual value of precipitation in the territory of the Republic of Armenia was 558 mm or 94% of the norm (592 mm) for the 1961–1990 period. The spatial distribution of precipitation is quite irregular. During 1935–2016, the climate in the northern (Vanadzor, Stepanavan), southern (Meghri), and central (Ararat valley) regions became more arid. Precipitation increased in the Shirak plain, in the Lake Sevan basin, and in Aparan-Hrazdan regions (Fourth… 2020).

1.4 Hydrography (Wetlands)

The wetlands in Armenia are represented by lakes, small lakes, rivers, reservoirs, water courses, areas temporarily covered by water, marshes, and peat areas. The rivers of Armenia are the tributaries of two large rivers of the Southern Caucasus – the Araks river and the Kura river. There is only one big river in

Armenia – Arax – but the river net is rather dense; in the whole of Armenia, there are about 9500 rivers and small rivers, with the total length of about 23 thousand km, including 215 rivers longer than 10 km (Fig. 1.3). The average annual flow of the rivers originating in Armenia makes 6.3 billion m^3, and the total volume of surface waters is 7.2 billion m^3. A part of these resources flows to the neighboring countries; this volume has significantly increased during this century after the sharp reduction of water use in Armenia after gaining independence.

In the structure of inner waters, the lakes and small lakes are of special importance. According to up-to-date hydrological studies on the territory of the Republic of Armenia, there are about 250 lakes; more than the half of them is of a temporary character, and they are periodically drying. Lake Sevan is the most famous and important for economy and biodiversity conservation. The other large lakes are Arpi Lake (regulated), Sev Lake, and Akna Lake. The total water resources of the lakes of Armenia are estimated as much as 39.8 million m^3. The water resources of Lake Sevan have been used for years for economic purposes resulting in gradual deterioration of ecological situation of the lake. By 1999, the lake water level had decreased

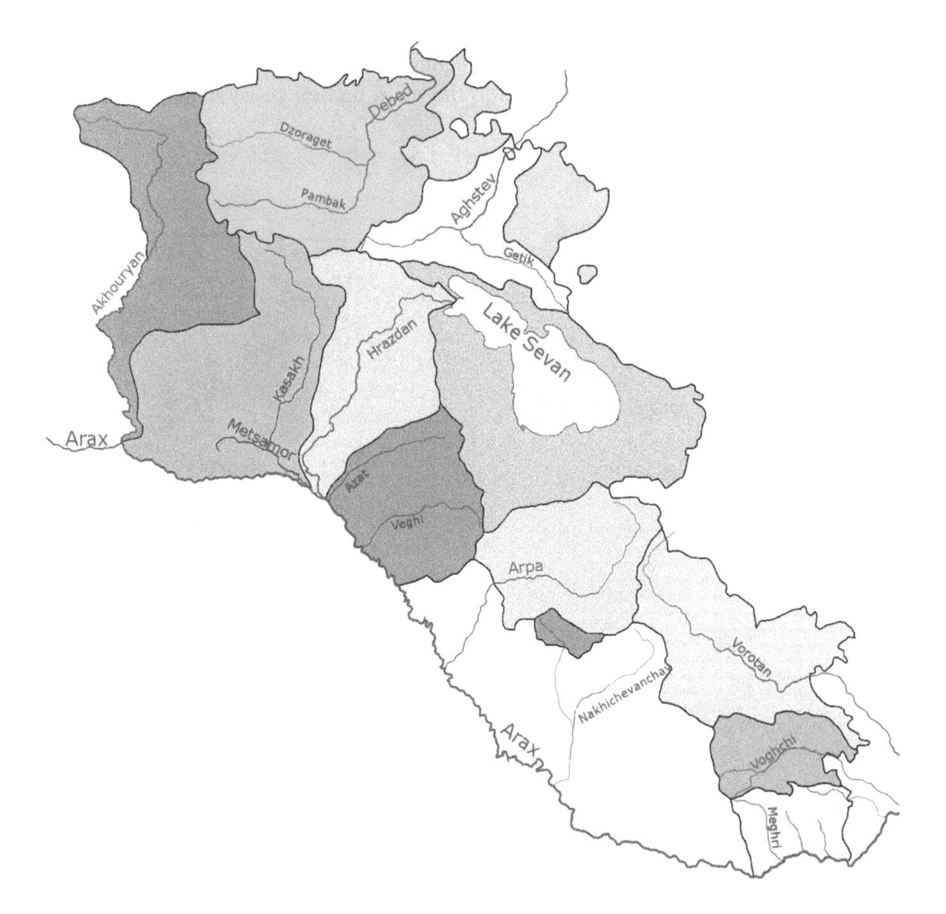

Fig. 1.3 Main rivers of Armenia

by 19.3 m in comparison with its initial mark. In order to prevent the lake euthrophication processes and improve the situation in the lake since 2002, the works to increase the water level have started.

The marsh formations located in the Araks river valley also belong to wetlands. In the Armenian side, the Khor Virap wetland is known, which is located in the place of Araks river old watercourse, as well as the system of Metsamor marsh formations, which includes Lake Ayghr, the River Sevjur and nearby marshes. On the territory of Armenia marshes occur also in the River Masrik valley and nearby Sevan peninsula, as well as in the relict marsh-lake formations in the meadows of Lori Plateau. In total, the marshes and peat areas in Armenia occupy 42 km^2. Lake Sevan, Lake Arpi, and Khor Virap wetland area are included in the list of wetlands of international importance under the Ramsar Convention (The fifth… 2014).

1.5 Soils

The diversity of climatic, geological, orographic, and historical conditions was also reflected in the diversity of the soils of Armenia. Currently, nine main soil types are distinguished here, five of which develop the main natural ecosystems (Table 1.1) (Melkonyan et al. 2004).

Table 1.1 Main soil types in Armenia

Landscape zones	Soil types	Area[a] Thousands ha	%	Elevation, m a.s.l.
Semidesert	Semideserts grey	152	5.8	850–1400
	Irrigated meadows grey	53	2.0	
	Paleo-hydromorph	2	0.1	
	Hydromorph salt-alkaline	29	1.1	
	Total	236	9.0	
Dry steppe	Mountain brown	242	9.2	1275–1950
Steppe	Black soil (Chernozem)	718	27.4	1350–2500
	Meadows' black soil	13	0.5	
	Ancient valley soil	48	1.8	
	Soils formed after the descent of the water level of Lake Sevan	18	0.7	
	Total	797	30.4	
Forest	Forest's grey	133	5.2	500–2400
	Forest carbonate soil	15	0.6	
	Forest brown soil	564	21.6	
	Total	712	27.4	
Mountain meadows	Mountain meadows	346	13.2	2250–4095
	Mountain meadow-steppe's	283	10.8	
	Total	629	24	

[a]In total, it was calculated for 2.616 thousand ha; water, urban areas, roads, bedrock outcrops, etc., were not taken into account

1.6 Vegetation

The diversity of landscapes and orography is an important determinant of Armenia's diverse vegetation. The lower mountain belt (480–1200 m) is covered by semidesert formations, gypsophilous or halophilous vegetation. There are salt marsh areas as well as the South Caucasian sand desert. The middle and upper mountain belts (1200–2200 m) are characterized by various kinds of steppe and forest vegetation, meadow steppes, shrub steppes, and thorny cushion (tragacanth) vegetation. The altitudinal span of the forest belt varies from 500 to 1500 (2000) m depending on the region and may be approaching to 2400 m when open park-like tree stands are included. The subalpine and alpine belts (2200–4000 m) are covered by meadows and carpets (Magakyan 1941; Takhtadjan 1941; Fayvush 2006) (Fig. 1.4). The main characters of different types of vegetation are presented in Table 1.2.

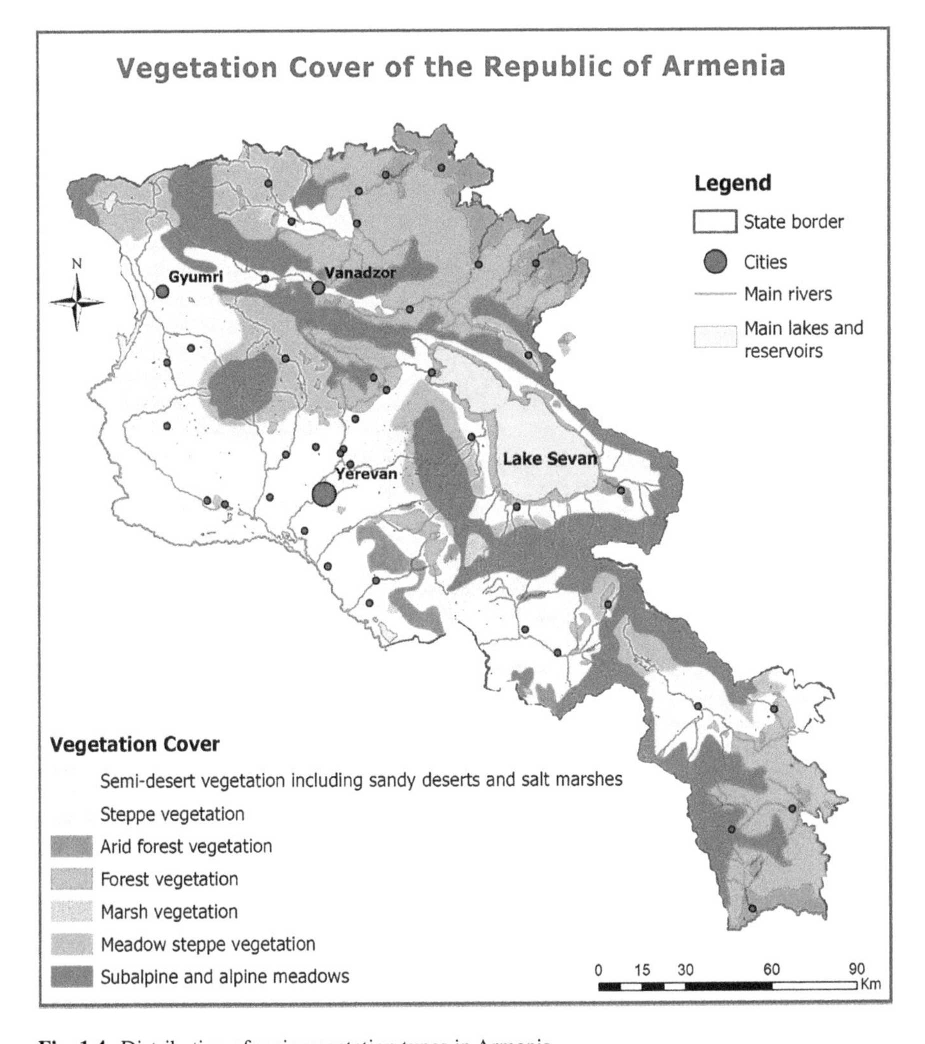

Fig. 1.4 Distribution of main vegetation types in Armenia

Table 1.2 Characters of main types of vegetation of Armenia

Type of vegetation	Altitude, m a.s.l.	Area, km^2	Number of species/endemics	Edificators	Endemics
Semidesert including sandy deserts and saline bodies ("solonchaks")	400–1250	4550	1400/16	*Artemisia fragrans*, *Capparis spinosa*, *Rhamnus pallasii*, *Kochia prostrata*, *Atraphaxis spinosa*, *Anisantha tectorum*, *Poa bulbosa*, *Centaurea squarrosa*, *Stipa capillata*, *Tanacetum chiliophyllum*, *Lepidium vesicarium*, *Euphorbia marschalliana*, *Taeniatherum crinitum*, *Calligonum polygonoides*, *Seidlitzia florida*, *Salsola ericoides*, *Halocnemum strobilaceum*, *Halostachys caspica*	*Allium schchianae*, *Centaurea alexandri*, *Centaurea arpensis*, *Cousinia daralaghezica*, *Isatis buschiorum*, *Allochrusa takhtajanii*, *Bufonia takhtajanii*, *Astragalus holophyllus*, *Papaver roseolum*, *Cotoneaster armenus*, *Verbascum gabrielianae*, *Scorzonera gorovanica*, *Papaver gorovanicum*, etc.
Steppes and meadow-steppes	1200–2200	8000	1600/46	*Stipa tirsa*, *Stipa pulcherrima*, *Stipa lessingiana*, *Festuca valesiaca*, *Xeranthemum squarrosum*, *Koeleria cristata*, *Bothriochloa ischaemum*, etc.	*Centaurea takhtajanii*, *Rhaponticoides hajastana*, *Rhaponticoides tamanianae*, *Centaurea vavilovii*, *Cousinia fedorovii*, *Cousinia takhtajanii*, *Scorzonera aragatzi*, *Tragopogon armeniacus*, *Isatis sevangensis*, *Merendera greuteri*, *Onobrychis takhtajanii*, *Alcea grossheimii*, *Polygala urartu*, etc.

Open arid forests	600–2200	2000	900/35	*Quercus araxina, Pistacia mutica, Paliurus spina-christi, Juniperus polycarpos, Juniperus foetidissima, Juniperus oblonga, Rosa spinosissima, Spiraea crenata, Rhamnus pallasii, Punica granatum, Amygdalus fenzliana,* etc.	*Smyrniopsis armena, Centaurea alexandri, Centaurea arpensis, Cousinia takhtajanii, Dianthus grossheimii, Amygdalus nairica, Cotoneaster armenus, Crataegus armena, Pyrus browiczii, Pyrus chosrovica, Pyrus daralaghezi, Pyrus gergerana, Rosa zangezura,* etc.
Forests	550–2400	3340	870/23	*Quercus macranthera, Quercus iberica, Fagus orientalis, Carpinus betulus, C. orientalis, Pinus kochiana, Taxus baccata, Platanus orientalis,* etc.	*Psephellus debedicus, Psephellus zangezuri, Colchicum goharae, Merendera mirzoevae, Pyrus complexa, Pyrus elata, Pyrus megrica, Rosa sosnovskyana, Rubus takhtadjanii, Linaria zangezura,* etc.
Subalpine and alpine meadows	2200–4000	4000	1100/24	*Dactylis glomerata, Phleum pratense, Agrostis lazica, Hordeum violaceum, Bromopsis variegata, Festuca varia, Anemone fasciculata, Doronicum oblongifolium, Cephalaria gigantea, Scabiosa caucasica, Taraxacum stevenii, Campanula tridentata, Pedicularis crassirostris, Carex tristis,* etc.	*Symphytum hajastanum, Colchicum ninae, Erodium sosnowskianum, Ornithogalum gabrielianae, Gladiolus hajastanicus, Papaver gabrielianae, Bromopsis zangezura, Poa greuteri, Trisetum geghamense, Alchemilla heteroschista, Alchemilla sevangensis, Alchemilla smirnovii, Verbascum sevanense, Agrostis trichantha,* etc.

(continued)

Table 1.2 (continued)

Type of vegetation	Altitude, m a.s.l.	Area, km^2	Number of species/endemics	Edificators	Endemics
Wetlands' vegetation (water-marsh vegetation)	400–3800	1774	630/3	*Phragmites australis, Typha latifolia, Juncus acutus, Cyperus fuscus, Hippuris vulgaris, Caltha polypetala, Deschampsia cespitosa, Glyceria plicata, Potamogeton natans, Groenlandia densa, Polygonum amphibium, Nymphoides peltatum, Nymphaea alba*	*Sonchus araraticus, Sonchus sosnowskyi, Linum barsegianii*
Petrophilous vegetation	400–4000	800	1100/27	*Cerasus incana, Sempervivum transcaucasicum, Nepeta mussinii, Astragalus microcephalus, Cystopteris fragilis, Cotoneaster integerrimus, Ephedra procera, Thalictrum minus, Saxifraga cartilaginea, Alopecurus tuscheticus, Campanula aucheri, Coluteocarpus vesicaria, Potentilla gelida, Potentilla porphyrantha*, etc.	*Allium struzlianum, Allium vasilevskajae, Seseli leptocladum, Sameraria odontophora, Thlaspi zangezuricum, Silene chustupica, Astragalus agasii, Astragalus sangezuricus, Ribes achurianii, Hypericum eleonorae, Acantholimon gabrielianae, Scrophularia olgae, Scrophularia takhtajianii*, etc.

1.6.1 Semideserts and Deserts

Semidesert vegetation formations are found in the lower montane belt at altitudes of 375–1200 m a.s.l. Main dominants of this vegetation type are *Artemisia fragrans*, *Capparis herbacea*, *Camphorosma lessingii*, *Seidlitzia florida*, and *Acantholimon* spp. This formation contains the main variation in gypsophilic and halophilic vegetation in Armenia. The semidesert vegetation of Armenia is very rich in endemic species, such as *Acantholimon vedicum*, *Allochrusa takhtadjanii*, *Astragalus holophylus*, *Bufonia takhtadjanii*, *Carthamus tamamschianae*, *Isatis buschiorum*, *Verbascum gabrielianae*, etc. The active speciation processes is one of the peculiarities of semidesert flora of Armenia. Over the last 25 years that have seen intensified processes of soil erosion and desertification, the expansion of the semidesert belt in the elevational profile went up by about 50 m, as observed in the occurrence of several edificators; in particular, the species *Artemisia fragrans*, *Capparis herbacea*, and *Rhamnus pallasii* have been recorded some 200–300 m above their previous altitudinal limits (The fifth… 2014). Around 80–90% of the semidesert belt is used for agricultural purposes, but often the rules of irrigation and soil cultivation are not followed, and that has resulted in soil erosion and secondary salination processes.

Very important ecosystems in the semidesert belt are ecosystems that develop on sandy soils. In this type of ecosystems, we included sand desert, volcanic slacks and sands, sandy patches, in the vegetation cover of which are represented *Calligonum polygonoides*, *Achillea tenuifolia*, *Aristida plumosa*, *Astragalus paradoxus*, *Allium materculae*, *Euphorbia marschalliana*, *Campanula propinqua*, *Nepeta trautvetteri*, *Taeniatherum crinitum*, *Noaea mucronata*, *Kochia prostrata*, *Haplophyllum villosum*, *Lepidium vesicarium*, etc.

1.6.2 Arid Open Woodlands and Tragacanth Vegetation

Arid open woodlands are one of the oldest types of vegetation and consist of both coniferous (juniper woodlands) and deciduous species (*Acer ibericum*, *Amygdalus fenzliana*, *Celtis glabrata*, *Pistacia mutica*, *Punica granatum*, *Pyrus salicifolia*, and others) and of shibliak (mainly spiny shrubs with *Paliurus spina-christi* as a dominant). Floristically, these communities are very rich. They occupy mainly the middle montane belt but also occur in the upper belt. Conspicuous are the tragacanth communities dominated by the spiny cushions of small shrubs. They occupy rather big areas in the middle and upper montane belts. Some species of *Astragalus*, *Acantholimon*, as well as *Onobrychis cornuta* dominate these communities (Fayvush and Aleksanyan 2016). Arid deciduous open forests are remarkably varied and are primarily distinguished by basic edificators; they are often polydominant communities in which several species predominate in the same stand (e.g., *Pistacia mutica* and *Punica granatum*, or *Amygdalus fenzliana* and *Celtis glabrata*, several species of *Pyrus*, etc.). There are more than 30 local endemic species growing in arid open forests, such as *Amygdalus nairica, Cousinia gabrielianae, Crataegus zangezura, Dianthus grossheimii, Pyrus chosrovica, Pyrus gergerana*, etc.

Juniper woodlands, dominated by various species of *Juniperus*, are very characteristic of Armenia from the lower to the subalpine belts. Dominants here are *Juniperus excelsa* and *J. foetidissima.*

Very characteristic ecosystems from the middle montane to the subalpine belt are tragacanth communities, in which cushion-shaped shrubs dominate, e.g., *Acantholimon bracteatum, Astragalus microcephalus, Astragalus lagurus, Astragalus uraniolimneus, Onobrychis cornuta, Gypsophila aretioides* and *Gundelia aragatsii.*

1.6.3 Steppes and Meadow Steppes

Mountain steppes in Armenia in the recent past were, apparently, the most widespread type of vegetation. They occupied all the mountain plateaus and treeless slopes in the middle montane belt. Researchers were strongly interested in the feather-grass steppes, which in their appearance are extremely similar to the steppes of Southern Russia. During the Soviet period, most steppe territories were plowed and used for agriculture. Presently, only small fragments of those steppes have been preserved on the steeper and stony slopes or in small patches between fields on the mountain plateaus. The most common communities are steppes dominated by *Stipa tirsa, S. pennata, S. lessingiana, S. arabica* and *Festuca valesiaca*, and also the so-called tragacanth steppes, dominated by *Astragalus microcephalus, A. lagurus, A. aureus,* and *Onobrychis cornuta*. Steppes are the richest vegetation types in terms of species, and they contain the largest number of endemic species in Armenia: More than 40 local endemics are known, e.g., *Alcea grossheimii, Centaurea takhtadjanii, Merendera greuteri, Rhaponticoides hajastana, Rhaponticoides tamanianae, Tragopogon armeniacus*, etc. In the upper montane belt, between the steppes and subalpine meadows, there is a relatively narrow belt of meadow-steppes, in which both steppe and meadow plants dominate.

1.6.4 Forests

The forest biodiversity of Armenia is evident from the many species of trees (125 species), shrubs (111), small shrubs (30), semishrubs (48), and woody lianas (9). Forest vegetation in the republic occurs mainly at altitudes of 500–2000 m a.s.l., while in some areas forests grow up to 2400 m a.s.l., forming the so-called park forests. The main forest areas of the republic are confined to the northern (62%) and southern (36%) regions, the central part of Armenia being much less afforested (2%). The main forest-forming species in Armenia are *Fagus orientalis, Quercus iberica,* and *Q. macranthera* and partly *Carpinus betulus* and *Carpinus orientalis*. In general, forest communities in Armenia occur in the foothills and the lower and middle mountain belts at slopes with inclinations of 20–25°. The timberline reaches up to 2300–2400 m a.s.l., though individual trees occur above the upper timberline till altitudes of 2700–2800 m a.s.l. Oak and beech forests are most dominant and are located at altitudes of 1300–2000 m a.s.l. (Makhatadze 1957). The stands dominated by *Pinus kochiana, Taxus baccata, Corylus colurna,* and other rare tree species decreased considerably in this area in historic time. At present they occur in patches or as sporadic trees. Moreover, *Taxus baccata, Corylus colurna,* and *Platanus orientalis*, which occur in small populations, are remarkable relict elements (Fayvush 2006; The fifth… 2014). The forest ecosystems are very diverse but occupy only about 10% of the territory of Armenia. We can distinguish riparian and gallery woodland dominated by *Populus* spp. and *Salix* spp., Irano-Anatolian mixed riverine forests (dominated by *Platanus orientalis* and *Populus euphratica*), *Fagus orientalis* forests, and non-riverine woodland with *Betula, Sorbus, Quercus iberica, Q. macranthera, Carpinus betulus, Fraxinus oxycarpa, Acer* spp., *Tilia cordata, T. caucasica, Ulmus* spp., *Pinus kochiana*, and *Taxus baccata*. About 20 locally endemic species are growing in the forests of Armenia, e.g., *Colchicum goharae, Merendera mirzoevae, Psephellus debedicus, Psephellus zangezuri,* and *Pyrus elata*. The open oak forests are important ecosystems in Armenia. *Quercus araxina* occurs only in the South Zangezur and Megri floristic regions in the lower montane belt, up to 1100 m a.s.l. The common species in these habitats are *Bothriochloa ischaemum, Carex humilis, Colutea cilicica, Cornus mas, Corylus avellana, Dactylis glomerata, Dictamnus albus, Genista transcaucasica, Jasminum fruticans, Ligustrum vulgare, Lonicera iberica, Melica transsilvanica, Origanum vulgare, Paliurus spina-christi, Rhamnus cathartica, Sambucus ebulus, Spiraea hypericifolia, Teucrium polium,* and *Thalictrum minus*. In some regions, the upper forest line is formed by scattered *Quercus macranthera* trees with an undergrowth of subalpine herbaceous species (shrubs are usually very small).

1.6.5 Subalpine and Alpine Vegetation

These ecosystems are typical for the upland zones and high altitudes (above 2200 m a.s.l.) in Armenia. These are subalpine meadows, subalpine tall grass vegetation, and alpine meadows and carpets.

1.6.6 Subalpine

The subalpine vegetation of Armenia is extremely diverse and is represented by a variety of ecosystems, including grasslands and lands dominated by forbs, mosses or lichens, heathland and scrub, and woody ecosystems. The grasslands comprise subalpine meadows and subalpine tall grass communities, and dominant species can be *Agrostis planifolia, Alopecurus armenus, Bromopsis variegata, Dactylis glomerata, Festuca ruprechtii, F. woronowii, Hordeum violaceum, Koeleria albovii, Phleum alpinum, Poa alpina, P. pratensis*. The dominant species in the subalpine forbs meadows can be *Anemone fasciculata, Astrantia trifida, Betonica macrantha, Carex brevicollis, Centaurea cheiranthifolia, Nepeta betonicifolia, Pimpinella saxifraga, Rhinanthus pectinatus, Scabiosa caucasica, Tanacetum coccineum, and Veratrum album.* Subalpine tallgrass communities are Ponto-Caucasian and dominated by *Aconitum orientale, Astrantia maxima, Campanula latifolia, Cephalaria gigantea, Dactylis glomerata, Delphinium flexuosum, Festuca gigantea, Galega orientalis, Lilium armenum, L. szovitsianum, Linum hypericifolium,* and *Thalictrum minus*. Subalpine heathland and scrub in Armenia include *Rhododendron caucasicum* heaths, and Southern Palaearctic mountain dwarf juniper scrub (*Juniperus hemisphaerica* and *Juniperus sabina*). Besides these, steppe scrub with *Spiraea* spp. is well represented in the upper montane and subalpine belt. In Northern Armenia, at the upper border of the forest, woody ecosystems are represented by the so-called subalpine crook stem forest (which are also called "Krummholz" or "Elfin Forest" in literature). *Acer trautvetteri, Betula litwinowii, Malus orientalis, Pyrus caucasica, Rubus idaeus, R. saxatilis, Sorbus aucuparia,* and *Viburnum lantana* are dominants in these communities.

1.6.7 Alpine

Alpine vegetation in Armenia is mainly dominated by alpine meadows and carpets, in which grasses dominate in meadows and herbs in carpets. Alpine meadows can be dominated by *Festuca* spp., *Bromopsis variegata, Carex* spp., and/or *Kobresia schoenoides*. The same species can act as dominants in most cases for the alpine carpets both on volcanic and limestone substrata. These ecosystems include carpets dominated by *Alchemilla grossheimii, Campanula tridentata, Cirsium rhizocepalum, Plantago atrata, Potentilla raddeana, Ranunculus dissectus* subsp. *aragazi, Sibbaldia parviflora*, or *Taraxacum stevenii*. About 20 local endemic species are found in the meadows and carpets of Armenia (*Colchicum ninae, Erodium sosnowskianum, Gladiolus hajastanicus, Ornithogalum gabrielianae, Symphytum hajastanum*, etc.).

1.6.8 Wetland Vegetation

These habitats are intrazonal and occur in all mountain belts. These are extremely diverse ecosystems occurring at lakes, rivers, streamlets, and marsh areas, which vary dependent on environmental conditions and elevations (Barsegyan 1990; Fayvush and Aleksanyan 2016). Over the last centuries, the changes due to economic activities have resulted in the reduction of wetland areas and the extinction of some wetland species. Over the last years, the following processes have caused serious threats: the construction of small hydropower plants on rivers with scarce water resources, which changed the ecosystem; fluctuation of the water level of

Lake Sevan, which resulted in the redistribution of water plants in the lake; and changes in the hydrological regime of rivers and lakes (e.g., on the Lori Plateau) (The fifth… 2014).

1.6.9 Vegetation on Screes and Rocks

These habitats are intrazonal and very characteristic for mountainous areas. Petrophilic ecosystems are characterized by an abundance of rare, stenochorous and stenotopic plant species. In this regard, they are of great floristic and environmental interest. In Armenia, these ecosystems are distinguished on the basis of their substrate – screes and rocks on a volcanic or limestone basis. These ecosystems are predominantly included in the category "inland unvegetated or sparsely vegetated habitats." Screes include temperate-montane acid siliceous screes (with *Amygdalus fenzliana, Cerasus incana, Coluteocarpus vesicaria, Euphorbia gerardiana, E. szovitsii, Fumana procumbens, Potentilla argentea, Stachys lavandulifolia, Stipa arabica*, etc.), temperate-montane calcareous and ultrabasic screes (with *Allium struzlianum, Helichrysum graveolens, Minuartia sclerantha, Tulipa biflora, T. julia*, etc.), acid siliceous screes of warm exposures (with *Amygdalus fenzliana, Atraphaxis spinosa, Bromus fibrosus, Cerasus incana, Ephedra procera, Nepeta mussinii, Poa bulbosa, Rhamnus pallasii*, etc.), and calcareous and ultrabasic screes of warm exposures (with *Allium materculae, Cleome ornithopodioides, Eremopyrum orientale, Eremostachys laciniata, Michauxia laevigata, Onosma sericea, Peganum harmala, Poa bulbosa, Rumex scutatus, Salvia dracocephaloides, Scrophularia thesioides, Serratula coriacea, Stachys inflata*, etc.). The category "inland cliffs, rock pavements and outcrops" includes acid siliceous inland cliffs (with *Artemisia splendens, Dianthus raddeana, Erigeron venustus, Helichrysum graveolens, Minuartia imbricata, Saxifraga moschata, S. juniperifolia, S. kolenatiana, Sedum pilosum, Symphyandra zangezura, Tanacetum zangezuricum*, etc.), basic and ultrabasic inland cliffs (with *Amygdalus fenzliana, Cerasus incana, Hypericum formosissimum, H. eleonorae, Parietaria elliptica, Potentilla porphyrantha, Rhamnus pallasii*, etc.), wet inland cliffs (with *Adiantum capillus-veneris, Lycopodium selago*, etc.), and almost bare rock pavements, including limestone pavements, weathered rock, and outcrop habitats (with species of *Sedum, Sempervivum, Parietaria*, and other genera).

1.7 Phytogeographical Location and Division

Armenia lies at the intersection of two phytogeographical subkingdoms: Boreal (Euro-Siberian) and Ancient Mediterranean (Takhtadjan 1986).

The ancient history of the formation of the flora and vegetation of Armenia, the diversity of natural conditions, the impact of various biotic and abiotic factors on them, and the different geological history of individual parts of the republic have led to a high degree of heterogeneity of its flora. In 1954, in preparation for the publication of the first volume of the Flora of Armenia, Takhtajan divided the entire territory of the republic into 12 floristic regions, reflecting the characteristics of their floristic composition and main ecosystems, and provided a corresponding map-scheme.

Based on numerous literature and herbarium data, as well as on the basis of long-term observations in nature, we made some adjustments to the floristic zoning of Armenia by A. L. Takhtajan (Tamanyan and Fayvush 2009) and in this work, we use this updated scheme to indicate the distribution of plant species and ecosystems throughout the republic. We accept 12 floristic regions within the following boundaries (Fig. 1.5).

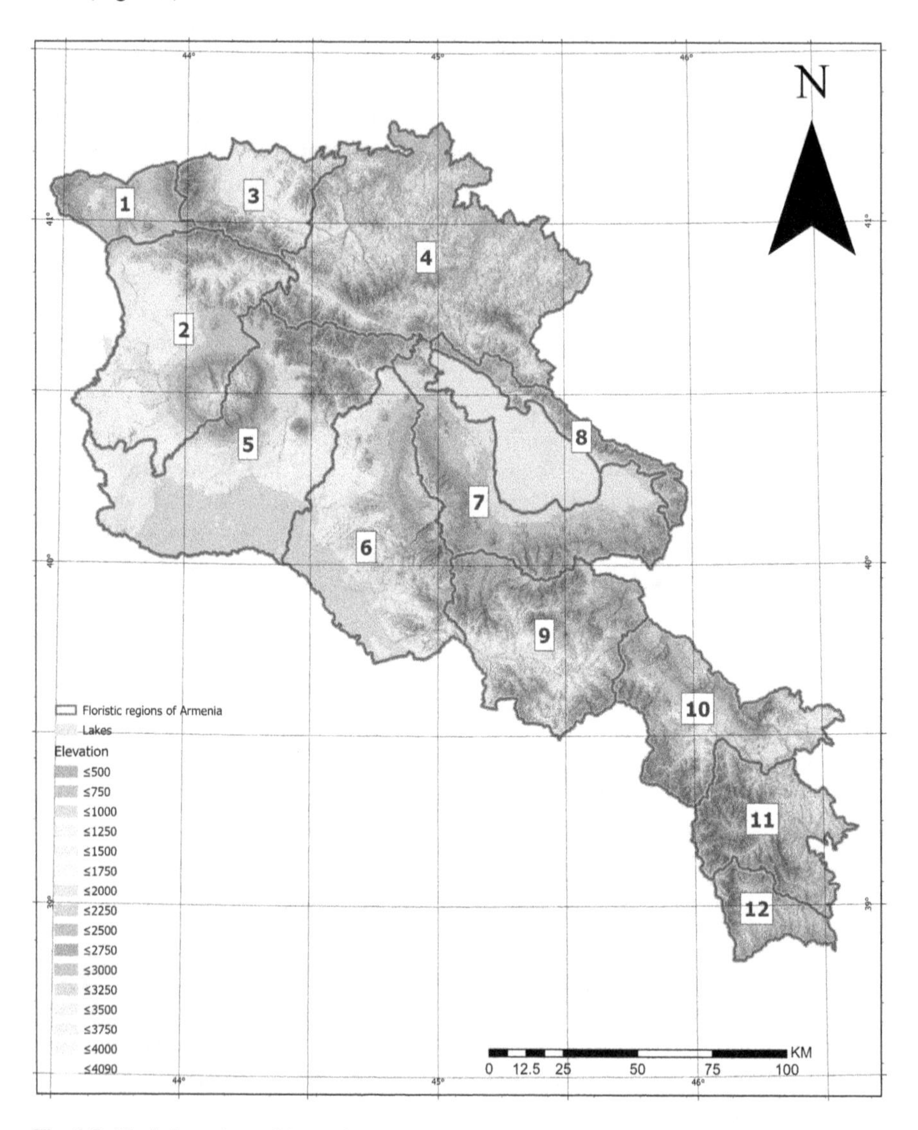

Fig. 1.5 Floristic regions of Armenia

1. Upper-Akhuryan floristic region. The boundaries of the region from the west and north are the state border of Armenia with Turkey and Georgia; from the east, the watershed of the Javakh range; and from the south, it runs a little south of the cities Ashotsk and Amasia, capturing the western part of the Bazum Range.
2. Shirak floristic region covers mainly the Shirak plateau, and in the west, it is limited by the state border with Turkey; in the north, it borders on the Upper Akhuryan floristic region, and it covers the entire Shirak ridge, the western part of the Pambak ridge to the watershed of the Kasakh river and the western part of the massif of Mount Aragats; and in the south, the region borders on the Yerevan floristic region along the physical border of the Shirak plateau, including Mount Arteni and vicinity of Talin town.
3. Lori floristic region covers the entire Lori plateau; the southern border runs along the watershed of the Bazum ridge, and the eastern border along the watershed of the Lalvar and Lejan mountains; in the west along the Javakh ridge, there is a border with the Upper Akhuryan floristic region, and the northern border coincides with the state border of Armenia with Georgia.
4. Ijevan floristic region covers the basins of the Aghstev and Debed rivers as well as the former Shamshadin administrative region north of the Sevan ridge.
5. Aparan floristic region covers the upper part of the Kasakh river basin and the entire basin of Marmarik river, Arailer volcano, the eastern part of the Aragats mountain massif, and the western part of the Ararat plain, and its border runs along the gorge of the Hrazdan river.
6. Sevan floristic region covers the entire basin of Lake Sevan, with the exception of the Sevan and Areguni ranges; the western border runs along the watershed of the Geghama highlands.
7. Areguni floristic region covers the Sevan and Areguni ranges and the coast of Lake Sevan, including the Ardnish peninsula.
8. Yerevan floristic region covers the eastern part of the Ararat plain, while its eastern border runs along the watershed of the Geghama Highlands and the western along the valley of the Hrazdan river.
9. Darelegis floristic region covers the basin of the Arpa river, and in the north along the Vardenis ridge, it borders on the Sevan floristic region; in the east, it is separated from the North Zangezur floristic region by the Zangezur ridge; in the west, it borders on the Yerevan floristic region, and in the south, the state border of Armenia is separated from the Nakhichevan Republic.
10. North Zangezur floristic region covers the basin of the Vorotan river, while its southern border runs along the Bargushat ridge and the left edge of the gorge of the Vorotan river.
11. South Zangezur floristic region covers the basin of the Voghji river; in the north, it borders on the North Zangezur floristic region; in the south, along the Meghri Range with Meghri; and in the west and east, the border of the region coincides with the state border of Armenia.
12. Meghri floristic region covers the middle reaches of the Araks river basin (within Armenia, the former Meghri administrative region) and is bounded from the north by the Meghri Range and from the south, west, and east by the state border of Armenia.

References

Aslanyan A (1958) Regionalnaja geologia Armenii (Regional geology of Armenia). Haypetrat, Yerevan

Aslanyan A (1985) O vozraste relief Armenii (osnovnye etapoy evolutsii) (On the age of the relief of Armenia) (main points of the evolution). Problems of Holocene geology, Yerevan, pp 14–19

Bagdasaryan A (1958) Climat Armjanskoj SSR (Climate of Armenian SSR). Yerevan

Barsegyan A (1990) Vodno-bolotnaja rastitelnosy' Armjanskoj SSR (Water-marsh vegetation of Armenian SSR). Yerevan

Dumitrashko NV (1979) Regional'naja geomorphologija Kavkaza (Regional geomorphology of the Caucasus). Moscow

Fayvush G (2006) Flora diversity of Armenia. Biodiversity of Armenia: from materials of the Third National Report. Yerevan, pp 9–12

Fayvush G, Aleksanyan A (2016) Mestoobitanija Armenii (Habitats of Armenia). Institute of Botany NAS RA, Yerevan

Fayvush G, Tamanyan K, Kalashyan M, Vitek E (2013) "Biodiversity hotspots" in Armenia. Ann Naturhist Mus Wien B115:11–20

Fourth National Communication on Climate Change (2020). Yerevan. UNDP Armenia

Gabrielian GK (ed) (1986) Fizicheskaja geographia Zakavkazja (Physical geography of the Transcaucasian). Yerevan State University, Yerevan

Gabrielyan GK (1962) Orography. In: Geology of Armenian SSR, vol 1. Yerevan, pp 25–30

Magakyan A (1941) Rastitelnost Armjanskoj SSR (Vegetation of Armenian SSR). Moscow-Leningrad

Makhatadze LB (1957) Dubravy Armenii (oak forests of Armenia). Yerevan

Melkonyan K, Ghazaryan H, Manukyan R (2004) Gjuhatntesakan nshanakutjan hogheri ecologiakan ardi vichaky, hoghogtagortsman makardaky, karavarman hamakargi katarelagortsumy ev ardjunavetutjan barelavman ughinery Hajastani Hanrapetutjunum (Current ecological status of agrarian soils, level of soil use, situation with management structure, and methods for improvement of soil productivity in the Republic of Armenia). Yerevan

Mittermeier RA, Turner WR, Larsen FW, Brooks TM, Gascon C (2011) Global biodiversity conservation: the critical role of hotspots. In: Zachos FE, Habel JC (eds) Biodiversity hotspots: distribution and protection of conservation priority areas. Springer, Heidelberg, pp 3–22

Takhtajan A (1941) Botaniko-geographicheskij ocherk Armenii (Phyto-geographical review of Armenia). Proc Inst Bot Armenian Branch USSR Acad Sci 2:3–156

Takhtajan AL (1954) Map of floristic regions of Armenia. Flora Armenia 1:3

Takhtajan A (1986) Floristic regions of the world. University of California Press, Berkley

Tamanyan KG, Fayvush GM (2009) K probleme floristicheskih rajonov Armenii (on the problems of floristic regions of Armenia). Flora, vegetation and plant resources of Armenia, Yerevan, 17, pp 73–78

The fifth national report to convention on biological diversity (2014) Yerevan

Third national communication on climate change under the United Nations Framework Convention on Climate change (2015) Yerevan

Chapter 2
Ecosystems of Armenia

George Fayvush, Alla Aleksanyan, and Vardan Asatryan

The richness of biodiversity is associated with the richness and diversity of natural ecosystems. All the main ecosystems of the Caucasus (with the exception of the humid subtropics) are represented in Armenia – deserts and semideserts, steppes, meadow-steppes, forests and open forests, subalpine and alpine vegetation, and intrazonal ecosystems. All of them are habitats for species of flora and fauna of Armenia.

Natural ecosystems and vegetation cover are extremely rapidly changing elements. Successional changes in vegetation caused by biogenic and abiogenic factors change both the general nature of vegetation and the distribution and ratio of habitats. In recent centuries, the anthropogenic factor has had a particularly serious impact on vegetation and ecosystems. It must be said that in the early stages of the development of human civilization, when primitive people were nomads and mainly gatherers, the effect of the anthropogenic factor on nature was minimal, on average at the level of the zoogenic factor. In the future, during the transition to a sedentary lifestyle, human impact increases, the transformation of natural ecosystems into agroecosystems, agrocoenoses, pastures, hayfields, and also into ecosystems of human settlements begins. Over time, this impact has intensified and intensified; in addition to the direct impact on natural ecosystems, an indirect one has also appeared – environmental pollution with greenhouse gases, waste, etc. All this leads to the change and destruction of existing ecosystems and the emergence of new

G. Fayvush (✉) · A. Aleksanyan
Institute of Botany after A. takhtadjan NAS RA, Yerevan, Armenia
e-mail: g.fayvush@botany.am; a.aleksanyan@botany.am

V. Asatryan
The Institute of Hydroecology and Ichthyology of Scientific Center of Zoology and Hydroecology of NAS of Armenia, Yerevan, Armenia
e-mail: vardan.asatryan@sczhe.sci.am

G. Fayvush (ed.), *Biodiversity of Armenia*,
https://doi.org/10.1007/978-3-031-34332-2_2

ones. This also leads to the emergence of new habitats, which are largely synanthropic. All this causes, on the one hand, a reduction in populations, up to complete extinction, of local rare species of plants and animals; on the other hand, it facilitates the penetration and spread of alien, often invasive, and undesirable species. From the point of view of nature conservation and natural biodiversity, of course, the most important are natural, preferably unmodified habitats, where the most important elements of the country's biodiversity, in particular endemic plants and animals, are represented. However, in the study of biodiversity in general and when working for purely utilitarian practical purposes, it is absolutely necessary to know about all the habitats that exist in the country.

In article 2 ("Use of terms") of the Convention of Biodiversity, an ecosystem is defined as "a dynamic complex of plant, animal and micro-organism communities and their non-living environment interacting as a functional unit." The concept of *habitat* is closely related but not identical to ecosystems. It is defined as "a location [area] in which a particular organism is able to conduct activities which contribute to survival and/or reproduction" (Stamps 2019). Thus, habitats are provided *by* ecosystems *for* individual species. In our work, we describe ecosystems of Armenia on the basis of habitat classification system EUNIS (Davies et al. 2004; Fayvush and Aleksanyan 2016).

In general, the scale selected for the EUNIS habitat classification is that occupied by small vertebrates, large invertebrates, and vascular plants. It is the same as that generally adopted by other European-scale typologies, for example, by the Palaearctic habitat classification (Devillers and Devillers-Terschuren 1996), and is comparable to the scale applied to the classification of vegetation in traditional phytosociology. All but the smallest EUNIS habitats occupy at least 100 m^2; there is no upper limit to the scale of the largest. At the smaller scale, "microhabitats" (features generally occupying less than 1 m^2 that are important for some smaller invertebrates and lower plants) can be described. Examples are decaying wood, found in mature forests and required by invertebrates whose function is decomposition, or animal dung in grassland environments. At the larger scale, habitats can be grouped as "habitat complexes," which are frequently occurring combinations or mosaics of individual habitat types, usually occupying at least 10 ha, which may be interdependent. Estuaries, combining tidal water, mud flats, saltmarshes, and other littoral habitats, are a good example.

The EUNIS habitat types are arranged in a hierarchy. Level 1 is the highest. There are 10 level 1 categories. Given that this classification is designed, first and foremost, for Europe, it shows the habitats that are not present in Armenia, and which we do not dwell on here. On the other hand, in Armenia, we present a number of ecosystems that are absent in Europe, and for which, we are forced to introduce new categories. Here, we give only a brief description of the ecosystems of Armenia; a more detailed description of them is given in a previously published monograph (Fayvush and Aleksanyan 2016). The classification is currently being reviewed, and some changes are being made to it. In general, the new version of the EUNIS classification can be found on the website https://www.eea.europa.eu/data-and-maps/data/eunis-habitat-classification-1.

All ecosystems of Armenia belong to terrestrial habitats, including inland surface waters. There are no (A) marine habitats and (B) coastal habitats in Armenia.

2.1 C. Inland Surface Waters

As described in EUNIS habitat classification, inland surface waters are non-coastal, aboveground, open, fresh, or brackish waterbodies (e.g., rivers, streams, lakes and pools, springs), including their littoral zones. Recently, it involves three categories, namely, "surface standing waters," "surface running waters," and "littoral zones of inland surface waterbodies" (EUNIS habitat classification 2022). Such classification includes constructed inland waterbodies of different salinity which support a seminatural plant and animal communities: intermittent and temporary waterbodies and banks and shores that are sufficiently frequently inundated to prevent the formation of closed terrestrial vegetation. The most up-to-date classification involves 236 habitat types of different levels throughout Europe. We distinguish between 131 inland surface water habitat types in Armenia. Among them, ten are missing from EUNIS classification.

2.1.1 C1. Surface Standing Waters

2.1.1.1 C1.1. Permanent Oligotrophic Lakes, Ponds, and Pools (Fig. 2.1)

These are low-nutrient-content waterbodies. The category includes oligotrophic waters of medium or high pH. Given the lack of sources of nutrient flux, such waterbodies are allocated mainly in the high-altitude locations where settlements are missing and grazing is very limited. If considering lakes higher than 2800 m a.s.l., there would be 313 such waterbodies in Armenia. Lakes Kari, Lessing, Tsaghkari, and Akna are among them.

2.1.1.2 C1.2. Permanent Mesotrophic Lakes, Ponds, and Pools (Fig. 2.2)

This class in general involves lakes and pools with waters fairly rich in nutrients (nitrogen and phosphorus) and dissolved bases (pH often 6–7). Many unpolluted lowland lakes and ponds are naturally mesotrophic and support dense beds of macrophytes. It should be noted that beds of charophytes can occur in mesotrophic (C1.25) lakes too. From the geographic perspective, this is widely distributed habitat class among Armenian lowland and mid-altitude lakes. The most typical are the

Fig. 2.1 Oligotrophic lake in Armenia (C1.1)

Fig. 2.2 Mesotrophic lake with *Nymphaea alba* (C1.24112)

lakes in Lori Plato (all lakes in Lori Plato are important botanical areas and a part of the Emerald Network), Lake Sevan (currently, it is closer to mesotrophic status), and some middle-sized lakes in the mid-altitude mountain belt.

2.1.1.3 C1.3. Permanent Eutrophic Lakes, Ponds, and Pools

Such lakes and pools are with mostly dirty grey to blue-green, more or less turbid, waters, particularly rich in nutrients (nitrogen and phosphorus) and dissolved bases (pH usually >7). Moderately eutrophic waters can support dense beds of macrophytes, but these disappear when pollution causes nutrient levels to rise further. In Armenia, these habitats are distributed in the lowland (Fig. 2.3) – in Ararat valley, Shirak plato, Vayots Dzor region, Meghri, and Idjevan floristic regions (Parzlich, Gosh).

2.1.1.4 C1.4. Permanent Dystrophic Lakes, Ponds, and Pools

This class involves lakes and pools with acidic waters of high humus content and often brown tinted (pH often 3–5). Such habitats are usually small artificial and semi-artificial ponds, pools, and reservoirs.

Fig. 2.3 Eutrophic lake in Armenia (C1.3)

2.1.1.5 C1.6. Temporary Lakes, Ponds, and Pools

This class involves freshwater lakes, ponds, pools, or parts of such freshwater bodies that become periodically dry, with their associated animal and algal pelagic and benthic communities. Habitats of the dry phase are listed under C3.5, C3.6, and C3.7.

2.2 C2. Surface Running Waters

These are running waters, including springs, streams, and temporary watercourses. There are 215 rivers of more than 10 km in the territory of Armenia. Besides, here are plenty of springs and streams and during spring season also many temporary watercourses.

2.2.1 C2.1. Springs, Spring Brooks, and Geysers

These are springs and resurgences, together with animal and plant communities dependent on the peculiar microclimatic and hydrological situation created by them. It involves also streams outflowing from springs and geysers or nearby springs, where the temperature regime is the same with the springs and different from the environment, and excludes vegetated spring mires (D2.2, D4.1), where springs emerge through a (usually small) expanse of vegetation with little or no open water. Areas of this habitat type are small but frequently encountered in almost all regions of Armenia in all altitudes. The exclusion are some volcanic structures, where groundwater can't reach the surface layer.

2.2.2 C2.2. Permanent Nontidal, Fast, Turbulent Watercourses

These are permanent watercourses with fast-flowing turbulent water and their associated animal and microscopic algal pelagic and benthic communities. Rivers, streams, brooks, rivulets, rills, torrents, waterfalls, cascades, and rapids are included. The bed is typically composed of rocks, stones, or gravel with only occasional sandy and silty patches. Features of the river bed, uncovered by low water or permanently emerging, such as gravel or rock islands and bars, are treated as the littoral zone (C3). Most of the rivers and streams in Armenia belong to this class. Usually, large rivers with prolonged watercourse in the upper course part also belong to this class. In the lower course part, the velocity decreases, and they are classified into C2.3 (Fig. 2.4).

Fig. 2.4 Permanent nontidal, fast, turbulent watercourses (C2.2 – Armenia, Debed River)

2.2.3 *C2.3. Permanent Nontidal, Smooth-Flowing Watercourses*

These are permanent watercourses with nonturbulent water and their associated animal and microscopic algal pelagic and benthic communities: slow-flowing rivers, streams, brooks, rivulets, and rills; and also fast-flowing rivers with laminar flow. The bed is typically composed of sand or mud. Features of the river bed, uncovered by low water or permanently emerging, such as sand or mud islands and bars, are treated as the littoral zone (C3). Concerning the mountain terrain of Armenia, such habitats are covering only small areas and can be encountered in the lower course parts of large rivers (Araks, Akhuryan, Arpa, Vorotan, Voghji, Aghstev, Debed, etc.) or in the middle or the upper course parts of the rivers that flow through mountain platos (Argichi, Akhuryan). Besides, some rivers where the velocity is artificially dropped due to dams (e.g., the Hrazdan River, the middle course part of the Vorotan River), are also involved in this class. Recently, many small- and medium-sized rivers where small hydropower plants are established (on the moment of 2019, there are 187 small HPPs) have been transformed into this class. Some major and medium-sized channels with natural bed and smooth flow also belongs to this class.

2.2.4 C2.5. Temporary Running Waters

These are watercourses that cease to flow for part of the year, leaving a dry bed or pools. This is a widely distributed class of habitats in the whole territory of Armenia but more typical for the Ararat valley, Shirak plato, Vayots Dzor, and Meghri regions. During spring season, these watercourses are fast-flowing and turbulent, leading to floods and mudflows, but in summer, the velocity drops, and till the mid-summer, they are drying out. However, after intense rainfalls, their flow could revive. Because soil moisture remains high, in the vegetation structure, hygrophytes or woody plants with deep root system are prevailing.

2.2.5 C2.6. Films of Water Flowing over Rocky Watercourse Margins

This class involves flowing water that is not contained within a channel but oozes over rocks. Such habitats are accompanying to waterfalls, where at the margins or on the rocks a film of water is flowing. These are very rare habitats in Armenia.

2.3 C3. Littoral Zone of Inland Surface Waterbodies

These are reedbeds and other water-fringing vegetation by lakes, rivers, and streams; exposed bottoms of dried up rivers and lakes; and rocks, gravel, sand, and mud beside or in the bed of rivers and lakes. This is a very widely distributed habitat class in Armenia, and dominant plant communities depend on the character of water reservoir.

2.3.1 C3.1. Species-Rich Helophyte Beds

These are the water-fringing stands of vegetation by lakes, rivers, and streams, with mixed species composition. These habitats are not so widely distributed, because most of the species of plants characteristic for this habitats are cenotically active and in most cases forming pure or near-pure thickets. But in this particular habitat class, floristic structure is rich, and community is multilayered and polydominant.

2.3.2 C3.2. Water-Fringing Reedbeds and Tall Helophytes Other Than Canes

These are water-fringing stands of tall vegetation by lakes (including brackish lakes), rivers, and brooks, usually species-poor and often dominated by one species. It includes stands of *Carex* spp., *Equisetum fluviatile, Hippuris vulgaris, Phragmites australis, Schoenoplectus* spp., *Sparganium* spp., and *Typha* spp. It excludes terrestrialized reed and sedge beds which are not at the water's edge (D5.1, D5.2). It is a very common habitat class in Armenia which encountered in all regions from the lower to the upper mountain belts.

2.3.3 C3.3. Water-Fringing Beds of Tall Canes

These are beds of tall canes lining permanent or temporary watercourses and water bodies. Included are beds of *Arundo donax* and *Erianthus ravennae.* They are very rare habitats in Armenia, known only in the banks of the Arax River.

2.3.4 C3.4. Species-Poor Beds of Low-Growing Water-Fringing or Amphibious Vegetation

This includes isoetids of the shores of oligotrophic lakes, *Nasturtium* by streams, Mediterranean dwarf *Scirpus* swards, and other species-poor but dissimilar types of vegetation. These habitats are not largely spread in Armenia. They are connected with oligo- and mesotrophic lakes, ditches, and channels. All habitats of this class are involved in the Resolution 4 of the Berne Convention.

2.3.5 C3.5. Periodically Inundated Shores with Pioneer and Ephemeral Vegetation

These are muddy, sandy, and gravelly shores and dried-up bottoms of lakes and rivers, with moderate cover of vascular plants. These include annuals (e.g., *Bidens* spp., *Cyperus* spp., *Persicaria* spp.), developing during the exposure phase as well as perennials tolerant of temporary total immersion.

2.3.6 C3.6. Unvegetated or Sparsely Vegetated Shores with Soft or Mobile Sediments

These are banks of sand, gravel, and mud in or by rivers, gravel by mountain streams, and mud bottoms of dried-up rivers and lakes, including saline lakes. They also involve exposed sand, gravel, and mud at the edge of lakes. More frequently such habitats can be found in the lower course part of the rivers and in the banks of lakes of lower and in some cases also middle mountain belt. Habitat types C3.62 (Unvegetated river gravel banks) are included in the Resolution 4 of the Berne Convention.

2.3.7 C3.7. Unvegetated or Sparsely Vegetated Shores with Nonmobile Substrates

These are periodically exposed rocks, pavements and blocks beside rivers and lakes, and in the draw-down zone of reservoirs. Rather widely distributed habitat type in Armenia but of small coverage.

2.3.8 C3.8. Inland Spray- and Steam-Dependent Habitats

These are spray-washed margins of pools below waterfalls and steamy margins of geysers and hot springs. Because waterfalls in Armenia are not large, such habitats are not occupying large areas too and typical only for permanent waterfalls. Geysers are missing in Armenia, and habitats related to wetted grounds, not because of evaporation or water dust, are formed around thermal springs. The most typical habitats are found near the largest waterfalls in Armenia – Shaki, Trchkan, etc.

2.4 D. Mires, Bogs, and Fens

These are wetlands, with the water table at or aboveground level for at least half of the year, dominated by herbaceous or ericoid vegetation. They include inland saltmarshes and excludes the water body and rock structure of springs and waterlogged habitats dominated by trees or large shrubs. Note that habitats that intimately combine waterlogged mires and vegetation rafts with pools of open water are considered as complexes.

Of the six categories of the second level according to the EUNIS classification, four are noted in Armenia. Ecosystems D1 (raised bogs, the source of water for which is only precipitation) and D3 (aapa, palsa, and polygonal bogs depending exclusively on glaciers) are absent in Armenia.

2.4.1 D2. Valley Mires, Poor Fens, and Transition Mires

These are weakly to strongly acid peatlands, flushes, and vegetated rafts formed in situations where they receive water from the surrounding landscape or are intermediate between land and water. Included are quaking bogs and vegetated noncalcareous springs. In general, peat habitats in Armenia are not very widespread; they are found mainly on the Lori plain, on the Pambak ridge, and in the basin of Lake Sevan (due to the draining of the lake, most of the peat bogs in this region turned out to be drained).

2.4.1.1 D2.1. Valley Mires

These are topogenous wetlands in which the peat-forming vegetation depends on water draining from the surrounding landscape. Most valley mires are habitat complexes including poor fens, transition mires, and pools. In Armenia, they are represented only by swamps of subcategory D2.11. Acid valley mires. The category includes peatlands and peat bogs of the middle and upper mountain belts (1400–2400 m a.s.l.). Presently, in Armenia, these habitats have been preserved only on the Lori plain, the previously existing habitats in the basin of Lake Sevan, after the descent of the lake waters, mostly dried up and the process of peat accumulation stopped.

2.4.1.2 D2.2. Poor Fens and Soft-Water Spring Mires

These are peatlands, flushes, and vegetated springs with moderately acid ground water, within valley mires or on hillsides. As in the rich fens, the water level is at or near the surface of the substratum, and peat formation depends on a permanently high water table. Poor-fen vegetation is typically dominated by small sedges (*Carex canescens, C. transcaucasica*), with pleurocarpous mosses (*Calliergonella cuspidata, Calliergon sarmentosum, Calliergon stramineum, Drepanocladus exannulatus, Drepanocladus fluitans*) or sphagna (*Sphagnum cuspidatum, Sphagnum papillosum, Sphagnum recurvum* agg., *Sphagnum russowii, Sphagnum subsecundum* agg.). Other characteristic vascular plants are *Agrostis canina and Juncus filiformis*. The habitats are inherent in the upper mountainous and subalpine belts, found in the Lori, Idjevan, Sevan, Aparan, and North Zangezur floristic regions.

2.4.1.3 D2.3. Transition Mires and Quaking Bogs

These are nonterrestrial waterlogged habitats with peat-forming vegetation and acidic groundwater or (for floating islands of vegetation) acidic lake waters. Characteristic species are *Carex diandra, C. lasiocarpa, C. rostrata,* and *Menyanthes trifoliata.* This category includes habitats that are found everywhere; the name itself says that these habitats are transitional between water and land, but they are still waterlogged habitats and are characterized by the presence of weak peat formation processes.

2.4.2 D4. Base-Rich Fens and Calcareous Spring Mires

These are peatlands, flushes, and vegetated springs with calcareous or eutrophic ground water, within river valleys, alluvial plains, or on hillsides. As in poor fens, the water level is at or near the surface of the substratum, and peat formation depends on a permanently high water table.

2.4.2.1 D4.1. Rich Fens, Including Eutrophic Tall-Herb Fens and Calcareous Flushes and Soaks

These are wetlands and spring-mires, seasonally or permanently waterlogged, with a soligenous or topogenous base-rich, often calcareous water supply. Peat formation, when it occurs, depends on a permanently high water table. Rich fens may be dominated by small or larger graminoids (*Carex* spp., *Eleocharis* spp., *Juncus* spp., *Molinia caerulea, Phragmites australis, Schoenus* spp., *Sesleria* spp.) or tall herbs (e.g., *Eupatorium cannabinum, Mentha longifolia, Caltha polypetala*). Fairly common ecosystems in Armenia are divided into a number of subcategories.

2.4.3 D5. Sedge and Reedbeds, Normally Without Free-Standing Water

These are sedge and reedbeds forming terrestrial mire habitats, not closely associated with open water. These are habitats common in Armenia, formed in areas with a high level of standing groundwater.

2.4.3.1 D5.1. Reedbeds Normally Without Free-Standing Water

These are terrestrialized stands of tall helophyte *Poaceae, Schoenoplectus* spp., *Typha* spp., horsetails, or forbs, usually species-poor and often dominated by one species, growing on waterlogged ground. They are classified according to dominant species which give them a distinctive appearance. They are very well represented in Armenia, found in all floristic regions (Fig. 2.5).

2.4.3.2 D5.2. Beds of Large Sedges Normally Without Free-Standing Water

These are terrestrialized stands of tall *Carex* and *Cyperus*, usually species-poor and often dominated by one species, growing on waterlogged ground. They are very well represented in Armenia. The category includes a number of subcategories. There is in this category one very interesting ecosystem – bladder sedge beds with high abundance of *Lychnis flos-cuculi* (D5.21421) – here, *Carex vesicaria* is a dominant, and very rare in Armenia, known only from one locality in the vicinity of Lermontovo village; *Lychnis flos-cuculi* is a co-dominant.

Fig. 2.5 Great reedmace beds (D5.131)

2.4.3.3 D5.3. Swamps and Marshes Dominated by Soft Rush or Other Large Rushes

These are stands of large *Juncus* spp. invading heavily grazed and trampled marshes or fens or (with *Juncus effusus*) eutrophicated poor fens and bogs, e.g., in the vicinity of bird colonies. Despite the fact that these ecosystems are widely distributed throughout Armenia, only two *Juncus* species, *J. effusus* (D5.31) and *J. buffonius* (D5.32), dominate in them.

2.4.4 D6. Inland Saline and Brackish Marshes and Reedbeds

These are saline wetlands, with closed or open vegetation, which are the non-coastal analogue of coastal saltmarshes and saline reedbeds.

2.4.4.1 D6.1. Inland Saltmarshes

These are salt meadows and swards of *Salicornia* and other *Chenopodiaceae* of inland salt basins of the nemoral zone. There are two subcategories in this category: interior Central European and Anatolian glasswort swards (D6.16), found on the Ararat valley, and in which, in addition to the dominant *Salicornia europaea*, very rare species *Microcnemum coralloides, Salsola soda, Halocnemum strobilaceum, Halostachys caspica*, etc., grow; and salt marshes with *Puccinellia gigantea and P. sevangensis* (D6.19).

2.4.4.2 D6.2. Inland Saline or Brackish Species-Poor Helophyte Beds Normally Without Free-Standing Water

These are terrestrialized stands of tall salt-tolerant helophytes, notably *Phragmites australis.* This category includes two subcategories: very common on Ararat valley dry halophile common reed – *Phragmites australis* – beds (D6.21) and unique salt marshes with *Juncus acutus* (D6.24), in which many very rare species included in the Red Data Book of plants of Armenia are growing here (*Sonchus araraticus, Linum barsegianii, Thesium compressum, Sphaerophysa salsula, Microcnemum coralloides, Frankenia pulverulenta,* etc.) (Fig. 2.6).

Fig. 2.6 Salt marshes with *Juncus acutus* (D6.24)

2.5 E. Grasslands and Lands Dominated by Forbs, Mosses, or Lichens

This is non-coastal land which is dry or only seasonally wet (with the water table at or aboveground level for less than half of the year) with greater than 30% vegetation cover. The vegetation is dominated by grasses and other nonwoody plants, including mosses, macrolichens, ferns, sedges, and herbs. This category includes semiarid steppes with scattered *Artemisia* scrub. It includes successional weedy vegetation and managed grasslands such as recreation fields and lawns.

2.5.1 E1. Dry Grasslands

These are well-drained or dry lands dominated by grass or herbs, mostly not fertilized and with low productivity. Included are *Artemisia* steppes.

2.5.1.1 E1.1. Inland Sand and Rock with Open Vegetation

This is an open, thermophile vegetation of sands or rock debris in the nemoral zone and, locally, in boreal or sub-Mediterranean lowland to montane areas of Europe. Included are open grasslands on strongly to slightly calcareous inland sands and vegetation formed mostly by annuals and succulents or semisucculents on decomposed rock surfaces of edges, ledges, or knolls, with calcareous or siliceous soils. In Armenia, this category is represented by only one subcategory Armenian flat rock debris swards (E1.116) – habitats are very widely distributed in Armenia from lower to alpine belt but do not occupy very big areas. They are common in all floristic regions. *Sedum* spp., *Saxifraga* spp., *Sempervivum transcaucasicum, Poa bulbosa, Erophila verna, Androsace maxima,* and *Androsace chamaejasme* usually dominate in these communities.

2.5.1.2 E1.2. Perennial Calcareous Grassland and Basic Steppes

These are perennial grasslands, often nutrient-poor and species-rich, on calcareous and other basic soils of the nemoral and steppe zones and of adjacent parts of the subboreal and sub-Mediterranean zones. Ecosystems of this category are very widely represented throughout Armenia; they are distinguished both by the diversity of dominants and the richness of the species composition. The classification is based on the dominant plant species. Subcategories Steppes with wild wheat species dominance (E1.2E22) – Unique habitats, *Triticum boeoticum, T. araraticum* and *T. urartu* are dominants in the communities; and Grass-forbs steppes (E1.2E24), in which endemic of Armenia *Smyrniopsis armena* is a dominant, should be highlighted (Fig. 2.7).

Fig. 2.7 Feather-grass steppes with *Stipa pennata* (E1.2E113)

2.5.1.3 E1.3. Mediterranean Xeric Grassland

These are meso- and thermo-Mediterranean xerophile, mostly open, short-grass perennial grasslands rich in therophytes: therophyte communities of oligotrophic soils on base-rich, often calcareous substrates. This category unites almost all ecosystems that are classified as semideserts in the Armenian geobotanical literature. They are very diverse ecosystems, widespread in Armenia, mainly in the lower mountain belt.

2.5.1.4 E1.4. Mediterranean Tall-Grass and Wormwood (*Artemisia*) Steppes

Traditionally, these habitats in Armenia are referred to as semideserts. The main edificator is usually *Artemisia fragrans*. According to EUNIS, ecosystems of this category are meso-, thermo-, and sometimes supra-Mediterranean formations of the Mediterranean basin, physiognomically dominated by tall grasses, between which may grow communities of annuals or sometimes chamaephytes. They include silicicolous as well as basiphile formations. In the semiarid regions between the Mediterranean and the deserts of western Asia, they dominate the landscape, forming a major steppe belt in which low scrub of *Artemisia* is prominent.

2.5.1.5 E1.8. Closed Mediterranean Dry Acid and Neutral Grassland

These are perennial grasslands on acid soils of the supra-Mediterranean zone, dominated by *Nardus stricta*. The only ecosystem E1.834 (Plant communities with *Nardus stricta* dominance) is represented in Armenia. The communities have rather poor floristic composition, and *Plantago atrata, Merendera raddeana, Tripleurospermum caucasicum, Alchemilla sevangensis, Hieracium pilosella,* and *Polygala alpicola* are more or less permanent in the composition. These are rather common habitats in Armenia on the altitudes 2000–3300 m a.s.l.

2.5.1.6 E1.A. Open Mediterranean Dry Acid and Neutral Grassland

It is a sandy open ground with vernal therophytes, not necessarily grasses, in the Mediterranean region. It includes open perennial grasslands and pastures on siliceous, usually skeletal, soils of the supra-Mediterranean zone. Only the subcategory Irano-Anatolian inland dunes (E1.A5), with ecosystem Psammophile communities with grasses and forbs (E1.A51) is known in Armenia. It is common in Ararat valley but rarer in lower mountain belt of other regions of Armenia. *Achillea tenuifolia, Astragalus paradoxus, Anisantha tectorum, Oligochaeta divaricata, Euphorbia marschalliana, Aphanopleura trachysperma, Erophila verna, Drabopsis nuda, Androsace maxima, Roemeria hybrida,* and *Koelpinia linearis* are common in the communities.

2.5.1.7 E1.C. Dry Mediterranean Lands with Unpalatable Non-vernal Herbaceous Vegetation

This includes dry land with less than 10% shrub cover and a large number of non-spring non-forage plants, including thorns (*Carthamus, Carlina, Centaurea, Onopordum*) and *Ferula* and *Phlomis* species. It is a very common habitat in Armenia, which is mainly the result of overgrazing of the lower and middle mountain belts pastures and is constantly expanding due to the ongoing degradation and destruction of ecosystems. Subcategories are distinguished here according to the main dominants of plant communities, which include most of the herbaceous expansive and invasive species of the flora of Armenia: *Cirsium* spp., *Onopordum* spp., *Carthamus* spp., Thistle *Centaurea* spp., *Silybum marianum*, *Cousinia* spp., *Carduus* spp., *Echinops* spp., *Verbascum* spp., as well as Giant hogweed (Heracleum spp.), *Astragalus galegiformis, Bupleurum exaltatum, Astrodaucus orientalis, Conium maculatum, Achillea filipendulina, Conyza canadensis, Tanacetum vulgare*, and *Sambucus ebulus* communities.

2.5.2 *E2. Mesic Grasslands*

These are lowland and montane mesotrophic and eutrophic pastures and hay meadows of the boreal, nemoral, warm, temperate, humid, and Mediterranean zones. They are generally more fertile than dry grasslands (E1) and include sports fields and agriculturally improved and reseeded pastures.

2.5.2.1 E2.1. Permanent Mesotrophic Pastures and Aftermath-Grazed Meadows

These are regularly grazed mesotrophic pastures of Europe, fertilized and on well-drained soils. Unlike Europe, in Armenia, they are typical for the middle and upper mountain belts, where meadow-steppe vegetation develops. Different ecosystems of meadow-steppe pastures are included in this category.

2.5.2.2 E2.2. Low- and Medium-Altitude Hay Meadows

These are mesotrophic hay meadows of low altitudes of Europe, fertilized and well-drained, with *Arrhenatherum elatius, Trisetum flavescens, Anthriscus sylvestris, Daucus carota, Knautia arvensis, Leucanthemum vulgare, Pimpinella saxifraga,* and *Geranium ruprechtii.* In Armenia, they are practically absent and very rarely formed in the river valleys; in most cases, the meadows of the lower and middle mountain belt are secondary post-forest, formed as result of deforestation.

2.5.2.3 E2.3. Mountain Hay Meadows

These are often species-rich hay meadows of the montane and subalpine levels of higher mountains. They are widespread ecosystems in Armenia, including subalpine meadows of the Ponto-Caucasian hay meadows subcategory (E2.32).

2.5.2.4 E2.5. Meadows of the Steppe Zone

These are lowland and montane mesotrophic pastures and hay meadows of the steppe zone of eastern Europe and Anatolia. They are rare habitats in Armenia. The conditions of the steppe belt do not favor the development of meadow vegetation here; therefore, these habitats occasionally develop in slightly more humid but not swampy places near springs or streams and occupy small areas.

2.5.2.5 E2.6. Agriculturally Improved, Reseeded, and Heavily Fertilized Grassland, Including Sports Fields and Grass Lawns

This category includes land occupied by heavily fertilized or reseeded permanent grasslands, sometimes treated by selective herbicides, with very impoverished flora and fauna, used for grazing, soil protection, and stabilization, landscaping, or recreation. In addition to artificial and improved pastures, this category includes sports fields (stadiums, golf courses, etc.), park and small lawns, and "green roofs."

2.5.2.6 E2.7. Unmanaged Mesic Grassland

This is a mesic grassland that is not currently mown or used for pasture. At present, due to the reduction in the number of livestock, such habitats have become quite widespread; pastures remote from settlements have begun to include them. Due to the lack of grazing, the state of the cover is improving, and they are at the succession stage of transition to the usual steppe, meadow-steppe, or meadow habitats.

2.5.2.7 E2.8. Trampled Mesophilous Grasslands with Annuals

These are herbage stands of low annuals in mesophilic, heavily uprooted habitats. Due to the economic crisis and the change in livestock management schemes, habitats of this type have become very common in the immediate vicinity of settlements and in the vicinity of livestock watering places. In the absence of the possibility of grazing on distant pastures or migration to summer pastures in the subalpine and alpine belts, pastures near settlements are subjected to intensive overgrazing; the same thing happens near watering places, and as a result, the herbage is knocked out, the floristic composition changes, and instead of perennial fodder grasses come annual weeds.

2.5.3 E3 Seasonally Wet and Wet Grasslands

This category includes unimproved or lightly improved wet meadows and tall herb communities of the boreal, nemoral, warm-temperate humid, steppic, and Mediterranean zones.

2.5.3.1 E3.3 . Sub-Mediterranean Humid Meadows

In Europe, these habitats usually include low-mountain, wet meadows rich in clovers (*Trifolium* spp.). In Armenia, this category of habitat is confined to high mountains – from the middle to subalpine belt.

2.5.3.2 E3.4. Moist or Wet Eutrophic and Mesotrophic Grassland

This category includes wet eutrophic and mesotrophic grasslands and flood meadows of the boreal and nemoral zones, dominated by grasses *Poaceae* or rushes *Juncus* spp. Besides tall rush pastures (*Juncus effusus, J. inflexus*) in this class, we include "Recently abandoned hay meadows," where thorny, poisonous, inedible plant species begin to dominate in the grass stand (*Polygonum bistorta, Filipendula ulmaria, F. vulgaris, Ranunculus* spp., *Alchemilla* spp., *Rumex* spp., *Anemone fasciculata, Veratrum album, Polygonum alpinum, and Leucanthemum vulgare.*

2.5.3.3 E3.5. Moist or Wet Oligotrophic Grassland

These are grasslands on wet, nutrient-poor, often peaty soils, of the boreal, nemoral, and steppe zones. In this category, only one subcategory of purple moor grass is represented in Armenia – Molinia – meadows and related communities. Ecosystems of this category are quite rare in Armenia and are found on the Lori plain and in the basin of the Lake Sevan. *Molinia coerulea* dominates here, and the vegetation includes *Deschampsia cespitosa, Betonica officinalis, Trollius europaeus, Carex panicea, C. pallescens, C. tomentosa, Festuca rubra*, and some species are common in Europe but rare in Armenia – *Potentilla erecta, Iris sibirica, Galium boreale,* and *Parnassia palustris.*

2.5.4 E4. Alpine and Subalpine Grasslands

These are primary and secondary grass- or sedge-dominated formations of the alpine and subalpine belts.

2.5.4.1 E4.1. Vegetated Snow Patch

These are vegetated areas that retain late-lying snow. Dominants may be mosses, liverworts, macrolichens, graminoids, ferns, and small herbs. In the highlands of Armenia, snow patches can persist for a long time; in some years, they persist throughout the summer. Here in places freed from melting snow, very characteristic open communities are usually formed, mainly from bulbous geophytes (*Puschkinia scilloides, Gagea glacialis, G. caroli-kochii, G. anisanthos, Merendera raddeana, Scilla armena, Colchicum szovitsii*, etc.) or representatives *Brassicaceae* and *Caryophyllaceae* families (*Draba bruniifolia, D. araratica, Cerastium cerastoides*, etc.), as well as *Primula algida, Gentiana verna ssp. pontica*, etc. Over time, as the vacated areas become overgrown, habitats of wet alpine carpets are formed.

2.5.4.2 E4.3. Acid Alpine and Subalpine Grassland

Alpine and subalpine grasslands developed over crystalline rocks and other lime-deficient substrates or on decalcified soils of mountains. In Armenia, these habitats include most of the high mountains of ridges and massifs of volcanic origin. The category includes diverse subcategories and ecosystems of alpine meadows (grasses and sedges dominate) and carpets (forbs dominate) developing on volcanic substrates (Fig. 2.8).

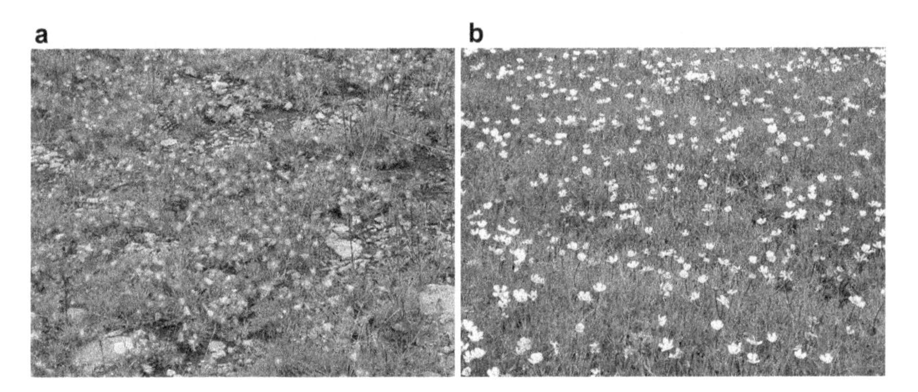

Fig. 2.8 Alpine carpets: A, *Campanula tridentata*-dominated carpets (E4.3A25); B, *Ranunculus dissectus* ssp. *aragazi*-dominated carpets (E4.3A29)

2.5.4.3 E4.4. Calcareous Alpine and Subalpine Grassland

This category includes alpine and subalpine grasslands of base-rich soils of the high mountains. They are not very widespread in Armenia, relatively well expressed in the Idjevan, Aparan, and Darelegis floristic regions. The category includes a variety of ecosystems that fall under the subcategory Caucasian alpine grassland. Representatives of herbs, grasses, and sedges dominate here. It should be taken into account that similar habitats in Armenia also develop on volcanic acidic substrates, where the same plant species often act as dominants in communities, but the floristic composition of ecosystems usually differs somewhat.

2.5.4.4 E4.5. Alpine and Subalpine Enriched Grassland

Presently, there are no artificially enriched (fertilized, with oversowing of seeds) pastures in Armenia. This category of pastures can be attributed only to those near the camps on summer pastures, naturally fertilized from a large number of animals, thanks to the nomadic system of animal husbandry. These pastures, on the one hand, are heavily overgrazed; on the other hand, they are oversaturated with organic fertilizers; as a result, the floristic composition is extremely depleted, and almost exclusively nitrophilous plant species grow. The most common communities are dominated by *Rumex alpinus*.

2.5.5 E5. Woodland Fringes and Clearings and Tall Forbs Stands

This category includes stands of tall herbs or ferns, occurring on disused urban or agricultural land, by watercourses, at the edge of woods, or invading pastures. They are stands of shorter herbs forming a distinct zone (seam) at the edge of woods.

2.5.5.1 E5.1. Anthropogenic Herb Stands

This category includes stands of herbs developing on abandoned urban or agricultural land, on land that has been reclaimed, on transport networks, or on land used for waste disposal. Generally, most of these habitats are ruderal, well represented in Armenia, and almost everywhere they stink naturally without additional human intervention.

2.5.5.2 E5.2. Thermophile Woodland Fringes

Woodland edge (seam) vegetation of the nemoral, boreo-nemoral, and sub-Mediterranean zones, composed of warmth-requiring drought-resistant herbaceous perennials and shrubs, which form a belt between dry or mesophile grasslands and the shrubby forest mantle, on the sunny side, where the nutrient supply is limited, or, sometimes, pioneering the woodland colonization into the grasslands. These habitats are common in Armenia, found mainly in the lower mountain belt.

2.5.5.3 E5.3. Bracken Fields

There are no typical habitats in Armenia, but one ecosystem is known in which the closely related species *Pteridium tauricum* dominates. Very rare ecosystem, it is known only on one very small site in the vicinity of N. Hand village in the South Zangezur floristic region.

2.5.5.4 E5.4. Moist or Wet Tall-Herb and Fern Fringes and Meadows

This category includes tall-herb and fern vegetation of the nemoral and boreal zones, including stands of tall herbs on hills and mountains below the montane level. Tall herbs are often dominant along watercourses, in wet meadows and in shade at the edge of woodlands. These are rather common habitats in Armenia, found both in forest and non-forest areas, usually in the middle and, less often, lower mountain belts.

2.5.5.5 E5.5. Subalpine Moist or Wet Tall-Herb and Fern Stands

This category includes luxuriant tall-herb formations of deep, humid soils in the middle to alpine, but mostly subalpine, levels of the higher mountains. Subalpine tall grass is a very characteristic type of vegetation for the Caucasus, best represented in the Greater Caucasus. It is less common in Armenia; the habitats are best represented in Northern Armenia, usually on gentle slopes above the forest line, but some areas are also found in Zangezur. The herbage in these habitats is high, often up to 1.5–2 m in height; the soil cover is usually 100%, but the turfing is very weakly expressed – no more than 10%–15%. Usually *Dactylis glomerata, Festuca gigantea, Lilium armenum, L. szovitsianum, Aconitum orientale, Cephalaria gigantea, Galega orientalis, Campanula latifolia, Delphinium flexuosum, Thalictrum minus, Linum hypericifolium, Astrantia maxima, A. trifida,* etc. are dominants here.

2.5.6 *E6. Inland Salt Steppes*

This category includes saline land with dominant salt-tolerant grasses and herbs. According to the EUNIS classification, these habitats are called steppes, while in Armenia, they traditionally mostly refer to semideserts.

2.5.6.1 E6.2. Continental Inland Salt Steppes

This category includes salt steppes and their associated salt-tolerant herbaceous communities outside the Mediterranean zone. This category is represented by subcategories "salt steppes" and "solonchaks grasslands". Ecosystems are very common in Ararat valley in lower mountain belt. They are close to salt marshes and differ by low level of underground water. A number of plant species which are very rare in Armenia are confined to these ecosystems (*Iris musulmanica, Linum barsegianii, Thesium compressum,* etc.). Ecosystem E6.251 was well represented in the Ararat valley (including Nakhichevan and Turkey), but its area has been significantly reduced due to desalinization and land cultivation activities. Simultaneously, this ecosystem is actually the only habitat for a rare invertebrate – the Armenian cochineal – for the conservation of which a special sanctuary has been organized in Armenia.

2.5.7 *E7. Sparsely Wooded Grasslands*

Grasslands with a wooded overstory that normally has less than 10% cover. In Armenia, these habitats are quite rare, since in most cases, the tree-shrub layer occupies more than 10% and usually such habitats belong to open or park forests of the subalpine belt. In fact, such habitats can be identified among subalpine grassland habitats, where individual trees grow above the upper border of the forest and up to the tree line. In addition, habitats with very rare trees are found on the plains (Ararat valley, Lori highland, Shirak and Ashots plateaus, Syunik highlands, and in the Lake Sevan basin), where these trees were planted artificially; in some cases, simply as individual trees; and in others, as a remnant of forest belts, where most of the trees died or were cut down.

2.6 F. Heathland, Scrub, and Tundra

This category includes non-coastal land which is dry or only seasonally inundated (with the water table at or aboveground level for less than half of the year) with greater than 30% vegetation cover. Heathland and scrub are defined as vegetation dominated by shrubs or dwarf shrubs of species that typically do not exceed 5 m maximum height. This category includes shrub orchards, vineyards, and hedges (which may have occasional tall trees). This also includes stands of climatically limited dwarf trees (krummholz) <3 m high, such as those occurring in extreme alpine conditions. It also includes *Salix* carrs.

2.6.1 F2. Arctic, Alpine, and Subalpine Scrub

This category includes scrub occurring north of or above the climatic tree limit but outside the permafrost zone. This includes scrub occurring close to but below the climatic tree limit, where trees are suppressed either by late-lying snow or by wind or repeated browsing.

2.6.1.1 F2.2. Evergreen Alpine and Subalpine Heath and Scrub

This category includes small, dwarf, or prostrate shrub formations of the alpine and subalpine zones of mountains, dominated by ericaceous species, dwarf junipers, brooms, or greenweeds. These are rather common habitats in Armenia, while habitats dominated by Ericaceae species are quite rare, but habitats with creeping junipers are very common, especially in Central and Southern Armenia. This category includes the subcategory Pontic alpenrose heaths, which is represented only by the ecosystem *Rhododendron caucasicum* heaths in Armenia. This ecosystem is distributed only in North Armenia; usually, it develops in the north slopes and occupies small areas in the subalpine belt. *Vaccinium myrtillus, Daphne glomerata, Poa nemoralis, P. longifolia, Anthoxanthum odoratum, Nardus stricta, Geranium sylvaticum, Pedicularis condensata, Scabiosa caucasica, Aetheopappus pulcherrimus, Coeloglossum viride, Myosotis alpestris,* and *Actaea spicata* are common species in these communities. This category also includes subcategories Southern Palaearctic mountain dwarf juniper scrub, represented by ecosystems dominated by *Juniperus sabina* and *Juniperus hemisphaerica*, as well as the subcategory alpine and subalpine meadows with *Vaccinium myrtillus* dominance (Fig. 2.9).

2.6.1.2 F2.3. Subalpine Deciduous Scrub

This category includes subalpine scrubs of *Alnus*, *Betula*, *Salix*, and Rosaceae (*Amelanchier*, *Rubus*, *Sorbus*, *Spiraea*), less than 5 m tall, often accompanied by tall herbs. In this category, we include steppe scrub, which is widespread in Armenia and well represented in Northern Armenia subalpine crook stem forest.

Fig. 2.9 *Rhododendron caucasicum* heaths in Armenia (F2.2261)

2.6.2 F3. Temperate and Mediterranean-Montane Scrub

These are shrub communities of nemoral affinities. They include deciduous and evergreen scrubs or brushes of the nemoral zone, and deciduous scrubs of the sub-Mediterranean and supra-Mediterranean zones.

2.6.2.1 F3.1. Temperate Thickets and Scrub

This category includes successional and plagioclimax scrub, mostly deciduous, of Atlantic, sub-Atlantic, or subcontinental affinities, characteristic of the nemoral zone, but also colonizing cool, moist, or disturbed stations of the Mediterranean evergreen forest zone. In general, this category is not typical for Armenia; we include only ecosystem *Hippophae rhamnoides* scrub in Sevan basin – artificial communities of *Hippophae rhamnoides,* which were created more than 60 years ago. Now, these communities became very dense, sometime impassable.

2.6.2.2 F3.2. Sub-Mediterranean Deciduous Thickets and Brushes

This category includes successional and plagioclimax scrub, mostly deciduous, of the sub-Mediterranean and supra-Mediterranean zones, but also colonizing cool, moist, or disturbed stations of the Mediterranean evergreen forest zone. This category includes two subcategories: Eastern Mediterranean deciduous thickets – edge ecosystem of oak forests (*Quercus iberica*) – and Ponto-Sarmatic deciduous thickets, so-called shibliak (*Paliurus spina-christi* thorn scrub) и Dew-berry dense scrub.

2.6.3 F5. Maquis, Arborescent Matorral, and Thermo-Mediterranean Brushes

This category includes evergreen sclerophyllous or lauriphyllous shrub vegetation, with a closed or nearly closed canopy structure, having nearly 100% cover of shrubs, with few annuals and some vernal geophytes; trees are nearly always present, some of which may be in shrub form. This includes pseudomaquis, in which the dominants are mixed deciduous and evergreen shrubs.

2.6.3.1 F5.1. Arborescent Matorral

This category includes successional and plagioclimax evergreen sclerophyllous or lauriphyllous vegetation of Mediterranean or warm-temperate humid affinities with a more or less dense, broken, or low arborescent cover and with a usually thick, high evergreen shrub stratum. This category in Armenia is only represented by the subcategory Juniper matorral, in which the dominants of individual ecosystems are *Juniperus excelsa* (=*J. polycarpos*) and *Juniperus foetidissima.*

2.6.3.2 F5.3. Pseudomaquis

This category includes mixed sclerophyllous evergreen and deciduous shrub thickets of the periphery of the range of Mediterranean sclerophyllous scrublands. In this category, we accept one subcategory, the Western Asian pseudomaquis. The habitats of Armenia, which we refer to this category, on the one hand, are physiognomically similar to the habitats indicated in its description; on the other hand, they differ greatly in their floristic composition. Here, we include arid open forests, widespread in Armenia, where edificators are most often *Pistacia mutica, Amygdalus fenzliana, Celtis glabrata, C. caucasica, Cerasus mahaleb, Pyrus* spp., *Punica granatum, Rhamnus pallasii, etc.*

2.6.4 F6. Garrigue

This category includes evergreen sclerophyllous or lauriphyllous shrub vegetation, with an open canopy structure and some bare ground, usually with many winter annuals and vernal geophytes. The typical garrigue, as a derivative of evergreen broad-leaved forests, is absent in Armenia; however, we refer some habitats to this category.

2.6.4.1 F6.8. Xero-halophile Scrubs

This category includes salt-tolerant shrub formations of dry ground in low-precipitation areas. In Armenia, these ecosystems include solonchaks with shrub vegetation. Ecosystems are very rare now because of irrigation and desalinization of soils in previous decades. They occupy small areas. *Halocnemum strobilaceum, Halostachys belangeriana, Bienertia cycloptera, Suaeda microphylla, Halimione verrucifera,* and *Tamarix octandra* are dominants in the communities (Fig. 2.10).

Fig. 2.10 Solonchaks on Ararat valley (F6.84)

2.6.5 *F7. Spiny Mediterranean Heaths: Phrygana, Hedgehog-Heaths, and Related Coastal Cliff Vegetation*

This category includes shrublands with dominant low spiny shrubs, widespread in Mediterranean and Anatolian regions with a summer-dry climate, occurring from sea level to high altitudes on dry mountains.

2.6.5.1 F7.3. East Mediterranean Phrygana

This category includes spiny shrublands, widespread at low and middle altitudes in the eastern Mediterranean and Anatolian regions. In Armenia, such habitats are very common and widespread, but they differ significantly from the typical habitats of the Eastern Mediterranean, so we call them Armenian phryganoids. Ecosystems in this category are dominated by *Acantholimon* species (*A. karelinii, A. vedicum, A. fedorovii, A. festucaceum, A. manakyanii*) and *Camphorosma lessingii.*

2.6.5.2 F7.4. Hedgehog-Heaths

This category includes primary cushion heaths of the high, dry mountains of the Mediterranean region and Anatolia, with low, cushion-forming, often spiny shrubs, in particular of genera *Acantholimon*, *Astragalus*, *Erinacea*, *Vella*, *Bupleurum*, *Ptilotrichum*, *Genista*, *Echinospartum*, and *Anthyllis* and various composites and labiates; secondary are zoogenic cushion heaths of the same regions, either downslope extensions of the high-altitude formations, and dominated by the same species. These habitats are very characteristic in Armenia, distributed from the middle mountainous to subalpine belt, and are found in all floristic regions; in the classical geobotanical literature of Armenia, they are known as tragacanths. In Armenia, one subcategory Central Anatolian hedgehog-heaths is presented, within which ecosystems are distinguished by the main dominants of the shrub layer: *Astragalus microcephalus, A. lagurus, A. aureus, A. uraniolimneus, A. sangezuricus, Onobrychis cornuta, Gypsophila aretioides, Acantholimon bracteatum, A, festucaceum, A. takhtadjanii, Atraphaxis spinosa, Gundelia aragatsi,* and *G. armeniaca.*

2.6.6 *F9. Riverine and Fen Scrubs*

This category includes riversides, lakesides, fens, and marshy floodplains dominated by woody vegetation less than 5 m high.

2.6.6.1 F9.1. Riverine Scrub

This category includes scrub of broad-leaved willows, e.g., *Salix caprea, S. pentandroides, S. triandra*, and *S. wilhelmsiana* (less than 5 m tall) beside rivers. In Armenia this includes riverside scrub of *Hippophae rhamnoides* and *Myricaria germanica as well.*

2.6.6.2 F9.2. Willow Carr and Fen Scrub

This category includes low woods and scrubs colonizing fens, marshy floodplains, and fringes of lakes and ponds, dominated by large- or medium-sized shrubby willows, generally *Salix cinerea, S. pentandroides,* and *S. caucasica*, alone or in association with *Rhamnus cathartica* or *Betula pubescens*, any of which may dominate the upper canopy.

2.6.6.3 F9.3. Southern Riparian Galleries and Thickets

This category includes tamarisk, oleander, chaste tree galleries and thickets, and similar low woody vegetation of permanent or temporary streams and wetlands of the thermo-Mediterranean zone. In Armenia, these ecosystems are found in the lower mountain belt, mainly on the Ararat plain, in the Darelegis, Meghri, and Idjevan floristic regions, with only *Tamarix* species participating in them. Ecosystems are rather common in Armenia but occupy small areas. Usually, they are developed near water streams drying in the summer season. *Tamarix ramosissima, T. smyrnensis,* and *T. octandra* are the most common dominants in these communities.

FA. Hedgerows

This category includes woody vegetation forming strips within a matrix of grassy or cultivated land or along roads, typically used for controlling livestock, marking boundaries, or providing shelter.

FB. Shrub Plantations

This category includes plantations of dwarf trees, shrubs, espaliers, or perennial woody climbers, mostly cultivated for fruit or flower production, either intended for permanent cover of woody plants when mature or else for wood or small tree production with a regular whole-plant harvesting regime. This subcategory includes shrub plantations for ornamental purposes or for fruit and vineyards.

2.7 G. Woodland, Forest, and Other Wooded Land

Woodland and recently cleared or burnt land where the dominant vegetation is, or was until very recently, trees with a canopy cover of at least 10%. Trees are defined as woody plants, typically single-stemmed, that can reach a height of 5 m at maturity unless stunted by poor climate or soil. This includes lines of trees, coppices, regularly tilled tree nurseries, and tree-crop plantations. This includes *Alnus* and *Populus* swamp woodland and riverine *Salix* woodland and excludes *Corylus avellana* scrub and *Salix* and *Frangula* carrs and stands of climatically limited dwarf trees (krummholz) < 3 m high, such as that occurring at the arctic or alpine tree limit.

2.7.1 G1. Broad-Leaved Deciduous Woodland

This category includes woodland, forest, and plantations dominated by summergreen non-coniferous trees that lose their leaves in the winter. This includes woodland with mixed evergreen and deciduous broad-leaved trees, provided that the deciduous cover exceeds that of evergreens.

2.7.1.1 G1.1. Riparian and Gallery Woodland, with Dominant Alder, Birch, Poplar, or Willow

This category includes riparian woods of the boreal, boreo-nemoral, nemoral, and sub-Mediterranean and steppe zones, with one or few dominant species, typically *Alnus*, *Betula*, *Populus,* or *Salix*. This includes woods dominated by narrow-leaved willows *Salix alba* and *Salix excelsa* in all zones including the Mediterranean. This category is represented by subcategory Armenian willow galleries; these ecosystems are common in Armenia, and they are represented near all big rivers in lower and middle mountain belts. *Salix alba, S. excelsa, S. caprea, S. armeno-rossica, S. triandra,* and *S. pseudomedemii* usually are dominants in these communities.

2.7.1.2 G1.2. Mixed Riparian Floodplain and Gallery Woodland

This category includes mixed riparian forests, sometimes structurally complex and species-rich, of floodplains and of galleries beside slow- and fast-flowing rivers of the nemoral, boreo-nemoral, steppe, and sub-Mediterranean zones. This includes gallery woods with *Acer*, *Fraxinus*, *Prunus,* or *Ulmus* and floodplain woodland characterized by mixtures of *Alnus*, *Fraxinus*, *Populus*, *Quercus*, *Ulmus*, and *Salix*. This subcategory represents the ecosystem mixed oak-elm-ash woodland of great rivers. Diverse riverine stands in areas flooded only during major floods. Not very common habitats are relatively well represented in the Aparan, Idjevan, and South Zangezur floristic regions along the banks of the Marmarik, Debed, Aghstev, and

Voghji rivers. In forest stands, *Quercus iberica, Acer campestre, Fraxinus excelsior,* and *F. oxycarpa* are represented with approximately equal abundance; *Ulmus carpinifolia* is less common; and *Salix caprea* is usually abundant closer to the river banks. In addition, we singled out the ecosystem as a separate subcategory Riparian woodland with invasive species, which is rather rare in Armenia. In some places in Ararat valley, there are communities with *Acer negundo* or *Gleditsia triacanthos* dominance.

2.7.1.3 G1.3. Mediterranean Riparian Woodland

This category includes alluvial forests and gallery woods of the Mediterranean region. Dominance may be of a single species, or few species or mixed with many species including *Fraxinus*, *Liquidambar*, *Platanus*, *Populus*, *Salix*, *and Ulmus*. This category includes one subcategory Irano-Anatolian mixed riverine forests (G1.37) with two ecosystems rare in Armenia, but very important in terms of biodiversity representation. Plane grove in Tsav river valley is the only ecosystem in the Caucasus where *Platanus orientalis* dominates in the woodland. It exists in the Tsav river valley in South Zangezur floristic region on altitude 650–750 m a.s.l. *Juglans regia, Celtis caucasica, Ficus carica, Rubus armeniacus, Punica granatum, Malus orientalis, Crataegus stevenii, C. pentagyna, Teucrium hircanicum, Euonymus velutina, Swida iberica, and Ranunculus cicutarius* grow here. And the second, also very rare ecosystem in Armenia Riverine forests with *Populus euphratica* dominance, it is more or less well represented in Megri floristic region. One small habitat is known in Yerevan floristic region in small gorge on Urts range (Fig. 2.11).

Fig. 2.11 Plane grove in Tsav river valley (G1.371)

2.7.1.4 G1.6. Beech Woodland

This category includes forests dominated by beech *Fagus sylvatica* in Western and Central Europe, and *Fagus orientalis* and other *Fagus* species in southeastern Europe and the Pontic region. The category is represented by a single subcategory Caucasian beech forests, in which numerous diverse ecosystems are grouped by habitat types (beech forests on dry sites, beech forests on mesic sites, beech forests on wet sites) (Fig. 2.12).

2.7.1.5 G1.7. Thermophilous Deciduous Woodland

This category of habitats is completely uncharacteristic for Armenia. We refer to it only intensively expanding habitats dominated by an invasive species – *Ailantus altissima* stands. *Ailanthus* was introduced into Armenia for use in urban greening, escaped from the culture, and forms monodominant communities in different conditions in lower and middle mountain belts. Ecosystems are very good represented in Lori, Idjevan, Yerevan, Darelegis, North and South Zangezur, and Megri floristic regions.

Fig. 2.12 Beech forests without forbs or grasses in soil cover (G1.6H25)

2.7.1.6 G1.9. Non-riverine Woodland with Birch, Aspen, or Rowan

Forests or woods, dominated by *Betula*, *Populus tremula,* or *Sorbus aucuparia,* are well represented in North Armenia. Three subcategories are represented here: Ponto-Caspian birch woods, whose ecosystems are represented along the upper forest line of Northern Armenia and are dominated by *Betula litwinowii* and *B. pendula*; Aspen groves of North Armenia are best represented in the Upper Akhuryan and Lori floristic regions; Rowan woodland is also commonly found along the upper forest line.

2.7.1.7 G1.A. Meso- and Eutrophic Oak, Hornbeam, Ash, Sycamore, Lime, Elm, and Related Woodland

This category includes woods, typically with mixed canopy composition, on rich and moderately rich soils. This includes woods dominated by *Acer*, *Carpinus*, *Fraxinus*, *Quercus*, *Tilia*, and *Ulmus*. This category includes the most important forest ecosystems of Armenia, in particular Armenian oak forests, ecosystems in which are grouped, first of all, by the dominant species of oak – *Quercus iberica, Q. macranthera,* or *Q. araxina*. This category also includes as subcategories hornbeam woodland (*Carpinus betulus*), lime woodland (*Tilia caucasica, T. cordata*), and Caucasian oak-hornbeam forests (Fig. 2.13).

Fig. 2.13 Oak forest with *Quercus macranthera* (G1.A1D2)

2.7.1.8 G1.C. Highly Artificial Broad-Leaved Deciduous Forestry Plantations

This category includes cultivated deciduous broad-leaved tree formations planted for the production of wood, composed of exotic species, of native species out of their natural range, or of native species planted in clearly unnatural stands, often as monocultures. In Armenia, such plantations are rare, some years ago, fast-growing poplars have been planted, and recently, work has begun on the establishment of *Paulownia* plantations.

2.7.1.9 G1.D. Fruit and Nut Tree Orchards

This category includes stands of trees cultivated for fruit or flower production, providing permanent tree cover once mature. These are very common ecosystems in Armenia, including walnut and orchards with tall trees.

2.7.2 *G2. Broad-Leaved Evergreen Woodland*

There are no natural stands of this category of habitats in Armenia. In recent years, attempts have been made to create olive plantations.

2.7.3 *G3. Coniferous Woodland*

This category includes woodland, forest, and plantations dominated by coniferous trees, mainly evergreen (*Abies*, *Cedrus*, *Picea*, *Pinus*, *Taxus*, *Cupressaceae*) but also deciduous *Larix*.

2.7.3.1 G3.4. Scots Pine Woodland South of the Taiga

This category includes forests of *Pinus sylvestris* ssp. *sylvestris* and *Pinus sylvestris* ssp. *hamata* of the nemoral and Mediterranean zones and of their transitions to the steppe zone. Included are, in particular, the forests of Scotland; of the alpine system, of the Mediterranean peninsulas; of the lowlands of Central Europe; of the East European Nemoral zone and its adjacent wooded steppes, formed by *Pinus sylvestris* ssp. *Sylvestris*; as well as those of Anatolia, of the Caucasus, and of Crimea, formed by *Pinus sylvestris* ssp. *hamata*. In Armenia, the only wild pine species is identified as *Pinus kochiana*, belonging to the *Pinus sylvestris* group, and therefore, the subcategory is represented only by the ecosystem Ponto-Caucasian Scots pine forests.

2.7.3.2 G3.9. Coniferous Woodland Dominated by *Cupressaceae* or *Taxaceae*

This category includes woods dominated by *Cupressus sempervirens*, *Juniperus* spp., or *Taxus baccata* of the nemoral and Mediterranean mountains and hills. In Armenia, only ecosystems dominated by *Juniperus* and *Taxus baccata* species are represented. Basically, we have ecosystems dominated by *Juniperus excelsa* (= *J. polycarpos*) and *Juniperus foetidissima*, as well as a rather rare ecosystem in Armenia Armenian yew groves. There are known three yew groves in Idjevan and one in South Zangezur floristic regions. In North Armenia, yew groves usually are placed on 2nd–3rd terraces along mountain rivers or on the slopes among beech forests, in South Armenia – on slopes among oak forest (Fig. 2.14).

2.7.3.3 G3.F. Highly Artificial Coniferous Plantations

This category includes plantations of exotic conifers or of European conifers out of their natural range or of native species planted in clearly unnatural stands, typically as monocultures in situations where other species would naturally dominate. In Armenia, they are widely distributed, since pine species (both local and introduced) have been widely used and are used in afforestation and the creation of protective forest belts and stripes.

Fig. 2.14 Anatolian Grecian (*Juniperus excelsa*) juniper woods (G3.935)

2.7.4 G4. Mixed Deciduous and Coniferous Woodland

This category includes forest and woodland of mixed broad-leaved deciduous or evergreen and coniferous trees of the nemoral, boreal, warm-temperate humid, and Mediterranean zones. They are mostly characteristic of the boreonemoral transition zone between taiga and temperate lowland deciduous forests and of the montane level of the major mountain ranges to the south. Neither coniferous nor broad-leaved species account for more than 75% of the crown cover.

2.7.4.1 G4.8. Mixed Non-riverine Deciduous and Coniferous Woodland

This category includes mixed non-riverine woodland without a significant *Pinus* component, comprising elements of *Fagus*, *Betula*, *Populus tremula,* or *Sorbus aucuparia.* In North Armenia, pine is quite often found in beech and oak forests, but very rarely its abundance increases to a significant one.

2.7.4.2 G4.9. Mixed Deciduous Woodland with *Cupressaceae* or *Taxaceae*

This category includes mixed non-riverine woodland without a significant *Pinus* component, comprising elements of meso- and eutrophic *Quercus*, *Carpinus*, *Fraxinus*, *Acer*, *Tilia*, *Ulmus,* and related woodland together with *Cupressaceae* or *Taxaceae* woodland. The category is represented by two subcategories. Mixed beech-yew forests are pretty usual ecosystems in North Armenia. Unlike yew groves, located mainly on river terraces, these habitats are confined to sloping beech forests and occupy small areas. Mixed open forests with *Juniperus* spp. and deciduous trees and shrubs are ecosystems which are rather common in Armenia; *Juniperus excelsa* is dominant in communities; and species from shibliak and arid deciduous open forests are included in the composition.

2.7.4.3 G4.F. Mixed Forestry Plantations

This category includes mixed plantations of coniferous and deciduous species where at least one constituent is exotic or outside its natural range, or if composed of native species, then planted in clearly unnatural stands. These are rare ecosystems in the conditions of Armenia, since almost everywhere created forest plantations were made in the form of monocultures. They are best represented in the basin of Lake Sevan, where during Soviet time, during the afforestation of soils freed from the waters of the lake, such plantations were often created. Currently, projects are being implemented to transform forest monocultures into more sustainable polydominant communities.

2.7.5 *G5. Lines of Trees, Small Anthropogenic Woodlands, Recently Felled Woodland, Early-Stage Woodland and Coppice*

This category includes stands of trees greater than 5 m in height or with the potential to achieve this height, either in more or less continuous narrow strips or in small (less than about 0.5 ha) plantations or small (less than about 0.5 ha) intensively managed woods. Woodland and coppice that is temporarily in a successional or non-woodland stage can be expected to develop into woodland in the future.

2.7.5.1 G5.1. Lines of Trees

This category includes more or less continuous lines of trees forming strips within a matrix of grassy or cultivated land or along roads, typically used for shelter or shading. These ecosystems are common in Armenia, created along most roads and railways as snow and wind protection plantations, as well as shelterbelts.

2.7.5.2 G5.2. Small Broad-Leaved Deciduous Anthropogenic Woodlands

This category includes plantations and small intensively managed woods of deciduous broad-leaved trees less than about 0.5 ha in area. These habitats are not often found in Armenia, since after planting for the purpose of afforestation or reforestation, maintenance is usually carried out only in the first years. In reality, these habitats exist in forest areas, where, after reforestation in small areas, care is taken by forestry employees.

2.7.5.3 G5.4. Small Coniferous Anthropogenic Woodlands

This category includes plantations and small intensively managed woods of coniferous trees less than about 0.5 ha in area. Like the previous subcategory, these ecosystems are not often found in Armenia, since after planting for the purpose of afforestation or reforestation, maintenance is usually carried out only in the first years.

2.7.5.4 G5.5. Small Mixed Broad-Leaved and Coniferous Anthropogenic Woodlands

This category includes plantations and small intensively managed woods less than about 0.5 ha in area, with mixed coniferous and broad-leaved trees. The proportion of conifers is in the range of 25–75%. Like the previous subcategory, ecosystems that are not often found in Armenia, since after planting for the purpose of afforestation or reforestation, maintenance is usually carried out only in the first years.

2.7.5.5 G5.6. Early-Stage Natural and Seminatural Woodlands and Regrowth

This category includes early stages of woodland regrowth or newly colonizing woodland composed predominantly of young individuals of high-forest species that are still less than 5 m in height. This includes young native woodland replanted with indigenous trees and naturally colonizing stands of non-native trees. Currently, in Armenia, these habitats include plantings in areas of afforestation of non-forest areas, carried out both within the framework of the state program, and in the course of implementation of international projects, as well as reforestation sites in areas with cut forests. Afforestation is usually carried out with native plant species.

2.7.5.6 G5.7. Coppice and early-stage plantations

This category includes woodland treated as coppice without standards. These are plantations with a dominant canopy of young trees that are still less than 5 m in height and plantations of dwarf trees or shrubs cultivated for wood or small-tree production, with a regular whole-plant harvesting regime, including short-rotation *Salix* beds for biomass production, Christmas tree crops, and tree nurseries.

2.7.5.7 G5.8. Recently Felled Areas

This category includes land that recently has supported deciduous or coniferous woodland after the trees have been clear-felled or burnt. This includes woodland with successional vegetation dominated by tall herbs, grasses, or shrubs, provided that these will soon be overtopped by a tree canopy. Currently, these habitats in Armenia do not occupy large areas. On the one hand, there have been no official passing cuttings in Armenia in recent decades; on the other hand, unsystematic cuttings carried out in the 1990s in the vicinity of settlements, which by their nature are close to passing cuttings, have now led to the formation of coppice stands almost everywhere, belonging to other categories of habitats. Forest fires are a fairly common occurrence in Armenia, and due to predicted climate change, their frequency may increase. In this regard, measures are being taken in Armenia to prevent these phenomena.

2.8 H. Inland Unvegetated and Sparsely Vegetated Habitats

This category includes non-coastal habitats with less than 30% vegetation cover (other than where the vegetation is chasmophytic or on scree and/or cliff) which are dry or only seasonally wet (with the water table at or aboveground level for less than half of the year). This includes subterranean nonmarine caves and passages including underground waters.

2.8.1 H1. Terrestrial Underground Caves, Cave Systems, Passages, and Waterbodies

This category includes natural caves, cave systems, underground waters, and subterranean interstitial spaces with associated communities of animals, fungi, and algae. Armenia is a mountainous country; therefore, these habitats are well represented, although in most cases, the caves are not deep, especially on volcanic massifs. On the other hand, there are caves with an extensive network of underground galleries and halls, about 1 km long, which are located mainly in karst rocks and are best represented in Vayots Dzor region. This category is subdivided into a number of subcategories with their characteristic biodiversity: cave entrances, cave interiors, dark underground passages, underground standing waterbodies, and underground running waterbodies. Disused underground and tunnels are included in this category as separate subcategory.

2.8.2 H2. Screes

This category includes accumulations of boulders, stones, rock fragments, pebbles, gravels, or finer material, of non-aeolian depositional origin, unvegetated, occupied by lichens or mosses, or colonized by sparse herbs or shrubs. Included are screes and scree slopes produced by slope processes, moraines, and drumlins originating from glacial deposition, sandar, eskers, and kames resulting from fluvio-glacial deposition, block slopes, block streams, and block fields constructed by periglacial depositional processes of downslope mass movement and ancient beach deposits constituted by former coastal constructional processes. High mountain, boreal, and Mediterranean unstable screes are colonized by highly specialized plant communities. They or their constituting species may also inhabit moraines and other depositional debris accumulations in the same areas. A very few communities form in lowland areas elsewhere.

2.8.2.1 H2.3. Temperate-Montane Acid Siliceous Screes

This category includes siliceous screes of high altitudes and cool sites in mountain ranges of the nemoral zone, including the alps, pyrenees, and Caucasus. This category includes habitats of volcanic mountain ranges and massifs from the middle mountain to the alpine belt. Here, we distinguish subcategories according to the altitudinal location of screes: acid siliceous screes of middle mountain belt of Armenia and subalpine and alpine screes on volcanic substrates and at the same time, mobile screes in alpine and subalpine mountain belts and fixed screes in alpine and subalpine belts with their characteristic floristic composition differ (Fig. 2.15).

2.8.2.2 H2.4. Temperate-Montane Calcareous and Ultrabasic Screes

This category includes calcareous and calcschist screes of high altitudes and cool sites in mountain ranges of the nemoral zone, including the alps, pyrenees, and Caucasus. In Armenia, they are characteristic of the central and western regions of the republic (Shirak, Yerevan, Darelegis floristic regions). The species composition is very rich and diverse; it includes *Helichrysum graveolens, Minuartia sclerantha, Allium struzlianum, Tulipa biflora, T. julia,* and many others.

Fig. 2.15 *Coluteocarpus vesicaria* on mobile screes in subalpine mountain belt (H2.351)

2.8.2.3 H2.5. Acid Siliceous Screes of Warm Exposures

Siliceous screes of warm exposures in mountain ranges of the nemoral zone, including the alps, pyrenees, and Caucasus, and of Mediterranean mountains, hills, and lowlands. Ecosystems are rather rare in Armenia, because in the lower mountain belt, steep slopes are almost absent, and screes are going to be overgrown. Habitats are presented in Yerevan, Darelegis, and Megri floristic regions. *Amygdalus fenzliana, Cerasus incana, Atraphaxis spinosa, Rhamnus pallasii, Ephedra procera, Poa bulbosa, Bromus fibrosus, and Nepeta mussinii* are common species in these ecosystems.

2.8.2.4 H2.6. Calcareous and Ultrabasic Screes of Warm Exposures

This category includes calcareous and calcschist screes of warm exposures in mountain ranges of the nemoral zone, including the alps, pyrenees, and caucasus, and of Mediterranean mountains, hills, and lowlands. Ecosystems are not very widely distributed in Armenia. They are common on not high ridges of Ararat valley and Darelegis floristic region. *Salvia dracocephaloides, Poa bulbosa, Onosma sericea, Peganum harmala, Stachys inflata, Eremostachys laciniata, Eremopyrum orientale, Allium materculae, Scrophularia thesioides, Cleome ornithopodioides, Serratula coriacea, Michauxia laevigata, and Rumex scutatus* are common species in these ecosystems.

2.8.3 H3. Inland Cliffs, Rock Pavements, and Outcrops

This category includes unvegetated, sparsely vegetated, and bryophyte- or lichen-vegetated cliffs, rock faces, and rock pavements, not presently adjacent to the sea, and not resulting from recent volcanic activity. Habitats are very typical for such a mountainous country as Armenia. They are found everywhere, while in the forest areas, these habitats are represented mainly by individual rocks or slopes in the gorges of mountain rivers; then, in other parts of the republic, large rock massifs are common. Rocky cliffs in the gorges and canyons of most mountain rivers are very characteristic.

2.8.3.1 H3.1. Acid Siliceous Inland Cliffs

This category includes dry noncalcareous inland cliffs. Specific plant associations colonize montane and Mediterranean cliffs. Most of the subdivisions refer to them. The habitats are well represented in Armenia; they are both rocky massifs and steep banks of canyons and gorges of mountain rivers in the zone of volcanic highlands and ridges. High-altitude and arctic siliceous cliffs stand out in this category, including the alpine rocks of the volcanic uplands of Armenia and the Araks mountain ranges. In the subcategory mountain siliceous cliffs, we single out rock ecosystems in forested and non-forested regions of Armenia. This category also includes disused siliceous quarries as a subcategory.

2.8.3.2 H3.2. Basic and Ultrabasic Inland Cliffs

This category includes dry, calcareous inland cliffs. Specific plant associations colonize montane and Mediterranean cliffs. Most of the subdivisions refer to them. These habitats are common in Armenia, confined to nonvolcanic mountainous areas. Here, in the subcategory bare limestone inland cliffs, we single out the high-altitude and arctic limestone cliffs and mountain limestone cliff ecosystems. This category also includes disused chalk and limestone quarries. In addition, the same category includes bare inland basaltic and ultrabasic cliffs, which are well represented in Armenia, especially in the canyons of mountain rivers (Fig. 2.16).

2.8.3.3 H3.4. Wet Inland Cliffs

This category includes very wet, dripping, overhanging, or vertical rocks of hills, mountains, and Mediterranean lowlands. Habitats are characteristic of Armenia, but occupying very small areas. The vegetation is represented mainly by moss species; there are also wet rocks on which *Adiantum capillus-veneris and Lycopodium selago* are growing.

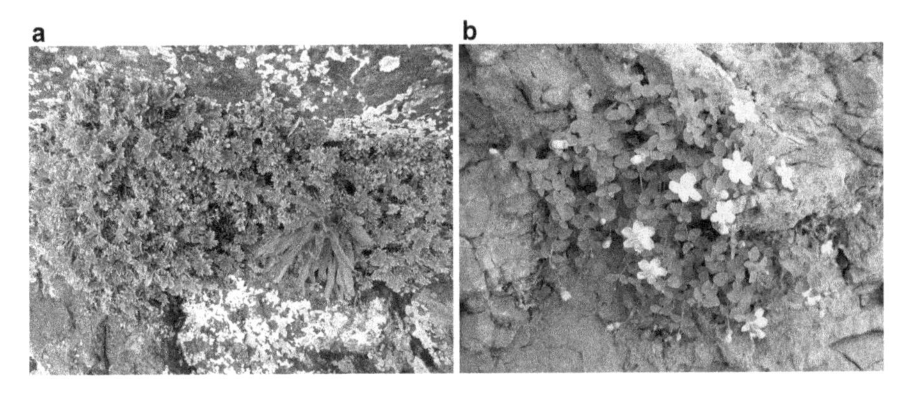

Fig. 2.16 (**a**) *Potentilla porphyrantha* in alpine cliffs on volcanic plateaus of Central Armenia (H3.1B12). (**b**) *Hypericum formosissimum* in mountain limestone cliffs (H3.2E2)

2.8.3.4 H3.5. Almost Bare Rock Pavements Including Limestone Pavements

This category includes more or less level surfaces of rock exposed by glacial erosion, by weathering processes, or by aeolian scouring, bare or colonized by mosses, algae, or lichens. The hard rock surface may be exposed or partially covered by erosional rock debris, in particular, those produced by frost weathering, heaving, thrusting, or cracking. Included are rock surfaces in karst landscapes, rock dome tops, whaleback, roche moutonné, flyggberg, and rock basin formations of periglacial areas, golec and felsenmeer formations, level surfaces of dykes, and old lava flows. Vascular plant communities may colonize cracks and weathered surfaces. In Armenia, these habitats are very rare; they are represented almost exclusively in the southern part of the Darelegis floristic region, where karst rocks are represented. They occupy small areas, since in most cases, they are largely eroded and overgrown with specific vegetation.

2.8.3.5 H3.6. Weathered Rock and Outcrop Habitats

This category includes rocks and outcrops colonized by pioneer communities, especially of *Crassulaceae*. The substrates are mostly siliceous, occurring in the alpine or montane levels of higher mountains of the nemoral zone. These habitats are closely related to the previous category; on separate weathered stone pillars, the first stage of succession appears, and overgrowing begins with the introduction of species *Sedum, Sempervivum, Parietaria* и др.

2.8.4 H4. Snow or Ice-Dominated Habitats

This category includes high mountain zones and high-latitude land masses occupied by glaciers or by perennial snow. Of all the diversity of habitats of this category, only small snow packs are found in Armenia, confined to the highest mountain ranges and ridges, most often located on the slopes of northern exposures. In Armenia, these are almost permanent snow patches, often persisting for several years, sometimes completely disappearing in the hottest years. They are typical only for the highest mountain ranges – Aragats, Geghama, Vardenis, Zangezur ridges, Ishkhanasar, Khustup Mountains, and some others.

2.8.5 H5. Miscellaneous Inland Habitats with Very Sparse or No Vegetation

This category includes miscellaneous bare habitats, including glacial moraines, freeze-thaw features, inland sand dunes, burnt ground, and trampled areas.

2.8.5.1 H5.3. Sparsely or Unvegetated Habitats on Mineral Substrates Not Resulting from Recent Ice Activity

This category includes accumulations of sand, boulders, stones, rock fragments, pebbles, or gravels, which are unvegetated, occupied by lichens or mosses, or colonized by sparse herbs or shrubs. Included are inland dunes, moraines, and drumlins originating from glacial deposition, sandar, eskers, and kames resulting from fluvio-glacial deposition, block slopes, block streams, and block fields constructed by peri-glacial depositional processes of downslope mass movement and ancient beach deposits constituted by former coastal constructional processes. In this category, we include the subcategories clay and silt with very sparse or no vegetation and stable sand with very sparse or no vegetation, which includes sand desert – the only ecosystem in Armenia with desert vegetation: *Calligonum polygonoides* is dominant here, and *Achillea tenuifolia, Aristida plumosa, Astragalus paradoxus, Allium materculae, Ceratocarpus arenarius, Euphorbia marschalliana, Oligochaeta divaricata, Verbascum suvorovianum, Koelpinia linearis, and Bellevalia albana* are common species – and sandy habitats including volcanic shlaks and sands, sandy patches in Ararat valley, and abandoned sandy quarries. This category also includes the subcategory inland non-lacustrine dunes, represented in Armenia by sandy hills and foothills in Ararat valley and artificial sandy hills. This category also includes the ecosystem Gypsaceous vegetation – Hammada, which is well represented in Yerevan floristic region, especially on Erah range. *Acantholimon hohenackeri, A. karelinii, Acanthophyllum squarrosum, Acantholepis orientalis, Salsola nodulosa, S. cana, Anisantha tectorum, Teucrium polium, Stachys inflata, Astragalus holophyllus, Convolvulus commutatus, Iris elegantissima, and Biebersteinia multifida* are common species here. In addition, we include in this category Armenian boulder fields – “chingils” – very common ecosystems are well represented on volcanic massives of Central and South Armenia. *Doronicum oblongifolium, Hieracium cymosum, Tanacetum chiliophyllum, Solenanthus stamineus, Minuartia dianthifolia, Delphinium foetidum, and Alchemilla sericata* are common species in these habitats.

2.8.5.2 H5.4. Dry Organic Substrates with Very Sparse or No Vegetation

This category includes unvegetated raw humus that is not the result of burning. It is characteristic, but not occupies large areas of habitat, usually formed near farms (dung dumps) or summer camps in the mountains, where manure storage areas are also allocated. At later stages, they begin to overgrow with various nitrophilic plant species (*Urtica dioica, Symphytum asperum, Anchusa azurea,* etc.).

2.8.5.3 H5.5. Burnt Areas with Very Sparse or No Vegetation

This category includes burnt ground that has not yet developed as a cover of vascular plants. In some regions of Armenia, the local population is convinced that mountain pastures must be burned in autumn in order to increase their productivity next year. As a result, habitats belonging to this category are formed. Currently, there is propaganda against this practice; as a result, the areas of these habitats are decreasing.

2.8.5.4 H5.6. Trampled Areas

This category includes bare ground resulting from trampling by humans or by other vertebrates including birds. В категорию включаются Unsurfaced pathways (H5.61) and habitats trampled by domesticated animals (H5.62), which are common for villages vicinities, summer camps, watering places, etc.

2.9 I. Regularly or Recently Cultivated Agricultural, Horticultural, and Domestic Habitats

This category includes habitats maintained solely by frequent tilling or arising from recent abandonment of previously tilled ground such as arable land and gardens. This includes tilled ground subject to inundation and excludes shrub orchards, tree nurseries, and tree-crop plantations.

2.9.1 I1. Arable Land and Market Gardens

This category includes croplands planted for annually or regularly harvested crops other than those that carry trees or shrubs. They include fields of cereals, of sunflowers and other oil seed plants, and of beets, legumes, fodder, potatoes, and other forbs. Croplands comprise intensively cultivated fields as well as traditionally and extensively cultivated crops with little or no chemical fertilization or pesticide application. Faunal and floral quality and diversity depend on the intensity of agricultural use and on the presence of borders of natural vegetation between fields.

2.9.1.1 I1.1. Intensive Unmixed Crops

This category includes cereal and other crops grown on large, unbroken surfaces in open-field landscapes. These habitats are quite common in Armenia, usually located on mountain plateaus. The main crops grown in Armenia are wheat, barley,

Fig. 2.17 Wheat field on Ararat valley (large-scale intensive unmixed crops – I1.11)

potatoes, corn; tobacco and Jerusalem artichoke are also grown in small areas. This category is subcategorized by field size: large-scale intensive unmixed crops, >25 ha; medium-scale intensive unmixed crops, 1–25 ha; and small-scale intensive unmixed crops, <1 ha (Fig. 2.17).

2.9.1.2 I1.2. Mixed Crops of Market Gardens and Horticulture

This category includes intensive cultivation of vegetables, flowers, and small fruits, usually in alternating strips of different crops. These ecosystems are very common in Armenia, occupying large areas. This category also includes large-scale market gardens and horticulture and small-scale market gardens and horticulture, including allotments.

2.9.1.3 I1.3. Arable Land with Unmixed Crops Grown by Low-Intensity Agricultural Methods

This category includes traditionally and extensively cultivated crops, in particular, of cereals, harboring a rich and threatened flora of field weeds including *Agrostemma githago, Centaurea depressa, Adonis aestivalis, A. flammea, Consolida orientalis, Papaver fugax, etc*. These habitats are very common in Armenia, as at present, due to the high cost of fertilizers and pesticides, most crops are cultivated extensively.

2.9.1.4 I1.5. Bare Tilled, Fallow, or Recently Abandoned Arable Land

This category includes fields abandoned or left to rest and other interstitial spaces on disturbed ground. This includes set-aside or abandoned arable land with forbs planted for purposes of soil protection, stabilization, fertilization, or reclamation. Abandoned fields are colonized by numerous pioneering, introduced, or nitrophilous plants. They sometimes provide habitats that can be used by animals in open spaces.

2.9.2 I2. Cultivated Areas of Gardens and Parks

This category includes cultivated areas of small-scale and large-scale gardens, including kitchen gardens, ornamental gardens, and small parks in city squares.

2.9.2.1 I2.1. Large-Scale Ornamental Garden Areas

This category includes cultivated areas of large-scale recreational gardens. The vegetation, usually composed mainly of introduced species or cultivars, can nevertheless include many native plants and supports a varied fauna when not intensively managed. This category of habitat is typical for all large- and medium-sized cities of Armenia, where, when planning, territories for city parks were necessarily allocated. There are subcategories within a category: park flower beds, arbors, and shrubbery and botanical gardens including dendroparks.

2.9.2.2 I2.2. Small-Scale Ornamental and Domestic Garden Areas

This category includes cultivated areas of ornamental gardens and small parks beside houses or in city squares. This includes kitchen gardens in the immediate vicinity of dwelling places.

2.9.2.3 I2.3. Recently Abandoned Garden Areas

Abandoned flowerbeds and vegetable plots in gardens are rapidly colonized by abundant weeds. These habitats regularly appear in Armenia, when city parks and squares are no longer paid attention to by city services. At present, these habitats are not numerous in Yerevan and are usually found in inconspicuous places.

2.10 J. Constructed, Industrial, and Other Artificial Habitats

This category includes primarily human settlements, buildings, industrial developments, the transport network, and waste dump sites. This includes highly artificial saline and nonsaline waters with wholly constructed beds or heavily contaminated water (such as industrial lagoons and saltworks) which are virtually devoid of plant and animal life.

2.10.1 J1. Buildings of Cities, Towns, and Villages

This category includes buildings in built-up areas where buildings, roads, and other impermeable surfaces occupy at least 30% of the land. This includes agricultural building complexes where the built area exceeds 1 ha.

2.10.1.1 J1.1. Residential Buildings of City and Town Centers

This category includes buildings in urban areas where buildings, roads, and other impermeable surfaces occupy at least 80% of the land, and with continuous or nearly continuous buildings, which may be houses, flats, or buildings occupied for only part of the day.

2.10.1.2 J1.2. Residential Buildings of Villages and Urban Peripheries

This category includes residential buildings in suburbs and villages where buildings and other impermeable surfaces occupy between 30% and 80% of the land area.

2.10.1.3 J1.3. Urban and Suburban Public Buildings

This category includes buildings with public access, such as hospitals, schools, churches, cinemas, government buildings, shopping complexes, and other places of public resort. Also in this category, old town walls and stony walls and roofs of ancient monasteries and castles are included.

2.10.1.4 J1.4. Urban and Suburban Industrial and Commercial Sites Still in Active Use

This category includes buildings in sites with current industrial or commercial use. This includes office blocks, factories, industrial units, large (greater than 1 ha) greenhouse complexes, large animal-rearing batteries, and large farm units.

2.10.1.5 J1.5. Disused Constructions of Cities, Towns, and Villages

This category includes disused factories, houses, offices, factories, or other buildings. These habitats are quite common in Armenia due to the closure of many industrial enterprises, abandoned and currently unused collective farms and state farms, etc.

2.10.1.6 J1.6. Urban and Suburban Construction and Demolition Sites

This category includes non-rural sites in which buildings are being constructed or demolished. These habitats are common in Armenia; construction sites are more common, and less often are sites where buildings are destroyed for subsequent new construction.

2.10.1.7 J1.7. High-Density Temporary Residential Units

This category includes residential buildings that are not intended to be present for more than 10 years. This includes temporary buildings – such as work camps for shift workers in areas remote from cities. These habitats are not very common in Armenia, most often found near large mining enterprises under construction.

2.10.2 J2. Low-Density Buildings

This category includes buildings in rural and built-up areas where buildings, roads, and other impermeable surfaces are at a low density, typically occupying less than 30% of the ground.

2.10.2.1 J2.1. Scattered Residential Buildings

This category includes houses or flats in areas where buildings, roads, and other impermeable surfaces are at a low density. These habitats are typical for small rural settlements, as well as for the outskirts of large villages. In addition, some holiday villages can be assigned to this category.

2.10.2.2 J2.2. Rural Public Buildings

This category includes rural buildings with public access, such as government buildings, schools, shops, or places of worship. This is a common habitat found in almost all rural settlements.

2.10.2.3 J2.3. Rural Industrial and Commercial Sites Still in Active Use

This category includes rural buildings used for industry, offices, warehousing, etc.

2.10.2.4 J2.4. Agricultural Constructions

This category includes structures dispersed within the rural or natural environment established for the purpose of agricultural activities; permanent or temporary residences; small-scale commercial, artisanal, or industrial activities; recreation; research; and environmental protection. They include isolated greenhouses, animal shelters, harvest-drying structures, sheds and huts, and field and pasture enclosures.

2.10.2.5 J2.5. Constructed Boundaries

This category includes walls and fences in areas where buildings are at low density. Ecosystems, in most cases separate individual farms, or separate them from other habitats. Ecosystems very common in Armenia, typical for most rural settlements.

2.10.2.6 J2.6. Disused Rural Constructions

This category includes disused constructions. Recently, there has been an increase in the number of habitats of this category due to the outflow of the rural population. Some buildings are used only during part of the year, and some are not used at all. This category also includes non-restored historical and architectural monuments.

2.10.2.7 J2.7. Rural Construction and Demolition Sites

This category includes rural sites in which buildings are being constructed or demolished. These habitats are common in Armenia, which is typical for most rural settlements.

2.10.3 J3. Extractive Industrial Sites

This category includes sites in which minerals are extracted, which includes quarries, open-cast mines, and active underground mines.

2.10.3.1 J3.1. Active Underground Mines and Tunnels

This category includes artificial underground spaces. They may constitute important substitution habitats for cave-dwelling bats and for significant subterranean invertebrates such as crustaceans, planarians, etc. These habitats are common in Armenia, including complexes of mines and operating tunnels (metro, road, railway).

2.10.3.2 J3.2. Active Open-Cast Mineral Extraction Sites, Including Quarries

This category includes areas used for open-sky mining and quarrying activities and presently in operation. These are very common habitats in Armenia, the areas of which have been intensively expanded in recent years.

2.10.3.3 J3.3. Recently Abandoned Aboveground Spaces of Extractive Industrial Sites

This category includes disused sites that were formerly quarries or open-cast mines. These are ecosystems that are not often found in Armenia, since older abandoned quarries where weeds have begun to grow, belong to other categories.

2.10.4 J4. Transport Networks and Other Constructed Hard-Surfaced Areas

This category includes roads, car parks, railways, paved footpaths, and hard-surfaced areas of airports, water ports, and recreational areas.

2.10.4.1 J4.1. Disused Road, Rail, and Other Constructed Hard-Surfaced Areas

This category includes unused areas that may be overgrown with grassy weeds or trees. These habitats are common in Armenia, although not occupying large areas. Asphalt pavement deteriorates over time and unused roads are gradually overgrown with weeds.

2.10.4.2 J4.2. Road Networks

This category includes road surfaces and car parks, together with the immediate highly disturbed environment adjacent to roads, which may consist of roadside banks or verges. These habitats are common in Armenia, and the road network in the republic is quite well developed. Road shoulders where clearing is carried out can be up to 2 m wide in some places.

2.10.4.3 J4.3. Rail Networks

This category includes railway tracks and the immediate highly disturbed environment adjacent to railways, which may consist of banks or verges. The railway network in Armenia is not dense and does not occupy large areas. Habitats are typical only for Central and Northern Armenia.

2.10.4.4 J4.4. Airport Runways and Aprons

This category includes hard surfaces in the airports other than buildings. In Armenia, this category actually includes very few habitats – two operating airports in Yerevan and one in Gyumri; the rest with concrete runways are practically nonfunctional (the airport in Kapan is under restoration). A few more inactive airports do not have concrete strips, are not used for their intended purpose, and belong to other categories of habitats.

2.10.4.5 J4.5. Hard-Surfaced Areas of Ports

This category includes hard surfaces in ports other than buildings. There are practically no ports in Armenia; with a big stretch, several small areas on the shores of Lake Sevan can be attributed to these habitats.

2.10.4.6 J4.6. Pavements and Recreation Areas

This category includes paved areas, city squares, and hard-surfaced recreation areas where the traffic is on foot or if wheeled then does not use the hard-surfaced area as a route. These habitats are common in Armenia, which are typical for large- and medium-sized cities as well as for some resorts and holiday homes.

2.10.4.7 J4.7. Constructed Parts of Cemeteries

This category includes hard-surfaced areas within cemeteries. These are very common habitats in Armenia. Most cemeteries in urban-type settlements belong to this category of habitats.

2.10.5 J5. Highly Artificial Man-Made Waters and Associated Structures

This category includes inland artificial waterbodies with wholly constructed beds or heavily contaminated water and their associated conduits and containers.

2.10.5.1 J5.3. Highly Artificial Nonsaline Standing Waters

This category includes artificial watercourses and basins, together with their associated containers, holding freshwater with no perceptible flow. This category includes ponds and lakes with completely man-made substrate, water storage tanks, intensively managed fish ponds, and standing waterbodies of extractive industries. This category includes subcategories specific to Armenia: intensively managed fish ponds and water storage tanks and small water reservoirs.

2.10.5.2 J5.4. Highly Artificial Nonsaline Running Waters

This category includes artificial watercourses and basins, together with their associated containers, carrying freshwater with perceptible flow. This includes sewers, running discharges from extractive industrial sites, subterranean artificial watercourses, and channels with completely man-made substrate. In addition to canals and underground artificial watercourses, it includes artificial fish ponds with running waters for trout farming.

2.10.5.3 J5.5. Highly Artificial Nonsaline Fountains and Cascades

This category includes artificial watercourses and basins, together with their associated containers, with freshwater that spurts or splashes. These habitats are very common in Armenia; in most large- and medium-sized cities, there are fountains and cascades of various shapes and sizes.

2.10.6 J6. Waste Deposits

This category includes tips, landfill sites, and slurries produced as byproducts, usually unwanted, of human activity. Unfortunately, due to the lack of waste processing or waste incineration enterprises, the territories of these habitats are very large.

2.10.6.1 J6.1. Waste Resulting from Building Construction or Demolition

This category includes dumps of building waste when not forming a part of construction or demolition sites or when it is so large as to constitute a separate habitat. These habitats occupied large areas in the disaster zone after the tragic earthquake of 1988; by now, they are practically overgrown and have moved into other categories. Today, habitats of this category do not occupy large areas and are usually located near large settlements. Near small settlements, landfills are usually not separated by the nature of the garbage and are mixed.

2.10.6.2 J6.2. Household Waste and Landfill Sites

This category includes sites used for disposal of household waste, including landfill sites that may be used for several types of waste. These habitats are very characteristic of Armenia occupying large areas, especially near large settlements. At the same time, there are both specially allocated landfills equipped with equipment and numerous illegal places for dumping garbage and household waste.

2.10.6.3 J6.3. Nonagricultural Organic Waste

This category includes sewage waste and sewage slurries. These habitats are not typical for Armenia, since treatment plants are few in number, their condition leaves much to be desired, and waste occupies small areas.

2.10.6.4 J6.4. Agricultural and Horticultural Waste

This category includes dung heaps, slurry lagoons, decaying straw, and dumps of unwanted produce. These very characteristic and frequent habitats are found in almost all rural settlements and towns. This category also includes the ecosystem areas for drying manure for pressed dung, which is very typical for the agricultural treeless regions of Armenia.

2.10.6.5 J6.5. Industrial Waste

This category includes heaps, tips, and mounds formed as byproducts of industrial activities. This includes slag heaps, mine waste, dumped quarry waste, and mineral wastes resulting from chemical processes. These ecosystems are common in Armenia, although large areas are occupied only by tailings and waste from the mining and processing industry.

2.11 Some Rare Ecosystems of Armenia

Among the huge variety of ecosystems in Armenia, rare ones are also presented, occupying very small areas but extremely important from the point of view of biodiversity conservation. The importance of these ecosystems is determined primarily by the fact that they are the habitat of the rarest species of plants and animals, and their destruction or serious change can lead to the extinction of these species. Below we provide brief descriptions of some of these ecosystems. All these ecosystems will be included in the Red Book of Ecosystems of Armenia.

2.12 C. Inland Surface Waters

2.12.1 C1. Surface Standing Waters

2.12.1.1 C1.24112. Northern *Nymphaea* Beds (Fig. 2.18)

This category includes aquatic plant communities with floating leaves dominated by *Nymphaea alba*. Currently, the state of the ecosystem is good; the water lily is slowly spreading over the surface of the lakes with an increase in the density of the

Fig. 2.18 Northern *Nymphaea* beds (C1.24112)

cover. *Juncus effusus, Carex leporina,* and *C. disticha* are represented in the coastal parts of the ecosystem. Currently, the ecosystem exists in three small lakes (total area less than 5 ha) on the Lori Plain and was recently (according to the oral report of I. Gabrielyan) found in Lake Chili in the Aparan floristic region. In the lakes where this ecosystem is represented, and along their shores, a number of rare species included in the Red Book of Plants of Armenia also grow: *Ranunculus lingua, Utricularia intermedia, Salvinia natans,* and *Potentilla erecta*. A threat to the existence of the ecosystem may be a change in the water regime of the lakes of the Lori Plain, associated with forecasted climate change, the use of lake waters for economic purposes and use of lakes as a watering place for domestic animals, as well as the "privatization" and "capture" of some lakes with the construction of mansions on their shores.

2.12.2 C3. Littoral Zone of Inland Surface Waterbodies

2.12.2.1 C3.2. Water-Fringing Reedbeds and Tall Helophytes Other Than Canes

C3.21111. Freshwater *Phragmites australis* and *Thelypteris palustris* Beds (Fig. 2.19)

This category is an extremely rare ecosystem in Armenia. The first layer of vegetation is completely dominated by *Phragmites australis*, while the second layer is dominated by *Thelypteris palustris*. *Carex acuta, C. diandra, C. pseudocyperus,* and *C. rostrata* are also abundant, and *Utricularia vulgaris* grows in the water nearby. Only one habitat is known – an island on Lake Chmoy in the Darelegis floristic region near the village Martiros. *Thelypteris palustris* is included in the Red Book of Plants of Armenia; in addition, *Menyanthes trifoliata*, included in the Red Book, also grows here. An existential threat to the ecosystem may be a change in water level caused by using the lake's water for irrigation and global climate change.

Fig. 2.19 Freshwater *Phragmites australis* and *Thelypteris palustris* beds (C3.21111)

2.12.3 D5. Sedge and Reed Beds, Normally Without Free-Standing Water

2.12.3.1 D5.2. Beds of Large Sedges Normally Without Free-Standing Water

D5.21212. Slender Tufted Sedge Beds with *Iris lazica* (Fig. 2.20)

This category is a very rare ecosystem in Armenia. Here, on moist alkaline soils, *Carex acuta* dominates, and *Iris lazica* is abundant. The floristic composition of the ecosystem is poor, except for the two abovementioned species; *Carex orbicularis* ssp. *kotschyanus, Polygonum bistortum, Alisma plantago-aquatica, Ranunculus sceleratus, Alchemilla sericata,* and *Sanguisorba officinalis* are registered. The ecosystem occupies a small area (less than 1 ha) in a swampy area in a relief depression. The ecosystem is known only from the environment of Darik village in the Upper Akhuryan floristic region. A threat to the existence of the ecosystem may be a change in the water regime due to climate change and the development of the territory for pasture.

Fig. 2.20 *Iris lazica* in slender tufted sedge beds (D5.21212)

Fig. 2.21 Bladder sedge beds with high abundance of *Lychnis flos-cuculi* (D5.21421)

D5.21421. Bladder Sedge Beds with High Abundance of *Lychnis flos-cuculi* (Fig. 2.21)

This category includes a very rare ecosystem, known only in the Ijevan floristic region in the vicinity of the Lermontovo village. In addition to the dominant species of *Carex vesicaria* and *Lychnis flos-cuculi*, in the ecosystem, the rare in Armenia *Potentilla erecta* is richly represented, and the more common are *Deschampsia cespitosa, Rhinanthus minor*, and others. *Lychnis flos-cuculi* and *Potentilla erecta* are included in the Red Book of Plants of Armenia. A threat to the existence of the ecosystem may be a change in the water regime due to climate change, the development of land for infrastructure facilities (restaurants, rest houses, hotels, and guest houses), as well as intensive grazing.

2.12.4 D6. Inland Saline and Brackish Marshes and Reedbeds

2.12.4.1 D6.2. Inland Saline or Brackish Species-Poor Helophyte Beds Normally Without Free-Standing Water

D6.24. Salt Marshes with *Juncus acutus* (Fig. 2.6)

This is not widespread, but very characteristic of the eastern part of the Ararat valley ecosystem. In fact, there is one fragmented habitat. Here, with the complete dominance of the *Juncus acutus* (included in the Red Book of Plants of Armenia), which forms large tussocks, a number of rare species are found in the community: *Sonchus araraticus, Linum barsegianii, Thesium compressum, Sphaerophysa salsula, Frankenia pulverulenta, Falcaria falcarioides, Cirsium alatum, Silene eremitica,*

Merendera sobolifera, Astragalus corrugatus, Iris musulmanica, and *Microcnemum coralloides*. Threats to the ecosystem may be a change in the water regime due to climate change, a decrease in the level of groundwater, and the impact of an anthropogenic factor (development of saline lands, imperfection of the irrigation system, burning of vegetation, intensive grazing, etc.).

2.13 E. Grasslands and Lands Dominated by Forbs, Mosses, or Lichens

2.13.1 E1. Dry Grasslands

2.13.1.1 E1.2. Perennial Calcareous Grassland and Basic Steppes

E1.2E22. Steppes with Wild Wheat Species Dominance (Fig. 2.22)

This category includes unique ecosystem where the vegetation is dominated by wild wheat species – *Triticum boeoticum, T. araraticum,* and *T. urartu.* The communities dominated by *T. boeoticum* are most often represented; the other two species usually act as companion species. About 292 species of vascular plants are registered in the ecosystem, including a number of wild relatives of cultivated plants: *Aegilops*

Fig. 2.22 Steppes with wild wheat species dominance in Erebuni reserve (E1.2E22)

columnaris, A. cylindrica, A. tauschii, A. triuncialis, Hordeum bulbosum, H. geniculatum, H. murinum, and *Secale vavilovii*. The ecosystem is represented in the Yerevan and Darelegis floristic regions. In the vegetation, a number of rare species included in the Red Book of Plants of Armenia species are represented: *Szovitsia callicarpa, Cichorium glandulosum, Astragalus guttatus, Salvia spinosa, S. suffruticosa, Triticum araraticum, Triticum urartu, Amblyopyrum muticum, Gundelia armeniaca, Lactuca takhtadzhianii, Amberboa moschata, Chardinia macrocarpa, Actinolema macrolema, Phalaris paradoxa,* and *Iris elegantissima*. The ecosystem may be threatened by irrigation of neighboring orchards and fields.

2.13.1.2 E1.3. Mediterranean Xeric Grassland

E1.3361. Semidesert with *Salsola dendroides* Dominance and High Abundance of *Cistanche armena* (Fig. 2.23)

This is a very rare ecosystem; the area of one (larger) of the two known habitats is less than 1 ha. This is known only in the Yerevan floristic region in the vicinity of the Khor Virab monastery. The vegetation cover is completely dominated by *Salsola dendroides*, which is the host of the highly ornamental parasite *Cistanche armena*. In the ecosystem, rare species included in the Red Book of Plants of Armenia are

Fig. 2.23 *Cistanche armena* in semidesert with *Salsola dendroides* dominance (E1.3361)

represented: *Cistanche armena* (=*Cistanche salsa*), *Amberboa amberboi, Cousinia tenella,* and *Nonea polychroma*. A threat to the ecosystem may be a change in the water regime in connection with the irrigation of the surrounding gardens and agricultural lands.

2.13.1.3 E1.4. Mediterranean Tall Grass and Wormwood – Artemisia – Steppes

E1.4511. Wormwood Semidesert with *Iris lycotis* (Fig. 2.24)

This is a rare ecosystem, which occupies a small area (about 2 ha) in the eastern part of the Ararat valley in the vicinity of the village Tigranashen. The vegetation cover is completely dominated by *Artemisia fragrans*, and Iris lycotis (included in the Red Book of Plants of Armenia) is very abundant. The vegetation also includes *Stipa arabica, Moltkia coerulea, Taeniatherum crinitum, Kochia prostrata,* and *Koelpinia linearis*. Threats to the ecosystem are intensive grazing and possible development of the territory for agricultural crops or as a result of road expansion.

Fig. 2.24 *Iris lycotis* in Wormwood semidesert (E1.4511)

2.13.2 E2. Mesic Grasslands

2.13.2.1 E2.1. Permanent Mesotrophic Pastures and Aftermath-Grazed Meadows

E2.1611. Grass-Meadow-Steppes with High Abundance of *Acanthus dioscoridis* (Fig. 2.25)

This is a very rare ecosystem, in which only one habitat of about 2 ha is known at the foot of Mount Hatis. *Festuca valesiaca, Koeleria macrantha,* and *Dactylis glomerata* act as dominants in the ecosystem; *Hordeum bulbosum, Eremopoa persica,* and *Stipa tirsa* are quite abundant; and on relatively more rocky areas, *Acanthus dioscoridis* (the rarest in Armenia, included in the Red Book of Plants) is abundantly represented. In the ecosystem, *Rosa spinosissima* and *Cerasus incana* are found as separate bushes, and *Scutellaria orientalis, Stachys atherocalyx, Phlomis tuberosa, Cerinthe minor, Crambe orientalis, Coronilla varia, Verbascum pyramidatum,* and *Vicia grossheimii* are represented in the grass cover. Ecosystem threats can be restricted areas of occupancy and degradation of habitat caused by the expansion of arable lands, grazing, haymaking.

Fig. 2.25 *Acanthus dioscoridis* in grass-meadow-steppe (E2.1611)

Fig. 2.26 *Nectaroscordum tripedale* in Ponto-Caucasian tall-herb communities (E5.5A1)

2.13.3 E5. Woodland Fringes and Clearings and Tall Forbs Stands

2.13.3.1 E5.5. Subalpine Moist or Wet Tall-Herb and Fern Stands

E5.5A1. Ponto-Caucasian Tall-Herb Communities with *Nectaroscordum tripedale* Dominance (Fig. 2.26)

This is rare in Armenia, very decorative, and highly fragmented ecosystem. *Hordeum bulbosum* and *Prangos ferulacea* are represented as co-edifiers in the composition of the vegetation; *Papaver orientale* is often found. It occupies small areas, each plot is usually less than 1 ha. The ecosystem is represented in the Aparan, Darelegis, and Meghri floristic regions. The reference site is located in the vicinity of the Amberd fortress on Mount Aragats. The main threats are climate aridization and recreational trampling.

2.14 F. Heathland, Scrub, and Tundra

2.14.1 F2. Arctic, Alpine, and Subalpine Scrub

2.14.1.1 F2.3. Subalpine Deciduous Scrub

F2.33711. Steppe Scrub with *Asphodeline taurica* (Fig. 2.27)

This is very rare in Armenia, with mixed communities of steppes and steppe shrubs on the Shirak Range (neighborhood of the Djadjur Pass). Of the shrubs, two species of *Spiraea* (*S. crenata, S. hypericifolia*) dominate; *Agropyron pectinatum, Koeleria macrantha,* and *Festuca valesiaca* currently dominate in the steppe areas; *Stipa pulcherrima* occurs in small spots; and *Asphodeline taurica* grows with a very high

Fig. 2.27 *Asphodeline taurica* in steppe scrub (F2.33711)

abundance on the entire slope. The ecosystem contains a number of rare species included in the Red Book of Plants of Armenia: *Asphodeline taurica, Rhaponticoides tamanianae, Asperula affinis, Paracarym laxiflorum, Valeriana eriophylla, Allium oltense, Allium rupestre, Allium struzlianum,* and *Dracocephalum austriacum.* The main threats may be climate change, intensive grazing, and forest plantations.

2.14.2 F3. Temperate and Mediterranean-Montane Scrub

2.14.2.1 F3.2. Sub-Mediterranean Deciduous Thickets and Brushes

F3.24761. Shibliak – *Paliurus spina-christi* Thorn Scrub with *Iris iberica* (Fig. 2.28)

This is a very rare ecosystem (only one locality of about 3 ha is known in the vicinity of the village of Bagratashen), in which, with the dominance of *Paliurus spina-christi*, the grass cover is abundantly represented by the rarest in Armenia *Iris iberica*. It also includes *Rosa spinosissima, Bothriochloa ischaemum, Cynodon dactylon, Medicago lupulina,* and *Teucrium polium*. The main threats may be the expansion of orchards and vineyards and intensive grazing.

Fig. 2.28 *Iris iberica* in Shibliak – *Paliurus spina-christi* thorn scrub (F3.24761)

Fig. 2.29 *Paeonia tenuifolia* in Shibliak – *Paliurus spina-christi* thorn scrub (F3.24762)

F3.24762. Shibliak – *Paliurus spina-christi* Thorn Scrub with *Paeonia tenuifolia* (Fig. 2.29)

This is an extremely rare ecosystem in Armenia dominated by *Paliurus spina-christi* and quite abundantly represented by *Paeonia tenuifolia*. The composition includes *Jasminum fruticans, Rosa spinosissima, Bothriochloa ischaemum, Tulipa sosnovskiy, Cousinia takhtadjanii, Ophrys oestrifera, Aegilops cylindrica, Teucrium polium,* and *Hordeum bulbosum*. The area is occupied by several very small plots of this ecosystem in the South Zangezur floristic region in the vicinity of Kapan town in the gorge of the Khaladzh River, which is no more than 2 ha. The main threats may be the expansion of mining activities.

2.14.3 F6. Garrigue

2.14.3.1 F6.8. Xero-halophile Scrubs

F6.84. Solonchaks on Ararat valley (Fig. 2.10)

This ecosystem is represented in the Ararat and Armavir regions of Armenia, and very small areas are found in the Meghri region. At present, in connection with the intensive work on land desalinization carried out in Soviet times, this ecosystem is represented in the Ararat valley in small fragments (1–2 ha). The ecosystem includes common semidesert species like *Salsola dendroides, S. crassa, Suaeda altissima, Halanthium rarifolium, Limonium meyeri, Hibiscus trionum, Seidlitzia florida*, etc. Also, very rare species included in the Red Book of Plants of Armenia grow here: *Halocnemum strobilaceum, Halostachys belangeriana, Bienertia cycloptera, Tamarix octandra, Nitraria schoberi, Kalidium caspicum,* and *Salsola soda*. The main threats are the development of land for agricultural purposes, desalinization of solonchaks, and changes in the level of groundwater. The ecosystem is included in Resolution 4 of the Berne Convention.

2.14.4 F7. Spiny Mediterranean Heaths (Phrygana, Hedgehog-Heaths, and Related Coastal Cliff Vegetation)

2.14.4.1 F7.4. Hedgehog-Heaths

F7.4I23. Traganth Communities with *Gypsophila aretioides* Dominance (Fig. 2.30)

This category is not widespread but is very typical for the Yerevan floristic region (especially for the Urts Range) ecosystems. The large cushions of *Gypsophila aretioides* (included in the Red Data Book of Plants of Armenia) are non-thorny, but similar in character to the prickly cushions of traganth astragals and *Onobrychis cornuta*, and usually develop on very stony places. The main threats may be the development of land for agricultural purposes, the expansion of highways, fires. The ecosystem is included in Resolution 4 of the Berne Convention.

Fig. 2.30 Traganth community with *Gypsophila aretioides* dominance (F7.4I23)

2.15 G. Woodland, Forest, and Other Wooded Land

2.15.1 G1. Broad-Leaved Deciduous Woodland

2.15.1.1 G1.3. Mediterranean Riparian Woodland

G1.371. Plane Grove in Tsav River Valley (Fig. 2.11)

This is the only ecosystem in the Caucasus where *Platanus orientalis* dominates, and the community includes *Juglans regia, Celtis caucasica, Ficus carica, Rubus armeniacus, Punica granatum, Malus orientalis, Crataegus stevenii,* and *C. pentagyna.* Here, rare species included in the Red Book of Plants of Armenia are growing: *Platanus orientalis, Teucrium hircanicum, Euonymus velutina, Swida iberica, Ranunculus cicutarius,* and *Carex pendula.* A number of animal species included in the Red Book of Animals of Armenia, as well as in Resolution 6 to the Berne Convention, also live here. The main threats may be climate change and changes in the water regime due to economic activities.

2.15.1.2 G1.9. Non-riverine Woodland with Birch, Aspen, or Rowan

G1.927. Aspen Groves of North Armenia (Fig. 2.31)

This is a rare ecosystem found in the Upper Akhuryan and Lori floristic regions. On the Ashotsk Plateau, it is represented in the form of small separate groves among meadow-steppe vegetation, with an almost pure stand of *Populus tremula.* In

Fig. 2.31 Aspen grove in Armenia (G1.927)

addition, it is found on the slopes of the gorge of the Akhuryan River, and in the Lori region, it is found as separate fragments among mountain oak forests. The forest stand includes single specimens of *Quercus macranthera, Salix caprea, Viburnum lantana, Lonicera caucasica, Rubus idaeus, Prunus divaricata*; among forbs *Allium victorialis, Primula amoena, Lamium album, Saxifraga cymbalaria, Erysimum froehneri, Geum rivale,* and *Thesium ramosum*. The main threat may be the impact of the anthropogenic factor, in particular illegal logging.

2.15.2 G3. Coniferous Woodland

2.15.2.1 G3.9. Coniferous Woodland Dominated by *Cupressaceae* or *Taxaceae*

G3.97B. Armenian Yew Groves (Fig. 2.32)

This is a rare ecosystem, wherein three relatively large groves of *Taxus baccata* are known in the Idjevan floristic region and one in the South Zangezur region. In Northern Armenia, yew groves are usually located on the 2nd or 3rd terraces along mountain rivers or on slopes among beech forests, in Southern Armenia – on slopes among oak forests. Groves are located in the middle mountain belt; the vegetation includes *Fagus orientalis, Quercus iberica, Carpinus betulus, Tilia cordata, Acer*

Fig. 2.32 *Taxus baccata* in Armenian yew groves (G3.97B)

campestre, Fraxinus excelsior; and in the grass-forbs layer *Dryopteris filix-mas, Asplenium scolopendrium, Impatiens noli-tangere, Asperula odorata, Arum orientale, Lamium album,* and *Geranium robertianum* are common. The main threat may be the impact of the anthropogenic factor, in particular, illegal logging.

2.16 H. Inland Unvegetated or Sparsely Vegetated Habitats

2.16.1 H2. Screes

2.16.1.1 H2.3. Temperate-Montane Acid Siliceous Screes

H2.3511. Mobile Screes in Alpine Belt of Aragats Mountain (Fig. 2.33)

This category includes young screes usually on steep slopes with a very characteristic floristic composition including *Alopecurus tuscheticus, A. textilis, Erysimum gelidum, Alchemilla sericea, Campanula saxifraga ssp. aucheri, Allium schoenoprasum, Coluteocarpus vesicaria, Catabrosella fibrosa, Veronica orientalis, Corydalis alpestris, Catabrosella araratica, Sibbaldia procumbens*, etc. In this ecosystem, rare species included in the Red Book of Plants of Armenia are growing (*Draba araratica, Draba hispida, Didymophysa aucheri, Isatis takhtadjanii, Pseudovesicaria digitata, Dracocephalum botryoides, Delphinium foetidum*). The main threat may be climate change. The ecosystem is included in Resolution 4 of the Berne Convention.

Fig. 2.33 Screes on Aragats mountain (H2.3511)

2.16.2 H3. Inland Cliffs, Rock Pavements, and Outcrops

2.16.2.1 H3.1. Acid Siliceous Inland Cliffs

H3.1B12. Alpine Cliffs on Volcanic Plateaus of Central Armenia (Fig. 2.16)

This category includes a rare ecosystem presented on the Syunik and Geghama highlands and on the massif of Mount Aragats. The most characteristic species here are *Saxifraga exarata, Sempervivum transcaucasicum, Tanacetum parthenifolium, Murbeckiella huetii, Aetheopappus pulcherrimus, and Campanula bayerniana*, as well as those included in the Red Book of Armenian Plants *Draba araratica, Potentilla porphyrantha,* and others. The main threat may be the development of the mining industry.

H3.1B221. Ancient Volcanic Cliffs on the Mount Arteni (Fig. 2.34)

The ecosystem consists of rocks and cliffs on Mount Arteni (Shirak floristic region), which is a left over from ancient volcanic activity and not covered by later lava strata from Mount Aragats. *Asplenium septentrionale, Cystopteris fragilis, Cheilanthes persica, Ephedra procera, Scariola orientalis, Onosma tenuiflora, Arabis caucasica, Campanula crispa, Nepeta mussinii, Scutellaria orientalis,*

Fig. 2.34 *Campanula massalskyi* on ancient volcanic cliffs on the Mount Arteni (H3.1B221)

Cotoneaster armena, Parietaria elliptica, Galium incanum, etc., grow on the rocks here, as well as those included in Red Book of Plants of Armenia: *Campanula massalskyi, Hieracium pannosum,* and *Bupleurum sosnovskyi.* The main threat may be the expansion of quarries for the extraction of stone and perlite.

2.16.2.2 H3.4. Wet Inland Cliffs

H3.411. Armenian Wet Inland Cliffs with *Adiantum capillus-veneris* (Fig. 2.35)

This category includes very wet, oozing, overhanging, or vertical rocks. These habitats are characteristic of Armenia but occupy very small areas. Various types of mosses grow on these rocks, and *Adiantum capillus-veneris* grows abundantly. The ecosystem is represented by separate localities on the rocks in the canyons of the Debed, Aghstev, Arpa, and Vorotan rivers and on the Urts ridge. The main threats are recreational trampling, road expansion, and climate change, which could cause rocks to dry out.

Fig. 2.35 *Adiantum capillus-veneris* on Armenian wet inland cliffs (H3.411)

2.16.3 H5. Miscellaneous Inland Habitats with Very Sparse or No Vegetation

2.16.3.1 H5.3. Sparsely or Unvegetated Habitats on Mineral Substrates Not Resulting from Recent Ice Activity

H5.321. Sand Desert with *Calligonum polygonoides* Dominance (Fig. 2.36)

This category includes the only habitat in Armenia with a desert type of vegetation. It occupies an area of about 100 ha in the vicinity of the village of Vedi; the "Goravan Sands" state sanctuary is distinguished here. In the vegetation cover, *Calligonum polygonoides* is dominant, and characteristic species are *Achillea tenuifolia, Aristida plumosa, Allium materculae, Ceratocarpus arenarius, Euphorbia marschalliana, Oligochaeta divaricata, Verbascum suvorovianum, Koelpinia linearis, Bellevalia albana,* and some others. Rare species included in the Red Book of Plants of Armenia such as *Calligonum polygonoides, Hohenackeria excapa, Minuartia sclerantha, Astragalus paradoxus, Silene arenosa, Astragalus massalskyi, Rhinopetalum gibbosum,* and *Verbascum nudicaule* grow here. The main threat may be sand mining and grazing.

Fig. 2.36 Sand desert with *Calligonum polygonoides* dominance (H5.321)

References

Davies CE, Moss D, Hill MO (2004) EUNIS Habitat classification revised 2004. European Environment Agency, European Topic Centre on Nature Protection and Biodiversity.

Devillers P, Devillers-Terschuren J (1996) A classification of Palaearctic habitats. Council of Europe, Strasbourg: Nature and environment, No 78.

Fayvush G, Aleksanyan A (2016) Habitats of Armenia. Yerevan, Institute of Botany NAS RA

Chapter 3
Flora of Armenia

George Fayvush, Lusine Hambaryan, Iren Shahazizyan, Arsen Gasparyan, Astghik Poghosyan, Siranush Nanagulyan, Anahit Ghukasyan, Alla Aleksanyan, Jacob Koopman, and Helena Więcław

3.1 Algae

Lusine Hambaryan and Iren Shahazizyan

Algae are an integral component of aquatic and terrestrial ecosystems. They carry out the process of converting solar energy into the chemical energy of photosynthesis products, which is necessary to maintain life and the circulation of substances and energy in the biosphere of our planet, and create an energy base for the existence of organisms of other trophic levels.

Currently, the algoflora of Armenia contains 497 species of land and water algae, which are included in 131 genera. Summarizing the literary sources available to us

G. Fayvush (✉) · A. Gasparyan · A. Ghukasyan · A. Aleksanyan
Institute of Botany after A. Takhtadjan NAS RA, Yerevan, Armenia
e-mail: gfayvush@yahoo.com; gasparyan.arsen@yahoo.com; anyaghukasyan@gmail.com; alla.alexanyan@gmail.com

L. Hambaryan
Yerevan State University, Yerevan, Armenia

Scientific Center of Zoology and Ichtyology NAS RA, Yerevan, Armenia
e-mail: lus-ham@yandex.ru

I. Shahazizyan · A. Poghosyan · S. Nanagulyan
Yerevan State University, Yerevan, Armenia
e-mail: ishahazizyan@ysu.am; astchik@ysu.am; snanagulyan@ysu.am

J. Koopman
Choszczno, Poland
e-mail: jackoopman@e-cho.pl

H. Więcław
Institute of Marine and Environmental Sciences, University of Szczecin, Szczecin, Poland
e-mail: helena.wieclaw@usz.edu.pl

G. Fayvush (ed.), *Biodiversity of Armenia*,
https://doi.org/10.1007/978-3-031-34332-2_3

and the research data of recent years, it was found that diatom (Bacillariophyta) algae prevail in different ecosystems of the territory of Armenia (Table 3.1). In the second place, blue-green (Cyanophyta) algae or cyanobacteria, green (Chlorophyta), and yellow-green (Xantophyta) algae are present in significant numbers. Euglenophyta, Dinophyta, and Chrysophyta are found in a small number of species and often have low indicators.

It should be noted that compared to higher plants, the floristic composition of Armenian algae is poorly studied. The first information about algoflora of Armenia was presented at the end of the nineteenth century by Schmidlei (1897), then in the twentieth century: Vladimirova (1944), Derzhavin (1940), Stroykina (1958), and Barseghyan (1959, 1969), as well as in the works of Tambian (1962, 1965, 1967, 1977), which did not fully include the algoflora of Armenia. Targeted studies of diatom algae were carried out by Golovenkina (1967), who investigated the south-eastern part of Armenia and recorded 115 species of diatom algae.

Lake Sevan and its catchment basin make up a large part of the water balance of the Republic of Armenia. That is why most of the studies of recent years have been devoted to the algoflora of this lake. Therefore, the phytoplankton community is one of the links that quickly responds to changes in the trophic status of aquatic ecosystems, which is why studies of phytoplankton in the catchment area of the lake are highly relevant today (Asatryan et al. 2022; Hovsepyan et al. 2013; Reynolds 2006; Rodrigues et al. 2015).

Khachikyan (2013) gave an up-to-date description of the phytoplankton of the rivers of the Sevan catchment basin. In recent years, due to anthropogenic fluctuations in the lake level, an increase in pollution indicators, a decrease in fish and crayfish stocks, and an increase in concentrations of biogenic elements have caused the phenomenon of "blooming" of planktonic algae, which was observed in the eutrophic phase of the lake in the 60–70s of the twentieth century (Legovich 1979; Parparov 1979; Hovsepyan et al. 2010), and are repeated.

As already noted, Lake Sevan is considered a unique natural ecosystem, an important water body, one of the national symbols, which plays an important role in the Armenian economy. Studies of the taxonomic composition of the algoflora of the Lake Sevan basin revealed that the composition of phytoplankton is represented by 219 species of algae from 86 genera.

The processes of eutrophication occurring in Lake Sevan were caused not only by the direct anthropogenic impact on the ecosystem of the lake but also by the environmentally unjustified development of various sectors of the economy in its catchment area. According to some researchers, the main role in the process of eutrophication of the lake belongs to deterioration of the ecological state in the drainage basin (Khudoyan 1994; Legovich 1979; Hovhannisyan et al. 2010).

The composition of the algal flora of Lake Sevan and the most important rivers of Armenia is given in Table 3.1. It should be noted that diatoms predominate in the composition of the algoflora.

The most important of the 28 small and large rivers flowing into Lake Sevan are the tributaries of the Dzknaget, Gavaraget, Lichk, Argichi, Vardenis, Makenis

Table 3.1 Composition of the algal flora of Lake Sevan and the most important rivers of Armenia

Water basin	Bacillariophyta		Cyanophyta		Chlorophyta		Xanthophyta		Euglenophyta		Dinophyta		Chrysophyta		Total species
	G[a]	S[a]	G	S	G	S	G	S	G	S	G	S	G	S	
Sevan Lake	31	101	14	36	32	64	4	8	2	6	1	2	2	2	219
River Masrik, Makenis, Gavaraget	4	8	2	6	1	2	2	2	219	4	–	–	–	–	198
River Pambak, Tandzut	31	97	13	27	15	26	3	4	–	–	–	–	–	–	154
River Argichi	26	85	12	17	15	19	4	4	3	5	–	–	–	–	130
River Vorotan	26	71	9	21	15	28	2	2	2	5	–	–	–	–	127
River Voghji	25	58	12	24	23	24	2	3	3	7	–	–	–	–	116
River Hrazdan	20	53	5	7	7	13	2	2	2	4	1	1	–	–	80
River Marmarik	19	47	5	6	14	17	–	–	1	3	–	–	–	–	73
Total	35	265	33	109	41	87	16	24	3	8	1	2	2	2	497

[a] *G* number of genera, *S* number of species

(Karchaghbyur), Masrik, and the Arpa-Sevan aqueduct, which make up the main part of the water flow.

In 2008–2011, the phytoplankton community coexistence of Masrik, Makenis, Gavaraget tributaries, and Arpa-Sevan aqueduct was thoroughly and purposefully studied, and a analysis was made compared with the results of the 1990s. On the whole, the species composition of the phytoplankton in the rivers remained almost unchanged compared to the data obtained in 1990–1991. However, it should be noted that an increase in the number of certain types of algae species, in particular, blue-green ones, was observed in the composition of phytoplankton. During the investigation period of the phytoplankton community of the main tributaries of Lake Sevan, 184 species belonging to five divisions, 35 families, and 61 genera of algae were found (Khachikyan 2013). This ratio is quite common for mountain rivers (Poretsky 1953; Kharitonov 1981; Ermolaev 1981; Getsen 1985; Nikulina 2005; Remigailo 2011). According to life form, phytoplankton was dominated by benthic forms of algae. Especially in the group of diatom algae, a predominance of benthic species was observed (62%), which is characteristic of the phytoplankton of mountain rivers (Kiselev 1969; Karimova 1972; Kulumbaeva 1982; Levadnaya 1992).

Further studies continued, and in total, during our studies, 198 species and 68 genus planktonic algae were found in the composition of the phytoplankton of the surveyed rivers (Table 3.1), among which 56% are diatoms, 17% are blue-green, 22% are green, 2% are yellow-green, and 3% are euglena (Hambaryan and Mamyan 2016). The predominance of the group of diatoms is a regularity for the rheoplankton of many of the studied rivers in Armenia (Stepanyan 2009; Danielyan 2009; Mamyan 2013).

Subsequently, research was conducted on the qualitative features of epiphytic algae of the Argichi River, one of the main tributaries flowing into Lake Sevan. About 130 species and 60 genera of algae were recorded, with diatoms predominating, and green and blue-green algae also being significant.

In recent years, studies of the algal flora of other river basins of Armenia have been intensified. Mamyan (2013) studied the effect of the anthropogenic factor on the phytoplankton community of Pambak and Tandzut rivers. The River Pambak is the largest tributary of the transboundary River Debed, which plays a major role in forming the water quality. Flowing through the industrial cities of Vanadzor and Spitak, the river includes not only domestic but also manufacturing waters. Mamyan discovered 154 species and 62 genera of planktonic algae (Table 3.1). During the studied period in 2009–2012, a gradual increase in the biodiversity of phytoplankton community was observed, which is due to the expansion of the species diversity of both diatoms and blue-green, green, and yellow-green algae (Fig. 3.1).

Only one river, the River Hrazdan, takes its start from Lake Sevan. The length of this river is 141 km. The river flows in a southwesterly direction, passes through the Gegharkunik and Kotayk regions, the city of Yerevan, then through the Ararat region, and flows into the River Araks. Flowing to the southeast, flowing into the Ararat valley, the river becomes calmer and irrigates the Ararat valley. It should be noted that gross anthropogenic interference and use of Lake Sevan has led to a significant deterioration in the state of the lake and a reduction in the unique endemic flora and fauna of the ecosystem.

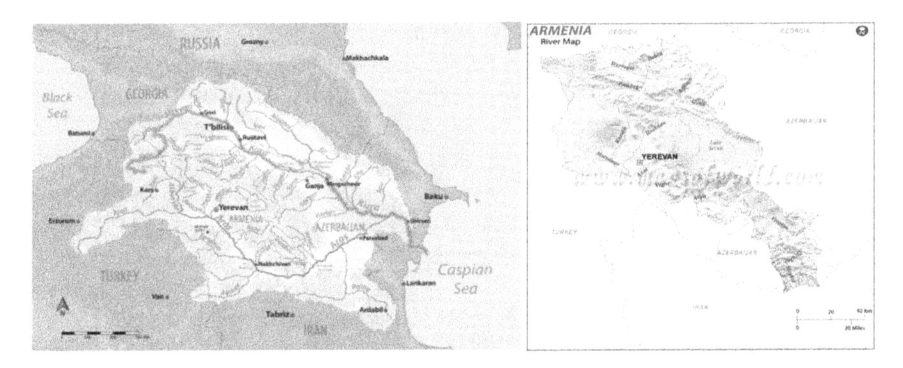

Fig. 3.1 Armenia river map

During the vegetation period of 2018, the phytoplankton community of the middle part of the River Hrazdan was studied, revealed about 80 species and 37 genera of algae belonging to six divisions were found. The phytoplankton community of the river was dominated by diatoms, and blue-green algae were subdominants (Stepanyan and Ghukasyan 2019). Eighty species belonging to six division were identified: Bacillariophyta, Chlorophyta, Cyanophyta, Euglenophyta, Dinophyta, and Xanthophyta (Table 3.1). In the studied part of the River Hrazdan, the most widely represented groups were Bacillariophyta and Cyanophyta. Species Chlorophyta, Euglenophyta, Dinophyta, and Xanthophyta were found in smaller numbers (Table 3.1).

Stepanyan (2019) studied the qualitative and quantitative indicators of the phytoplankton community of the River Marmarik, determined commonly occurring algae species, and gave the ecogeographic characteristics. The quality of the river water was assessed based on bioindicative methods were about 73 species and 39 genera of algae were recorded at the studied observation points of the River Marmarik (Table 3.1). The dominant composition of the phytoplankton of this river is represented by blue-green (Cyanophyta), diatom (Bacillariophyta), green (Chlorophyta), and euglenophyta (Euglenophyta) algae. Diatom algae dominated in terms of qualitative and quantitative values. According to species composition, green algae were subdominant, blue-green algae according to number, euglena algae according to biomass. The high values of the quantity and biomass of diatom algae in the River Marmarik were due to the chain *Fragilaria capucina,* large unicellular *Amphora ovalis, Cocconeis placentula, Pinnularia viridis, Ceratoneis arcus, Rhoicosphenia curvata*, colonial *Diatoma hiemale*, and microcellular, centric *Stephanodiscus astraea* species. Among the diatom algae, the genera *Navicula*, *Diatoma*, and *Surirella* (six species each) were found with the largest species composition. The dominant composition of blue-green algae was mainly represented by *Aphanothece clathrata, Microcystis aeruginosa,* and *M. wessenbergii. Oscillatoria limnetica* and *Phormidium foveolarum* were recorded in autumn with large values. The dominant composition of green algae was represented by *Ankistrodesmus acicularis* and *Lagerheimia genevensis*; the latter is not a species typical for the water ecosystems of Armenia and has been found in the River Marmarik for the first time (Badalyan 2006; Stepanyan and Hambaryan 2014, 2016; Khachikyan and Stepanyan 2018).

In 2018–2020, the study of phytoplankton of the River Voghji and Vorotan was carried out. As a result, 116 species and 65 genera of algae were found from the monitoring stations of the River Voghji (Table 3.1). The algoflora of the River Vorotan is represented by 127 species and 54 genera of algae.

Kalantaryan (2020) studied the prevalence and biodiversity of green unicellular microalgae and cyanobacteria in the water ecosystems of different regions in Armenia (mineral springs, fresh waters, lakes, rivers) and also assessed the biotechnological potential of the most promising strains of microalgae cultures, in particular the microalgae *Parachlorella kessleri*.

Thus, the basis of the species richness of the algal flora of Armenia is formed by green (Chlorophyta), diatoms (Bacillariophyta), and blue-green (Cyanophyta) algae, which combine more than half (about 58%) of the species composition of the algal flora of Armenia.

Currently, the study of the specifics of the typological distribution of algae in the water bodies of Armenia remains relevant. In historical perspective, one should note the unevenness of such studies and a certain dependence on the regional and practical significance of individual types of water bodies. It should be noted that during the entire study period of algae, various types of reservoirs of lotic and lentic systems and their habitats were covered – freshwater, soil, and aerophytic.

Of great interest for spatial dynamics, and as reserves for the conservation of original biodiversity, are areas with mountain landscapes, where a variety of algae also develops in the changing climate of the Earth (Barinova et al. 2011). The specificity and diversity of Armenian algae are due to a complex of hydrological, hydrochemical, and genesis factors, as well as the high mountain relief of water bodies.

Recently, in the phytoplankton of lakes in the whole world (Dokulil and Teubner 2000; Lürling et al. 2018; Molot et al. 2021; Salmaso et al. 2015; Sterner et al. 2020; Winter et al. 2011), there is a tendency to increase the number of cyanoprokaryotes, which are characterized by rapid growth leading to harmful water blooms. They reduce the ecosystem and economic opportunities of water resources as they lead to secondary pollution of water bodies and complicate the overall environmental situation both in specific water bodies and in the region as a whole. Apparently, such processes are due to a change in environmental conditions, an increase in anthropogenic influence, and an increase in the level of organic pollution (anthropogenic pressure of various recreation areas and human economic activity) of these water bodies, which will undoubtedly contribute to an increase in the degree of limnoecosystem trophicity.

One of the relatively well-studied aquatic ecosystems in Armenia is Lake Sevan. Unfortunately, despite the organization of the Sevan National Park in 1978, the ecological state of the lake and its drainage basin has not improved; on the contrary, an increase in anthropogenic pressure and pollution is observed in all zones of the protected area. Starting in 2018, an intensive flowering of toxic cyanobacteria was observed in the lake; work was carried out to identify species; and for the first time, a quantitative assessment of cyanotoxins in the littoral of Lake Sevan, in particular anatoxin A, was given (the study was carried out with the financial support of the ANSEF program http://ansef.org/molbio-5292AWARD).

Phytoplankton of Lake Sevan, as an important indicator of the state of aquatic ecosystems, has been studied from the 1930s to the present by numerous authors (Vladimirova 1947; Oganesyan 1994; Stroykina 1952; Mnatsakanyan 1984; Meshkova 1962; Legovich 1979; Nikulina and Mnatsakanyan 1984; Parparov 1979; Vardanyan 1993; Mikaelyan 1996; Khudoyan 1994). According to the available data, the main divisions of phytoplankton in Lake Sevan (Hambaryan 2001; Hambaryan et al. 2007, 2020a, b; Hovsepyan 2013, 2017; Hovsepyan et al. 2010, 2014; Hovsepyan and Khachikyan, 2016a, b), during long-term monitoring in different periods of trophicity, are: Bacillariophyta (diatoms), Cyanophyta (blue-green, cyanobacteria), Chlorophyta (green), and Euglenophyta (euglena), Xanthophyta (yellow-green) – these groups are of great importance for the formation of water quality and composition indicators characterizing changes in the lake ecosystem.

As is known, as a result of an artificial decrease in the water level, Lake Sevan, starting in the 1930s, as a result of eutrophication processes, went through the stages from an oligotrophic reservoir to a meso-eutrophic.

Representatives of the Charophyta (Characeae) and Dinophyta (Dinophyta) groups, in the period after the increase and the period of fluctuations in the level of 2015–2019, were rare and had average quantitative indicators. It has been established that in aquatic ecosystems subject to anthropogenic influence, as a rule, species diversity indicators decrease, phytoplankton biomass and quantity increase, and the proportion of diatoms decreases. Long-term studies of the dynamics of the dominant species in the phytoplankton composition of Lake Sevan revealed that in comparison with the oligotrophic period, there was an increase in the number of species at different stages of ecosystem development (Fig. 3.2).

With eutrophication, under the conditions of anthropogenic level fluctuations, an increase in the quantitative average annual indicators of the phytoplankton biomass was observed, which was accompanied by a deterioration in the lake water quality parameters (Fig. 3.3).

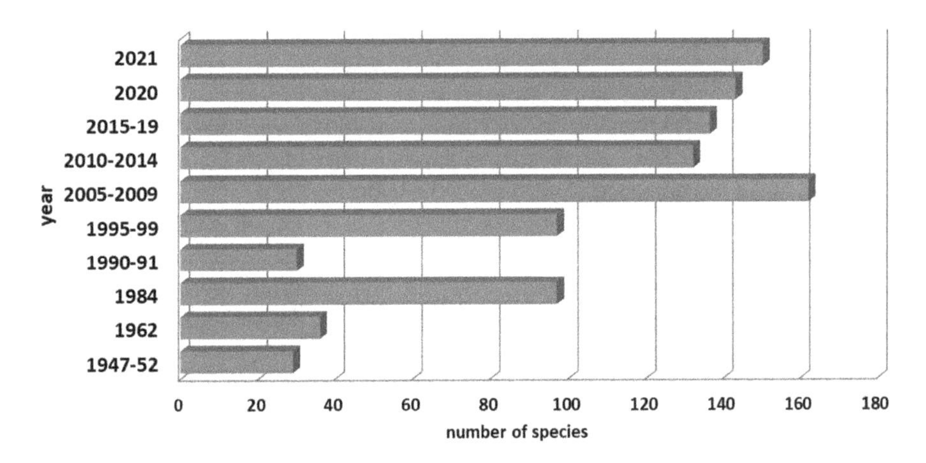

Fig. 3.2 The number of main species in the phytoplankton composition of Lake Sevan in a long-term aspect

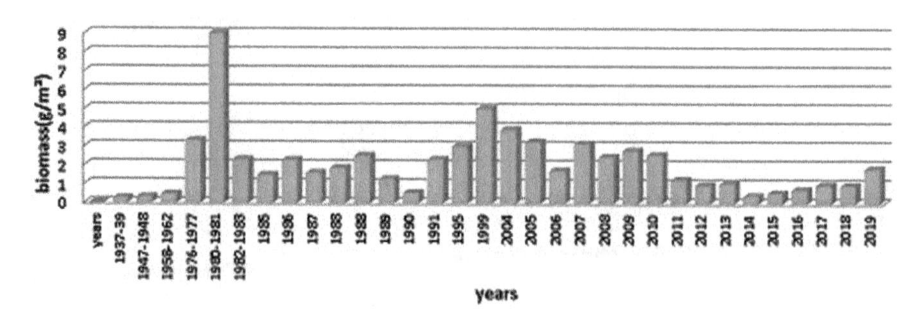

Fig. 3.3 Changes in the phytoplankton biomass of Lake Sevan over a long-term period

According to the studies of phytoplankton of the oligotrophic period 1939–1947 quantitatively (greens predominated qualitatively), diatoms were the predominant group of algae (Vladimirova 1947; Stroykina 1952). The biomass indicators were 0.32 g/m^3, the species composition was relatively poor, and peculiarities in the development of phytoplankton were noted: regular fluctuations in the number of individual species and the presence of their stable succession. Such a trend in the development of phytoplankton of Lake Sevan, characteristic of the prelaunch period and until the 1950s, had similar features with other high-mountain oligotrophic lakes in the world. In the period from 1956 to 1958, the level of the lake was lowered by 11 m, which caused sharp changes in the species composition of phytoplankton, and already in 1964–1965, the onset of eutrophication manifested itself, expressed by the flowering of species of the genus Anabaena (*Anabaena flos-aquae, A. lemmermanii*), which had not previously been noted by researchers in the community. In the period from 1983 to 1992 (the period of level stabilization, water diversion of the River Arpa) in Lake Sevan, an increase in the role of green algae and the appearance of a new dominant – *Binuclearia lauterbornii*, which became a monodominant species in the plankton of the period 1995–1999, was observed (period of repeated lowering of the water level) (Mnatsakanyan 1984; Hambaryan 2001). These facts testify to the partial reversibility of eutrophication processes in the lake as the level stabilizes, a decrease in the number of types of eutrophicity indicators was noted (Mikaelyan 1996; Hambaryan 2001). After 2001, the observed increase in the water level contributed to unpredictable succession in the autotrophic link, under the influence of changes in the hydrological regime and the formation of new ecotones in the littoral part of the lake. In the period 2005–2009, in the phytoplankton community of Lake Sevan, significant changes were observed due to an increase in the diversity of diatoms. After the flooding of the shores, the qualitative composition of diatoms was replenished by some periphyton and benthic genera: *Navicula*, *Cymbella*, *Achnantes*, and *Pinnularia*, and a number of species: *Ceratoneis arcus, Amphora coffeaeformis, Aphanothece stagnina, Cyclotella stelligera, Fragilaria capucina, F. construens*, as well as euglena *Phacus* sp. and *Trachelomonas hispida*, which in the period 2005–2014 became permanent representatives of the Sevan phytoplankton. After 2016, along with a gradual increase in the quantitative indicators of phytoplankton, more frequent harmful water blooms were periodically observed in the summer-autumn periods (Hovsepyan and

Khachikyan 2016a, b). In the spring of 2018, the general quantitative indicators of phytoplankton increased, in particular, the number of blue-green algae amounted to 40% of the total number of the community. The dominant complex included the following species: *Aphanothece clathrata, A. stagnina, Microcystis aeruginosa, M. wessenbergii, Gomphosphaeria lacustris, Oscillatoria lacustris, Merismopedia elegans, Spirulina abbreviata, Phormidium foveolarum*. At the end of June, against the background of an anomalous increase in water temperature, which at different points was 18–21 °C, a massive bloom was observed in the lake caused by cyanobacteria of the *Dolichospermum* (*Anabaena*) genus: *D. flos-aquae, D. lemmermanii,* and *D. spiroides* (Fig. 3.4).

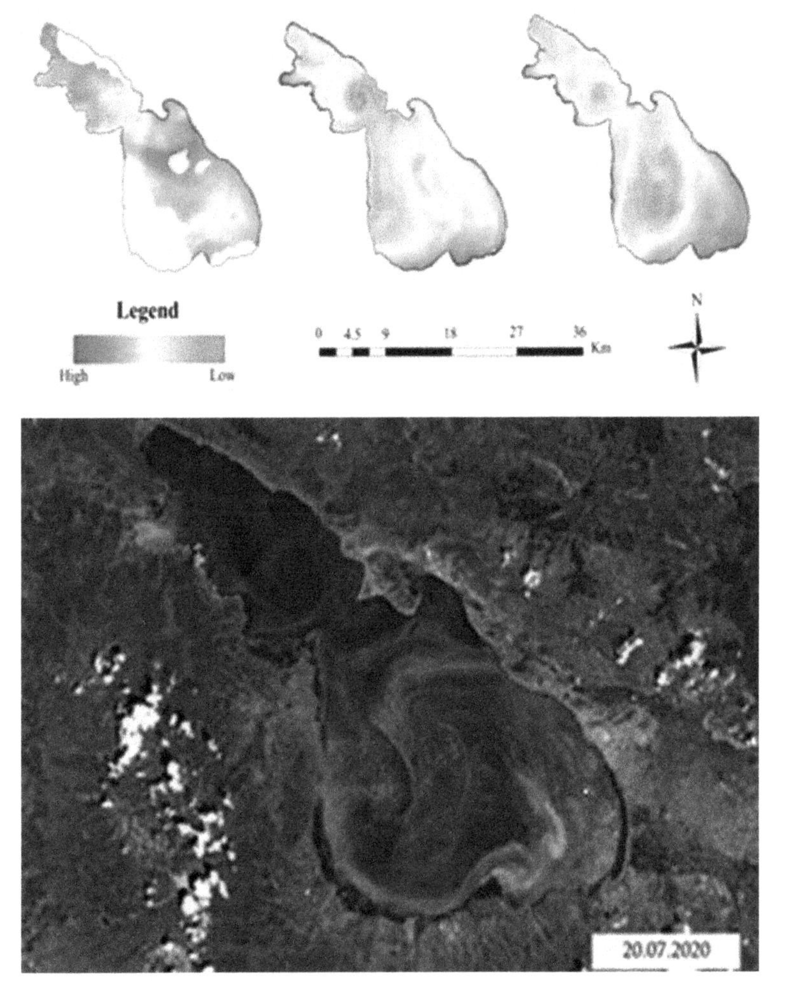

Fig. 3.4 Space images of temperature regime changes (high-low) according to a Landsat 8 OLI/TIRS satellite image and the spread of harmful cyanobacteria blooms in the waters of Lake Sevan, July 2020

It should be noted that the "bloom" captured the entire water area of the lake, and even according to visual observations, it was very intense (Figs. 3.5 and 3.6).

In the period from June to July 2018, the average quantity and biomass of cyanobacteria of the genus *Dolichospermum/Anabaena* were 3,576,000 cells/l and 14.3 g/m^3. Maximum values of 16,528,000 cells/l and 66 g/m^3 were recorded in the area of Shorzha Bay. In the lake, there was a change in the ecological conditions of transparency, indicators of dissolved oxygen, the accumulation of decaying algae biomass, and the presence of an unpleasant odor. It is known that blooming phytoplankton, especially cyanobacteria, causes a disturbance in the development of all links in the food chain and contribute to unpredictable changes in quantitative and qualitative parameters (Paerl et al. 2001, 2011). During the period of prevalence of cyanobacteria, the quantitative indicators of diatoms decreased in plankton, which led to a decrease in their role in phytoplankton. The role of diatoms increased in autumn (63% and 76%), when the development of the eutrophication species *Melosira granulata* (8,448,000 cells/l and 42 g/m^3) and *Cyclotella kuetzingiana* (1,392,000 cells/l and 3.06 g/m^3) reached the state flowering in different parts of Small Sevan (deeper part of Lake Sevan). In autumn, phytoplankton revealed species of *Zygnema* and *Nephrocitium agardhianum* which are indicators of eutrophication and have not been noted in previous studies.

During the active growing season of 2019 in Lake Sevan, three quantitative maxima were observed in the development of phytoplankton in the lake. In the

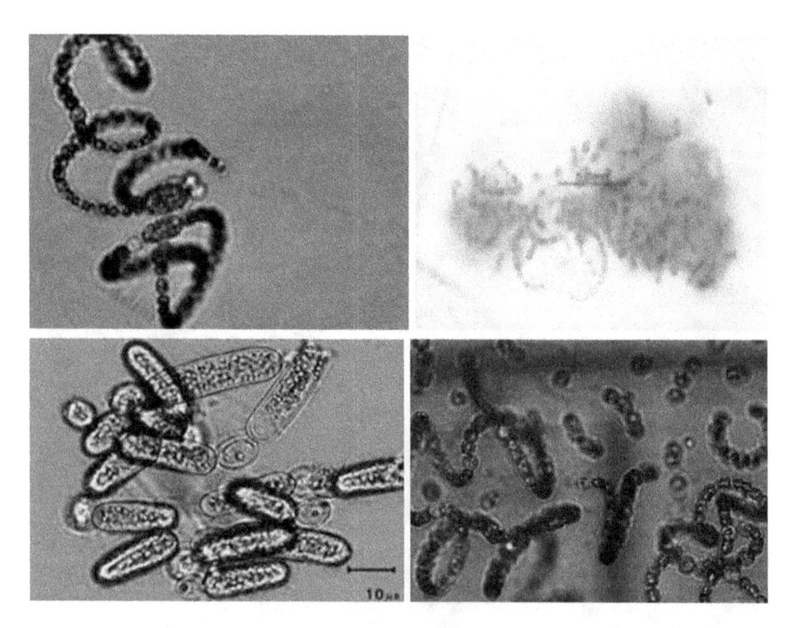

Fig. 3.5 Species of the genus *Dolichospermum/Anabaena* in water samples of Lake Sevan in 2019–2021

Fig. 3.6 Separate sections of the shore of Lake Sevan during the summer period 2019–2021 of flowering of cyanobacteria

spring, green algae reached the level of blooming (88%); in the summer, cyanobacteria (90%); and in the fall, diatoms (80%). Frequent algal blooms in the littoral zone indicate the presence of pollution and lead to nutrient enrichment, changing climatic conditions. Blooming of water in summer period with toxic species of the genus *Dolichospermum*/*Anabaena* causes changes in the habitat. CyanoHABs (cyanobacterial harmful algal blooms) is a danger of the transfer of blooms to the deep-water areas of the lake, as it was observed in 2018–2019 (Hambaryan et al. 2020a, b). In the summer of 2018 and 2019, environmental risks increased due to the intensive flowering of Cyanoprokaryota in the lake, which is an indicator of negative changes. An increase in the biomass of cyanobacteria (75%), dominance in summer plankton, and observed frequent water blooms indicate the eutrophication of Lake Sevan, which has been clearly manifested in recent years. During the flowering period (which lasted more than a month), by the beginning of July, a sharp increase in the quantitative indicators of cyanobacteria was observed. For the first time in lake water samples, the presence of secondary metabolites of cyanobacteria was recorded: microcystin and anatoxin A. The amount of anatoxin A in water was 0.147 μg/l. Thus, abrupt, unpredictable changes in community dynamics and the dominance of toxic species that release allelochemical substances that inhibit the development of nontoxic species are a key factor that reveals the unstable state and activation of eutrophication processes in aquatic ecosystems (Chakraborty et al. 2015). The detected toxins are widely known as substances that negatively affect the ecological conditions of the habitat and contain risks for animals and human health (Carmichael 2001; Hambaryan et al. 2019, 2020a, b; Gevorgyan et al. 2020).

The blooming of the lake water continued in 2020–2021; however, it was on a smaller scale and was mostly localized in the coastal zone. Studies of the 2021 flowering period revealed that along with *Dolichospermum/Anabaena flos-aquae* (β), another toxic alga, *Aphanizomenon f-a* (β), also developed, the indicators of which at some points reached 732,000 cells/l and 5.12 g/m^3.

The ecosystem of Lake Sevan is in an unstable state and needs detailed and ongoing research, as well as the creation of a lake model to predict ecosystem development scenarios.

3.2 Lichens

Arsen Gasparyan

Lichens are composite organisms traditionally known as an association of fungus and algae and/or cyanobacteria. At the same time, recent studies have showed the presence of other organisms, such as multiple algal species, viruses, and basidiomycete yeasts in lichen thalli (Morillas et al. 2022). Although lichens are represented in a wide range of natural (Fig. 3.7) and anthropogenic habitats, covering about 8% of the Earth's land surface (Asplund and Wardle 2017), they are often overlooked because of their small stature and slow growth. According to Lücking et al. (2017), 19,387 lichen-forming fungi names have been accepted globally yet. From the Caucasus region, more than 1840 lichens have been documented by Urbanavichus (2015). The recent revisions of lichen checklists from the neighboring countries of the South Caucasus, Georgia (Inashvili et al. 2022) and Azerbaijan (Alverdieva 2018) communicate about 713 and 811 species, respectively.

According to the latest lichenological studies, 619 lichenized fungi taxa have been reported from the Republic of Armenia so far (Gasparyan and Sipman 2016, 2020; Gasparyan et al. 2017). The major lichenological studies in Armenia have been initiated in the first decades of the twentieth century (Kara-Murza 1931a, b; Pakhunova 1933). Then, in the 1960s, Nikoghosyan (1963, 1964a, b, 1965a, b, 1966) explored the diversity of some genera of macrolichens, such as, *Parmelia* Ach., *Physcia* (Schreb.) Michaux, *Ramalina* Ach., *Xanthoria* (Fr.) Th.Fr. etc. (Fig. 3.8). Later, Barkhalov (1983) published the first comprehensive publication

Fig. 3.7 *Rusavskia elegans* (Link) S.Y. Kondr. & Kaernefelt on a rock

Fig. 3.8 *Xanthoria parietina* (L.) Th. Fr. on a bark of branch of deciduous tree

about the Caucasus lichens, where additions from the lichen biota of Armenia were also presented. At the same time, the lichen diversity of the Lake Sevan basin was extensively explored by Abrahamyan (1984, 1996). The end of the 2000s stood out by the publications of Harutyunyan and Mayrhofer (2009), where 114 new species for the country were reported.

Finally, the latest decade has been marked by the intensive studies of Gasparyan and colleagues (Gasparyan and Aptroot 2016; Gasparyan and Sipman 2013, 2014, 2016, 2020; Gasparyan et al. 2015, 2016, 2017; Zakeri et al. 2016). As a result, the species diversity has significantly increased by 205 species (49.5%) since the publication of the first catalogue of lichens of Armenia, where Harutyunyan et al. (2011) summarized available data about the 414 species of lichens from Armenia.

In Armenia, the largest lichen genus would be considered *Caloplaca* Th.Fr., but since the recent changes in the classification of the family Teloschistaceae (Arup et al. 2013), a large number of the *Caloplaca* spp. was accommodated within the newly proposed genera *Athallia* Arup, Søchting, and Frödén; *Calogaya* Arup, Søchting, and Frödén; *Blastenia* A. Massal, *Pyrenodesmia* A. Massal, *Rufoplaca* Arup, Søchting, and Frödén; *Variospora* Arup, Søchting, and Frödén; etc. Therefore, the largest lichen genus in Armenia is now *Lecanora* Ach. (excluding species which have recently been moved to the genera *Myriolecis* Clem. and *Protoparmeliopsis* M. Choisy), followed by *Acarospora* A. Massal. (23 taxa). Both genera are also among the largest lichen genera in the world (Lücking et al. 2017). The most diverse lichen genus globally is *Xanthoparmelia* (Vain.) Hale, comprising 820 species (Lücking et al. 2017). Locally, the genus *Xanthoparmelia* is represented by 13 species. The list of the largest genera of Armenian lichens is presented in Table 3.2.

Among 619 lichen-forming fungi growing in Armenia, 219 taxa are specialized epiphytic species (Fig. 3.9), which are well studied, whereas saxicolous, terricolous, and other groups of lichens remain sparingly known (Gasparyan and Sipman 2016). The highest diversity of epiphytic lichens is concentrated in the northern provinces of Armenia, especially Tavush and Lori, where relatively high levels of humidity and extensive forest cover, predominantly composed of oak (*Quercus* spp.), beech (*Fagus orientalis*), hornbeam (*Carpinus* spp.) and pine (*Pinus* spp.), and mixed and pure stands. For the saxicolous lichens, the available literature data suggest a greater species diversity in the Aragatsotn, Gegharkunik, Kotayk, and Vayots Dzor provinces of Armenia.

Table 3.2 The list of lichen genera with the largest number of species reported from Armenia

NN	Genus	Number of taxa
1.	*Lecanora* Ach.	28
2.	*Acarospora* A. Massal.	23
3.	*Cladonia* P. Browne	21
4.	*Rinodina* (Ach.) Gray	21
5.	*Aspicilia* A. Massal.	14

Fig. 3.9 Epiphytic lichen (*Ramalina sinensis* Jatta) on a bark of branch of deciduous tree

Overall, the abovementioned explorations indicate the presence of hidden diversity of the lichen biota in the country and the necessity of further research focusing on unexplored groups and habitats. Eventually, one of the important factors for the further boost of systematic lichenological research in Armenia has been the establishment of the Lichen Research and Conservation Group at the Institute of Botany after A. Takhtajan of the National Academy of Sciences of the Republic of Armenia. The group focuses on studies of lichen diversity, taxonomy, and phylogenetics.

3.2.1 Endemic Lichens

In comparison with vascular plants of Armenia, the level of endemism of lichens is low. Up to now, only three new species for science, *Megaspora cretacea* Gasparyan, Zakeri, and Aptroot (Zakeri et al. 2016), *Verrucaria juglandis* Gasparyan and Aptroot (Gasparyan and Aptroot 2016), and *Dermatocarpon cineritectum* Nikoghosyan (1965b) are known only from the current territory of Armenia. The *Megaspora cretacea* and *Verrucaria juglandis* were found during the OPTIMA's lichenological excursion to Armenia organized by A. Gasparyan. The first species grows on *Juniperus*. and is known only from a few localities in the "Khosrov Forest" State Reserve. In the southern part of the country, two other species of the genus *Megaspora*, *M. rimisorediata* Valadb. and A. Nordin and *M. verrucosa* (Ach.) Arcadia and A. Nordin can also be found. Another species, *V. juglandis*, was only collected from the bark of tree roots of *Juglans regia* in the riparian forest in the province Vayots Dzor. There are no fresh specimens and/or observations of *Dermatocarpon cineritectum* Nikoghosyan, which is known only from the publications of Nikoghosyan (1965b) and Barkhalov (1983).

3.3 Mosses

Astghik Poghosyan and Siranush Nanagulyan

Bryophytes are the most ancient and isolated group of higher plants. Leafy mosses represent the highest number of species among all mosses and are widely distributed from the Arctic to Antarctica. As highly adaptive organisms, mosses are extremely vital. They play an essential role in the formation of vegetation cover, especially at the initial stages of the formation of plant communities. Unlike vascular plants, mosses utilize the gametophyte as the main photosynthetic form that develops the sporophyte (sporogon) on it. Such a unique structure for plants raises much interest among researchers.

Mosses play a significant role in the biosphere, mainly in the regulation of the water balance of the continents; have an impact on the natural environment: increase soil moisture, affect its thermal and gas regime, physical and chemical properties, are pioneers in the process of colonization of new territories by plants, and ensure the accumulation of organic matter in the soil and the formation of primary humus. Along with lichens, bacteria, fungi, and algae, mosses are involved in soil-forming processes. Due to the group form of growth, mosses create dense or loose tufts that accumulate moisture. While actively participating in the formation of powerful moisture receivers in the form of swamps and mossy forests, they have a significant impact on the provision of ground with moisture. But mosses also slow down soil erosion, by partially absorbing surface runoff and transferring it underground, thereby strengthening the soil. Mosses are an important component of many plant communities in a variety of ecosystems. Moss colonies become small econiches with sufficient moisture, which capture, preserve, and facilitate the germination of spores and seeds of a variety of vascular plants. Furthermore, mosses are actively involved in the process of peat formation, contributing to the accumulation and storage of carbon from the atmosphere, which is very important at present, when climate change processes are taking place on the Earth.

Bryophytes are resistant to prolonged drying and are able to grow in places of uneven moisture distribution, even if very short-term seasonal moisture is present. Bacteria, fungi, and insects do almost no damage to mosses. They are being used in agriculture, in the form of natural fertilizers or as a source of energy (peat). Some species are used in medicine, as they have bactericidal properties, and pressed moss tiles can be used in construction. Mosses can assimilate radioactive substances and accumulate them for a long time and can serve as indicators of the degree of environmental pollution (Pogosyan 2003a; Nanagulyan et al. 2017).

3.3.1 History of the Study of Mosses in Armenia

The collection of mosses in the territory of Armenia to study their species composition began in the nineteenth century. Many botanists have collected mosses so far. The first publications on the mosses of Armenia are associated with the works of Belanger (1833–1837) and Boissier and Buhse (1860). In the first half of the twentieth century, information about new collections of mosses in Armenia was published by Kara-Murza (1931a, b), Zedelmeyer (1931, 1933), Magakyan (1941), Takhtadjan (1941), Yaroshenko (1946), Tonakanyan (1948), and Dolukhanov (1949), but these were mostly reports of the occasionally collected bryological specimens (Manakyan 1995; Pogosyan 2003a).

A consistent study of the bryoflora of Armenia began in the 50s of the twentieth century. In 1959, Abramova and Abramov published a list of 51 moss species. The same article noted the presence of rich collections of mosses by Shelkovnikov, which were identified by Broterus, but the list of these species, drawn up in the form of an article by Voronikhin, disappeared during the Second World War. Later, data referring to the mosses of Armenia were published by various authors: Dylevskaya (1949, 1959), Dylevskaya and Barseghyan (1971, 1975), Abramova and Abramov (1950, 1977), Dildaryan (1979 and others), Manakyan (1969, 1975 and others),

Dildaryan and Avetisyan (1981, 1983, 1987), Dildaryan and Petrosyan (1987), and Melikyan and Dildarian (1975). These articles provide information on the species composition of mosses in different regions of Armenia, ecological groups, the participation of the bryoflora in the local flora, as well as the patterns of distribution of mosses in altitudinal zones, and the confinement of moss synusia to different plant communities, the ratios of bryoflora and flora of higher vascular plants are indicated. All this contributed to the elucidation of the floristic relationships between the bryofloras of Armenia, the Caucasus, and Europe and supplements the data on the history of the floras of Armenia and the Caucasus (Pogosyan 2003a).

Nowadays, the bryoflora of Armenia remains unevenly studied. The flora of leafy mosses was mainly studied in the forest regions of the northeast (Dildaryan 1969, 1971, 1972) and southeast (Manakyan 1970, 1983, 1989a, b) of Armenia. The regions of the central part of Armenia remain poorly explored. Only little information is known about the Shirak region, Tsaghkunyats and Teheniss ridges, and Sevan region (Avetisyan and Dildaryan 1978, 1983, 1986, 1987, etc.; Dylevskaya and Barseghyan 1971, 1975; Avakyan and Dildaryan 1989, 1990; Barsegyan 1990). During 1998–2002, A.V. Poghosyan studied the bryoflora of the Arailer volcanic massif located in central Armenia. She carried out a taxonomic analysis of leafy mosses in this massif, identified ecological groups, analyzed the coenotic confinement, and for the first time in Armenia studied the ultrastructure of gametophyte stems of some species of leafy mosses (Pogosyan 2003a; Oganezova and Pohosyan 2004).

We have taken as a basis for the classification the checklist of Eastern Europe and North Asia mosses (Ignatov et al. 2006). Some changes and additions given in an annually reviewed species checklist of mosses from the Plant List (http://www.theplantlist.org/1.1/browse/B/) are followed for the convenience of the material presented.

3.3.2 *The Current State of Knowledge and Protection of the Bryoflora of Armenia*

Over the years, studies of the bryoflora of Armenia documented 433 species of mosses belonging to 168 genera and 65 families. Of these, 50 species (28 genera, 20 families) are from liverworts, and 383 species (140 genera, 45 families), from mosses (Manakyan 1995; Manakyan et al. 1999; Pogosyan 2003b).

Literary information about the liverworts of Armenia is scarce since those have been collected occasionally rather than systematically. Liverworts were mainly collected by Manakayan and identified by foreign authorities. Due to the lack of relevant experts, the study of liverworts in Armenia is currently insufficient. For Armenia, the statement "Information about the liverworts Caucasus in our literature is so scarce that all sorts of new data about them are interesting, no matter how modest they are," made back in 1914 by Voronov is actual even today (Manakyan et al. 1999).

The leading families of liverworts are Lophoziaceae, seven species (five genera); Scapaniaceae five (2); Aneuraseae, Lophocoleaceae, Marchantiaceae, and Ricciaceae, four species each (two genera each); and Pelliaceae and Plagiochilaceae, three each (1).

The leading families of Mosses are: Pottiaceae, 58 species (19 genera); Brachytheciaceae, 33 (12); Grimmiaceae, 29 (5); Amblystegiaceae, 25 (15); Bryaceae, 25 (3); Orthotrichaceae, 19 (2); Mniaceae, 15 (5); Dicranaceae, 10 (2); Polytrichaceae, 10 (4); and Sphagnaceae, 10 (1). The 10 main families include 234 species, which is about 62.7% of the species composition of leaf mosses in Armenia. The same families include 68 genera (48.2%). The remaining families are represented by 1–9 species, and they account for 73 genera and 139 species.

The genera *Bryum* (22 species), *Grimmia* (21), *Orthotrichum* (17), *Tortula* (12), *Didymodon* (10), *Sphagnum* (10), *Syntrichia* (10), *Brachythecium* (9), *Dicranum*, *Pohlia* (8 each), *Encalypta*, *Plagiomnium*, *Weissia* (7 each), *Fissidens*, and *Philonotis* (6 each). The leading 15 genera include 166 species (44.5%). The remaining 126 genera are represented by 1–5 species.

Geographical distribution in mosses is similar to higher plants. They are characterized by both disjunctively and fragmentation of habitats, relictness, and endemism. However, no endemic moss species have been found in Armenia so far. The bryoflora of Armenia consists of Holarctic species, many of which have a disjunctive range within the Caucasus (Manakyan 1986). When considering local floras within the Holarctic, moss endemism was found to be lower compared to other plants, and it increases significantly in the tropics.

A comparative analysis of the bryofloras of Central Europe, Bulgaria, Crimea, and neighboring states with the bryoflora of Armenia showed that the species composition of the moss flora of Armenia repeats the species composition of these regions by 45–81% and allows us to assume that the bryoflora of Armenia and the Caucasus as a whole, in its genesis, tends to broad-leaved forests of Europe (Manakyan 1986; Oganezova et al. 2005).

The composition of the leading families reflects the natural and climatic conditions of the republic, where the entire range of climatic conditions is represented from humid and moderate to arid, sharply continental. The most favorable for the growth of mosses are the forest and subalpine zones. The greatest diversity of taxa of all ranks is noted here. Under the conditions of Armenia, most leafy mosses and liverworts are confined to forests (308 and 29 species, respectively) and subalpine (150 and 29 species) ecosystems. According to the altitudinal distribution, the following can be stated: in the lower mountain belt 166 and 17 species, on average – 285 and 25 species; in the upper – 232 and 37 species, respectively. Table 3.3 shows the number of mosses species by vertical zones.

The main share in the diversity in Armenia falls on representatives of the abovementioned main families of bryoflora. However, the degree of their representation in different coenosis is not the same and depends on the habitat conditions.

Table 3.3 Diversity of bryophytes across vertical zones

Species	Mountain belts						
	L	M	U	LM	MU	LU	LMU
Liverworts+mosses	17 + 166	25 + 285	37 + 232	1 + 34	7 + 65	1 + 8	7 + 106
Total	183	310	269	35	72	9	113

L lower, *M* middle, *U* upper

It is known that the species of the Brachytheciaceae and Amblystegiaceae families are the most common and numerous in temperate and high latitudes. In Armenia, they grow in all altitudinal belts but the most abundant and diverse in the middle mountain belt, somewhat poorer in the upper, and sometimes in the lower.

Representatives of the family Bryaceae are distributed all over the globe. They grow in all elevation zones of Armenia, the highest taxonomic diversity falls on the upper, somewhat less in the middle, and insignificant in the lower zone, but they do not play a significant role in plant communities, except for some wetland formations, where they often dominate.

Representatives of the family Hypnaceae are also distributed all over the globe. In Armenia, predominantly terrestrial species are noted and grow at the entire altitude range, but most of all in the middle and upper mountain zones.

Representatives of the family Polytrichaceae are distributed in many areas of the globe. In Armenia, they are noted in small numbers and play an insignificant role in the composition of the vegetation cover. The highest species richness occurs at middle and high elevations.

Species of the family Grimmiaceae grow in subtropical and temperate zones and southern latitudes – in the mountain and high mountain zones. They do not play a significant role in the composition of the vegetation cover of Armenia, but they are of no small importance in the development of stony substrates and are evenly distributed in all mountain zones.

Representatives of the Trichostomaceae and Pottiaceae families are mostly adapted to arid conditions and are common in regions with a Mediterranean climate. In Armenia, they are found in all altitudinal zones, while the highest diversity of species occurs at mid-mountain zones; it is not much different from that at lower and upper elevations.

Representatives of the family Orthotrichaceae are found mainly on woody and shrubby plants. They do not play a significant role in the composition of the moss cover. Like the species of the family Grimmiaceae, these species contribute to the accumulation of air dust. However, unlike Grimmiaceae species, the dust produced by Orthotrichaceae settles on the bark of trees and prevents normal breathing of the trunks.

Species of the family Dicranaceae occur predominantly at northern and temperate latitudes and southern latitudes restricted to high mountains. In Armenia, this family is represented by a significant number of genera, but the species diversity is low. These species are concentrated at middle and higher elevations, mainly in the forest biome.

The family Mniaceae is found predominantly in the forest communities and mountain regions of the Northern Hemisphere. In Armenia, they form significant covers both on the soil and on separate blocks, stones, and rocky areas in the forests. The main species diversity is limited to the middle and upper mountain belts.

Species of the family Thuidiaceae are common at the temperate latitudes of the Northern Hemisphere, especially in East Asia. In Armenia, it is represented by a few species that play a significant role in the overgrowth of stones, trunk bases, and open areas in the forest zone.

Representatives of the family Entodontaceae are widespread in temperate and warm regions of the globe. In Armenia, they do not play a significant role in the vegetation cover and are rare, except for *Entodon concinnus* (De Not.) Par.

Representatives of the family Leskeaceae are distributed mainly in the temperate and warm temperate zones of the Northern Hemisphere, especially in East Asia, with focus on North and South America. In Armenia, the number of species is small, but they take an active part in the fouling of stones, rocks, and bases of tree trunks in the forest zone.

The family Bartramiaceae is represented in Armenia by a small number of species found mainly in the middle and high elevations. *Philonotis fontana* (Hedw.) Brid. plays an essential role in some wetland vegetation communities.

The representatives of the family Neckeraceae are predominantly distributed in tropics. In Armenia, the main number of species is collected in the relatively mesophytic forests of Tavush region, and only two species, *Neckera besseri* (Lobarz.) Jur. and *Thamnobryum alopecurum* (Hedw.) Gang., are found in forests of the Syunik region.

The remaining families are represented unevenly both in individual floristic regions and in various communities. Of these, the families Sphagnaceae, Meesiaceae, Andreaceae, Tetraphidaceae, Lembophyllaceae, Theliaceae, and Fabroniaceae should be highlighted; the generic diversity of which is not high, but from the botanical and geographical point of view, their presence in Armenia is of great interest.

The peat moss (*Sphagnum*) species *S. fimbriatum* Wils., *S. fuscum* (Schimp.) Klinggr., *S. girgensohnii* Russ., *S. riparium* Aongst., and *S. squarrosum* Crome are found in Armenia in relatively remote places, mainly in high mountains (Manakyan 1986).

Thus, another confirmation of the links between the bryoflora of the Republic of Armenia, Caucasus, and Europe is the presence of representatives of common moss genera.

As a result of considering the distribution of moss species in the territory of Armenia, intriguing results were obtained. Northeastern Armenia covers the Lori and Ijevan floristic regions, which are part of the Tavush and Lori administrative regions of Armenia. They border Georgia and Azerbaijan. The main type of vegetation cover is forests. For this region, 214 species of bryophytes are described, 20 species of them are liverworts, and 194 species are leafy mosses. The liverworts species belong to 14 families and 16 genera, and the leafy mosses species belong to 36 families and 89 genera.

The distribution of mosses in the forests of Northeastern Armenia depends on climatic and soil conditions. The vast majority of mosses are forest species: *Plagiomnium cuspidatum* (Hedw.) T. Cor., *Brachytheciastrum velutinum* (Hedw.) Ignatov and Huttunen, and *Dicranum scoparium* Hedw. Fewer species occur in treeless habitats, like *Tortula acaulon* (With.) R.H. Zander and *Tortula protobryoides* R.H. Zander. The region is also characterized by a small group of mosses from swampy and highly humid habitats: *Paludella squarrosa* (Hedw.) Brid., *Meesia triquetra* (Jolycl.) Ångstr., *Aulacomnium palustre* (Hedw.) Schwaegr., *Drepanocladus aduncus* (Hedw.) Warnst., and *Tomentypnum nitens* (Hedw.) Loeske.

Southeastern Armenia covers the Darelegis, Zangezur, and Meghri floristic regions, which are included in the Vayots Dzor and Syunik provinces of the Republic. The vegetation of Southeastern Armenia is very diverse and is represented by almost all the main vegetation types in Armenia, except for beech forests. For this region, 264 species of bryophytes are listed, of which 26 species are liverworts, and 238 species are leafy mosses. The liverworts species belong to 15 families and 19 genera. The leafy mosses are classified into 34 families, 107 genera, and 241 species.

The bryoflora of Southeastern Armenia, as well as Northeastern Armenia, is predominantly concentrated in forests. However, in dry habitats such as rocks, stones, and gravelly soils rich with moss, species can be found.

The degree of knowledge of the bryoflora in the floristic regions of Armenia is displayed in Table 3.4. For liverworts, 50 families, 105 genera, and 214 species are known, and for leafy mosses, 49, 126, and 264, respectively. The main diversity of the species composition of the bryoflora of Armenia falls on two thoroughly studied floristic regions – Ijevan and Zangezur – where forests prevail. The bryoflora of

Table 3.4 Taxonomic analysis of mosses by floristic regions of Armenia

Floristic regions of Armenia	Liverworts			Mosses			Bryophytes		
	Species	Genera	Family	Species	Genera	Family	Species	Genera	Family
Upper Akhuryan	–	–	–	2	2	2	2	2	2
Shirak	–	–	–	26	17	14	26	17	14
Aragats	3	2	2	49	31	19	52	33	21
Lori	6	6	6	97	59	32	103	65	38
Idjevan	18	15	13	206	99	40	224	114	53
Aparan	10	10	8	85	50	28	95	60	36
Sevan	21	15	11	129	65	29	150	80	40
Gegham	2	2	2	12	9	8	14	11	10
Yerevan	4	4	4	94	51	22	98	55	26
Darelegis	2	2	2	57	32	19	59	34	21
Zangezur	21	17	14	229	106	40	250	123	54
Meghri	13	10	8	120	61	30	133	71	38

Ijevan lacks representatives of the families of Cinclidotaceae, Plagiotheciaceae, Theliaceae, and Fabroniaceae, whereas representatives of the families of Andreaeaceae, Cinclidotaceae, Pterigynandraceae, Tetraphidaceae, Theliaceae, and Meesiaceae are lacking in Zangezur region. The distinctive features of the bryoflora in these areas are associated with natural conditions. In the Ijevan region, relatively mesophilic beech, beech-oak, and oak forests occupy almost 28% of the entire area of the region; in Zangezur, more xerophilic oak and oak-hornbeam forests occupy only around 14% of the area. In the Zangezur region, absolute heights reach almost 4000 m, and in Ijevan, they do not exceed 3100 m. A high number of species (138 species of leafy mosses from 73 genera, 31 families) are common to the northeast and southeast of Armenia. These species are also more or less common throughout Armenia (Manakyan 1986).

In the central and western parts of Armenia (Upper Akhuryan, Shirak, Aragats, Aparan, Sevan, Gegham, Yerevan floristic regions), special bryological studies similar to the works of Manakyan and Dildaryan were not conducted. For central Armenia, 201 species of leafy mosses from 83 genera and 32 families have been identified (Oganezova et al. 2005).

The wetland bryoflora of Armenia is very diverse. Wetland mosses are confined to high-mountain swamps, peat bogs, glacial and low-lying lakes, spring outlets, hollows, gorge bottoms, river valleys, etc., and represented by more than 80 species, belonging to 32 genera and 19 families, which present considerable interest. The most richly represented families are Amblystegiaceae, Cratoneuraceae, Bryaceae, and Brachytheciaceae (Dylevskaya and Barseghyan 1975; Pogosyan 2003a). Near the village of Gorayk in the Sisian region, at an altitude of 2300–2400 m above sea level, Manakyan (1986) discovered for Armenia a completely atypical *Sphagnum* bog, where *Sphagnum fuscum* and *S. squarrosum* predominate.

Table 3.4 presents the taxonomic analysis of mosses in the floristic regions of Armenia.

The rarest moss species, known from only one locality in Armenia, are *Crumia latifolia* (Kindb.) W.B.Schofield, *Orthotrichum urnaceum* Müll.Hal., *Haplocladium virginianum* (Brid.) Broth., *Sphagnum riparium* Ångstr, *Fissidens grandifrons* Brid., *Grimmia ramondii* (Lam. & DC.) Margad., *Grimmia decipiens* (Schultz) Lindb., Codriophorus *acicularis* (Hedw.) P.Beauv., *Cinclidium arcticum* (Bruch et al.) Schimp., *Aulacomnium turgidum* (Wahlenb.) Schwaegr, and *Paludella squarrosa* (Hedw.) Brid.

Manakyan (1995) analyzed the bryoflora of Armenia and identified 127 rare species of leafy mosses, which constitute 36.4% of the total number of species. Such a high percentage of rare species can be explained, on the one hand, by the relatively poor knowledge of the bryoflora of Armenia, especially its central and northern parts; on the other hand, there is a large variety of natural conditions, but at the same time, the presence of a large number of small areas with a diversity in combinations of climatic, hydrological, edaphic conditions, and others.

The ecological classification of mosses is not trivial, since their ability to grow in various conditions is wider than that in other groups of vascular plants. Mosses are small plants. Their size allows them to grow on different substrates and to develop

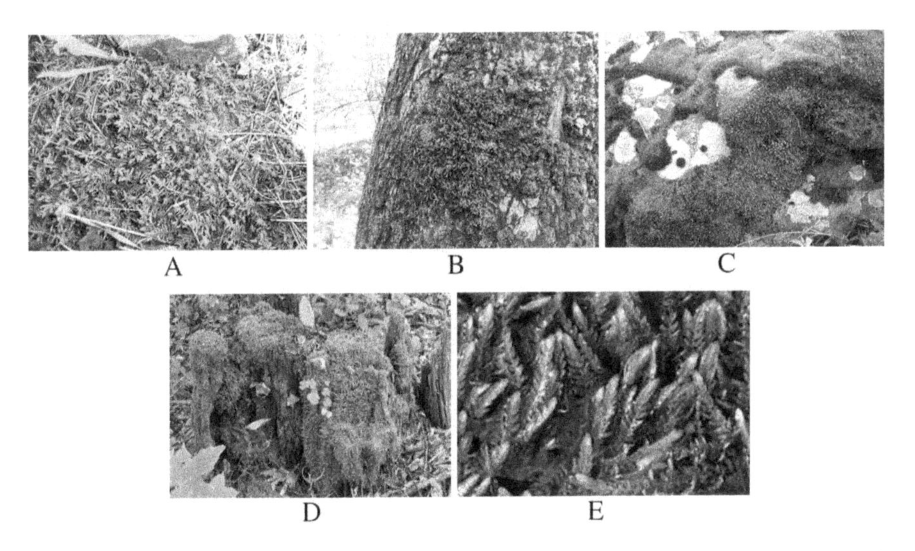

Fig. 3.10 Examples of different ecological groups of mosses in relation to occupied substrates: (**a**) epigeoids; (**b**) epiphytes; (**c**) epiliths; (**d**) epixyles; (**e**) hydrophytes

different habitats. Depending on the substrate, mosses are classified into the following groups (Fig. 3.10): epigeoids (A), epiphytes (B), epiliths (C), epixyles (D), and hydrophytes (E). In each of the noted ecological groups, only some of the mosses are strictly confined to the corresponding ecotope (Poghosyan et al. 2015).

Many mosses have a wide range of adaptability and are equally common on several types of substrates, so it is difficult to attribute them particularly to one ecological group. Such mosses include the species *Schistidium apocarpum* (Hedw.) Bruch et al., *Bryum capillare* Hedw., *Orthotrichum anomalum* Hedw., *Amblystegium serpens* (Hedw.) Bruch et al., *Hypnum cupressiforme* Hedw., etc.

Since mosses retain dust and small substrate particles, they contribute to soil accumulation. Consequently, epilithic mosses sometimes are accompanied by soil mosses. For example, on the surface of stones, one can observe the development of both epilithic and typically ground mosses – *Rhytidium rugosum* (Hedw.) Kindb., *Homalothecium lutescens* Bruch et al., and *Brachythecium albicans* (Hedw.) Bruch et al. Epigeoids, such as *Anomodon rugelii* (Müll.Hal.) Keissl., *Brachytheciastrum velutinum* (Hedw.) Ignatov and Huttunen, and *Hypnum cupressiforme*, often settle on the bases of trunks and protruding roots of trees. This is certainly because the soil is accumulating here. The ability to maximize the use of the most insignificant changes of a substrate, even under rather extreme arid conditions, indicates the extremely high adaptability of mosses and their considerable tolerance.

On the other hand, some species do have a narrow ecological specialization. G.F. Rykovsky (1989) revealed a tendency for the transition of the epiphytic genus *Orthotrichum* to an epilithic way of life. As a result of the global aridization of the climate, an increasing number of species of this genus have become epiliths (Fig. 3.11), although the most specialized species retain their epiphytic way of life

Fig. 3.11 *Orthotrichum* – epiliths

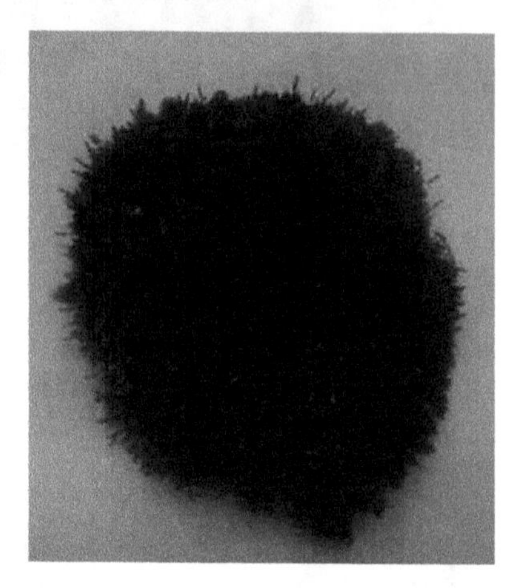

Fig. 3.12 *Orthotrichum* – epiphytes

(Fig. 3.12.). Air humidity is the main factor that determines species diversity and biomass accumulation of epiphytic mosses. Epiphytic mosses are particularly affected by deforestation. That leads to a progressive depletion of the epiphytic bryoflora. "Deforestation disrupts the development of the epiphytic component of ecosystems and narrows the possibilities of natural selection as a regulating and guiding force of evolution" (Rykovsky 1989).

In connection with the peculiarities of the ecology of mosses, the anatomical structure of their conductive tissues is of particular interest. The degree of development of the conductive system of leafy mosses is closely related to their ecology (Melikyan and Dildaryan 1975). The anatomical structure of moss gametophyte reflects the ecological group it belongs to (epigeoids, epiphytes, etc.). Thus, different ecological groups of mosses differ precisely in the structure of the conducting system. Pogosyan (2003a) studied the ultrastructure of the gametophyte stems of some species of mosses of the Arailer mountain. The results of that study confirm the data of Melikyan and Dildaryan (1975), Farra (1991), and Dildaryan et al. (2002) about the presence of some specialization in the structure of epigeoid stems, which enables them to use soil moisture as a way to regulate the water balance. Epiphytes and epiliths are either not or poorly differentiated by the structure of conducting tissues of the stems and are satisfied with the absorption of atmospheric moisture only (Pogosyan 2003a; Oganezova and Pogosyan 2004; Pogosyan et al. 2015).

3.4 Vascular Plants

George Fayvush, Anahit Ghukasyan, Alla Aleksanyan, Jacob Koopman and Helena Więcław

The flora of Armenia comprises about 3800 species of vascular plants (The Fifth National report to Convention on Biodiversity 2014). A characteristic feature of the flora is the absolute predominance of both the number of species and genera of angiosperms – about 97%. Club mosses, horsetails, and ferns are represented by 39 species, and gymnosperms, only nine species. Among angiosperms, dicots completely predominate – about 80% species. In general, such a distribution of the number of species by large taxonomic units is quite common both for the flora of the Northern Hemisphere and for the general terrestrial flora, in which there are only 12% of monocots, 83% of dicots, and only 0.2% of gymnosperms and ferns, 2%; mosses and liverworts, 3.3%; and hornworts, 0.05% (Pimm and Joppa 2015; Ranker and Sundue 2015; Von Konrat et al. 2010).

The largest families of the Armenian flora are listed in Table 3.5 and are shown in Fig. 3.13. As a matter of fact, the 30 largest families comprise more than 80% of all species of the flora of Armenia.

The largest genera of the Armenian flora are listed in Table 3.6.

The current flora of Armenia has a long geological history. As elsewhere, our flora results from speciation in situ and the migration of species from other (often very distant) areas. According to our calculations for the territory of Armenia, the influence of speciation and migration processes on the florogenesis was more or less balanced (Fayvush 1990). The Armenian flora contains widespread, polychorous plant species (e.g. *Phragmites australis* (Cav.) Trin. ex Steud., *Lythrum salicaria* L., *Chenopodium album* L.), as well as many species that originated in the Mediterranean, Asia Minor, and Irano-Turanian regions. However, at the same time, the territory contained powerful foci of speciation of some genera (Gabrielian and Fayvush 1986, 1989; Tamanyan and Fayvush 1987, etc.) resulting in 146 local endemic species (The Fifth… 2014).

In the present work, we have tried to combine classical floristic analysis with karyological analysis, namely, with the representation of various cytorases in Armenia (Fayvush and Ghukasyan 2023).

Determining the number of chromosomes of different species is the first step in karyological studies. The importance of such studies lies in their value in clarifying speciation and phylogenetic relationships. Counting chromosomes is the first step toward a better understanding of karyotype evolution and the role of chromosome evolution in species diversification within the genus.

Table 3.5 Largest and medium-sized families of the flora of Armenia

NN	Family	Number of genera	Number of species	Number of Armenian endemics	Number of karyologically investigated species
1.	*Asteraceae*	90	442	27	207
2.	*Poaceae*	102	336	13	142
3.	*Fabaceae*	33	324	15	64
4.	*Rosaceae*	29	214	31	29
5.	*Brassicaceae*	68	203	7	18
6.	*Caryophyllaceae*	32	183	10	38
7.	*Lamiaceae*	33	153	1	22
8.	*Scrophulariaceae*	20	150	8	17
9.	*Apiaceae*	61	140	3	25
10.	*Cyperaceae*	16	108	0	13
11.	*Chenopodiaceae*	30	87	0	8
12.	*Boraginaceae*	23	78	2	9
13.	*Ranunculaceae*	17	68	1	19
14.	*Polygonaceae*	7	50	0	6
15.	*Orchidaceae*	17	44	0	3
16.	*Rubiaceae*	9	44	0	2
17.	*Euphorbiaceae*	4	43	1	0
18.	*Alliaceae*	2	42	3	31
19.	*Liliaceae*	5	38	0	17
20.	*Orobanchaceae*	4	36	6	0
21.	*Hyacinthaceae*	6	34	1	27
22.	*Campanulaceae*	4	32	1	6
23.	*Iridaceae*	3	29	2	17
24.	*Geraniaceae*	2	28	1	14
25.	*Papaveraceae*	4	28	5	2
26.	*Plumbaginaceae*	5	26	1	1
27.	*Dipsacaceae*	5	25	0	1
28.	*Primulaceae*	8	25	0	6
29.	*Onagraceae*	5	23	0	2
30.	*Malvaceae*	7	22	1	0

Data on the chromosome numbers of Armenian flora are presented in the book *Number of Chromosomes of Flowering Plants of Armenia* (Nazarova and Ghukasyan 2004). Furthermore, for this study, we also analyzed chromosome data from other, more recent, studies (Nazarova 2009, 2011; Kotseruba et al. 2010, 2012; Ghukasyan 2010, 2011; Ghukasyan and Janjughazyan 2015; Ghukasyan and Akopian 2018; Ghukasyan et al. 2022; Hayrapetyan and Ghukasyan 2015, 2021; Oganezova 2013, Więcław et al. 2020). Of the 3800 species of vascular plants having been recorded in Armenia, 798 (21%) were included in our analysis, and these species allowed us to draw certain, relatively reliable conclusions, especially since among the studied species, all large- and medium-sized families known in Armenia were represented

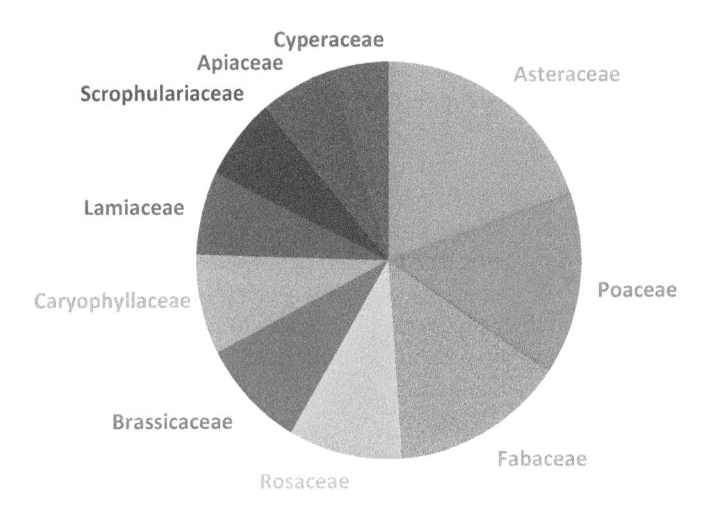

Fig. 3.13 Ten largest families of Armenian flora

Table 3.6 Largest genera of the flora of Armenia

NN	Genus	Number of species
1.	*Astragalus*	134
2.	*Carex*	71
3.	*Centaurea* s.str.	46
4.	*Allium*	41
5.	*Vicia*	41
6.	*Silene*	38
7.	*Euphorbia*	37
8.	*Verbascum*	37
9.	*Veronica*	34
10.	*Pyrus*	32
11.	*Trifolium*	31
12.	*Potentilla*	29
13.	*Rosa*	28
14.	*Alchemilla*	26
15.	*Campanula*	25
16.	*Onobrychis*	25
17.	*Ranunculus*	25
18.	*Cousinia*	24
19.	*Dianthus*	23
20.	*Lathyrus*	23

(Annex 1). Most karyologically investigated samples were diploid (585 races), with fewer tetraploids (178), hexaploids (76), and octaploids (13). There were also races (11) with high ploidy (ten or more), aneuploids (18), triploids (nine), and pentaploids (three). Interestingly, 49 species are represented by two or more cytoraces: 25

species are represented by di- and tetraploid races; four species have di-, tetra-, and hexaploid races; six species are di- and hexaploid races; three species are di- and octaploid races; six species are represented by tetra- and hexaploid races; and two, by tetra- and octaploid races. In addition, diploid and tetraploid species were represented by races with different base numbers. B chromosomes were found in the karyotypes of 24 species; of particular interest is the Armenian population of *Crepis pannonica*, in which there are five diploid cytoraces with $x = 4, 5, 6, 7$, and 8, as well as triploid and aneuploid cytoraces. Overall, this diverse karyological spectrum is indicative of the ongoing speciation and morphogenesis in modern Armenian plants.

In Armenia, the first three places in the spectrum of large families (Table 3.5), as in most Holarctic floras, are occupied by the families *Asteraceae, Poaceae*, and *Fabaceae*, which are very widespread all over the globe. This is a common phenomenon; however, it should be noted that the polymorphism of *Fabaceae* is due to polymorphism in a few large genera (especially *Astragalus*), which is very typical for the Middle and Central Asian floras. *Asteraceae* and *Poaceae* families are represented by a large number of genera with different numbers of species (from monotype to large). The fourth place in the spectrum of the flora of Armenia is occupied by the *Rosaceae* family, which is also widespread throughout the globe but especially richly represented in the temperate warm regions of the Northern Hemisphere. At the same time, it is characteristic that in the Irano-Turanian floras, the role of this family decreases from forest to nonforest flora. In this family, it is necessary to note the richness of the genus *Pyrus* (32 species), indicating intensive processes of speciation and morphogenesis in the South Caucasus. The next four families (*Brassicaceae, Caryophyllaceae, Lamiaceae, Scrophulariaceae*) are also widespread in the temperate regions of both hemispheres, but they are particularly diverse in the Mediterranean up to Central Asia, that is, in the region of the Ancient Mediterranean. The next two families, *Apiaceae* and *Cyperaceae*, are best represented in the temperate regions of the Northern Hemisphere and in the mountains of the tropics; in our case, they reflect the Palearctic features of the Armenian flora.

The most ubiquitous genera within the context of Armenian flora are *Astragalus, Centaurea* s.l., *Carex,* and *Allium*.

The largest genus in the Armenian flora is *Astragalus* (approximately 140 species). This genus is very typical for the entire Irano-Turan region. However, there are relatively few narrow endemic species (only 12) in Armenia, and the distribution ranges of most species are not limited to one country or a floristic region. Only 14 species have been karyologically studied, but among them, there is one tetraploid, one hexaploid, one octaploid, one pentaploid, and one species (*Astragalus sevangensis* Grossh.) for which tetra- and octaploid cytoraces have been registered (Annex 1).

The second largest genus in the flora of Armenia is *Carex*, which currently comprises 72 taxa, occurring mainly in wetlands, less often in dry habitats (Koopman et al. 2021). The genus *Carex* originated in the late Eocene in East Asia, with the exception of the subgenera *Psyllophorae* and *Uncinia*, which arose in the West Palearctic and America, respectively (Waterway et al. 2009; Starr et al. 2015; Martín-Bravo et al. 2019). Modern distribution of this genus is primarily a product of recent diversification on the Northern Hemisphere, as the great majority of dispersal events have taken place between Nearctic, East Asia, and West Palearctic regions (Martín-Bravo et al. 2019). The *Carex* taxa occurring in Armenia for this work were classified by Koopman and Więcław into eight different phytogeographic elements (Table 3.7; Fig. 3.14). The classification of phytogeographic regions is based on the concept of Takhtajan (1986), Grubov (2010) and Carta et al. (2022). When classifying *Carex* taxa to the appropriate phytogeographic element, their general distribution given in the POWO (2022) database was taken into account. Ultimately, six phytogeographic regions were identified: Circumboreal (CB), Eurasiatic (EA), Mediterranean-Iranian (MI), Central Asian (CA), Central European (CE), and Irano-Turanian (ITR). *Carex* taxa with a range covering two phytogeographic units were treated as connective elements: Eurasiatic-Mediterranean-Iranian (EA-MI) and Central Europe-Mediterranean-Iranian (CE-MI).

The highest percentage of taxa, 32.4% (23 taxa), belongs to the Eurasiatic element (EA), followed by Irano-Turanian (IRT: 16.9%, 12 taxa). The smallest number of taxa is associated with the Central European (CE: 1.4%, 1 taxon) and Mediterranean-Iranian (MI: 4.2%, 3 taxa) regions. However, 16 (22%) of the *Carex* taxa, classified as connecting elements, were related to the Mediterranean-Iranian region (Fig. 3.14).

There are no local *Carex* endemics in the flora of Armenia; however, several very rare species have been reported from Armenia, with a range related to Irano-Turan region (Khandjian 2001; Koopman et al. 2021). *Carex cilicica* currently occurs at only one site in the province of Vayots Dzor, known since 1965 from a collection by Barsegian (ERE) (Egorova 1999; Koopman et al. 2021). *Carex capitellata* (two localities in Tavush and Syunik prov.), *C. oligantha* (one site in Mt. Kaputjugh, Syunik prov.), and *C. pyrenaica* subsp. *micropodioides* (one site Kotayk prov.) were found in Armenia in the twentieth century and are known only from the herbarium collections in ERE.

Of the 72 *Carex* species growing in Armenia, only 12 have been karyologically studied; all of them are diploids and are widely distributed either in the Boreal or Ancient Mediterranean subkingdoms. The number of chromosomes (n) in karyologically investigated species of *Carex* varies from 19 to 42, and these species show different cytotypes. Polyploidy is most likely to be rare in species of the genus *Carex*. It seems that the evolution of karyotypes in this genus, which is important for species diversification, is driven by the fusion and fission of the holokinetic chromosomes inherent in species of *Carex* (Więcław et al. 2020).

It is extremely interesting to look at the distribution of the species of the genus *Centaurea* s.l. in the Armenian Highlands. In one of our previous works (Gabrielian and Fajvush 1989), we continued the analysis of Wagenitz (1986) of the distribution of species of this genus in Southeast Asia. For convenience, Wagenitz used the grid system by Davis (1965) in the flora of Turkey, dividing the entire area into squares each occupying 2° of latitude and longitude.

According to Wagenitz (1975, 1986), a great variety of species and sections of the genus *Centaurea* is concentrated in eastern Anatolia and especially where the three countries – Iran, Iraq, and Turkey – meet, where 35 species from 17 sections

Table 3.7 Phytogeographic elements in the genus *Carex* in Armenia

Phytogeographic elements	Definitions	Taxa
CB Circumboreal	Plants usually distributed in temperate and cold regions of North America, Asia, and Europe	*Carex atherodes*, *C. canescens*, *C. capillaris*, *C. diandra*, *C. lasiocarpa*, *C. obtusata*, *C. pseudocyperus*, *C. rostrata*, *C. supina*, *C. vesicaria*
EA Eurasiatic	Elements mostly distributed across the temperate zone of Europe and Asia	*Carex acuta*, *C. appropinquata*, *C. bohemica*, *C. capillifolia*, *C. cespitosa*, *C. depauperata*, *C. digitata*, *C. distans*, *C. disticha*, *C. elata*, *C. hartmaniorum*, *C. humilis*, *C. leersii*, *C. melanostachya*, *C. muricata* subsp. *ashokae*, *C. otomana*, *C. otrubae*, *C. pallescens*, *C. pediformis*, *C. praecox*, *C. rhizina*, *C. secalina*, *C. tomentosa*
MI Mediterranean-Iranian	Elements which are distributed across the Mediterranean regions, Southern Europe, in Southwest Asia, and North Africa	*Carex halleriana*, *C. phyllostachys*, *C. umbrosa* subsp. *huetiana*
CA Central Asian	Taxa distributed in temperate regions of Central Asia, Tien-Shan, Caucasus, Siberia, and Western Asia	*Carex alatauensis*, *C. caucasica*, *C. deasyi*, *C. diluta*, *C. songorica*, *C. stenophylla* subsp. *stenophylloides*
CE Central European	Taxa distributed in temperate regions of Central Europe to Caucasus	*Carex depressa* subsp. *transsilvanica*
IRT Irano-Turanian	Centre of diversity of Irano-Turanian is Western Asia: Anatolia and Irano-Armenia	*Carex aterrima* subsp. *medwedewii*, *C. capitellata*, *C. cilicica*, *C. liparocarpos* subsp. *bordzilowskii*, *C. nigra* subsp. *transcaucasica*, *C. oligantha*, *C. orbicularis* subsp. *kotschyana*, *C. oreophila*, *C. pachystylis*, *C. pseudofoetida* subsp. *acrifolia*, *C. pyrenaica* subsp. *micropodioides*, *C. tristis*
Connective elements		
EA-MI Eurasiatic – Mediterranean-Iranian	Elements distributed in Eurasiatic and Mediterranean-Iranian regions	*Carex acutiformis*, *C. caryophyllea*, *C. divisa*, *C. hirta*, *C. leporina*, *C. panicea*, *C. remota*, *C. riparia*, *C. spicata*, *C. vulpina*
CE-MI Central European – Mediterranean-Iranian	Elements distributed in Central European and Mediterranean-Iranian regions	*Carex brevicollis*, *C. divulsa*, *C. hordeistichos*, *C. michelii*, *C. pendula*, *C. sylvatica*

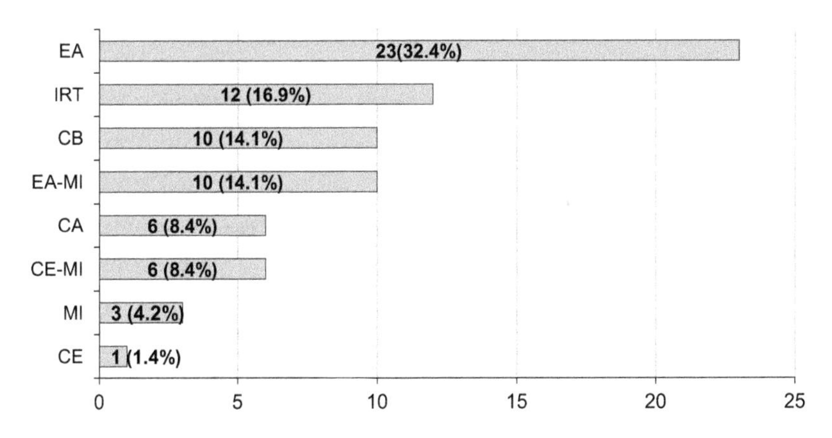

Fig. 3.14 Number and percentage of phytogeographic elements in the genus *Carex* in Armenia

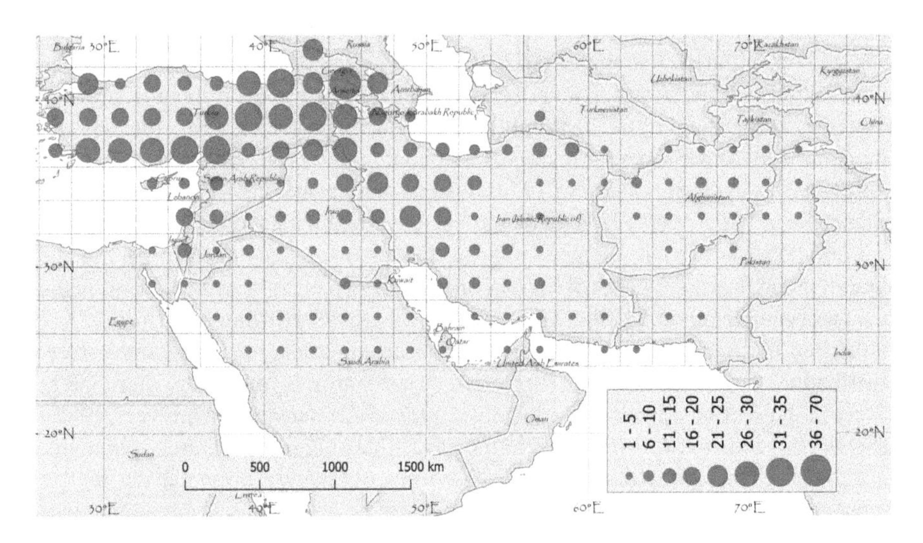

Fig. 3.15 *Centaurea* in SW Asia: species numbers per grid cell. After Wagenitz (1986), with additions by Gabrielian and Fajvush (1989)

grow. In the parts of eastern Anatolia bordering Armenia, 18–34 species from 13 to 19 sections are found. Wagenitz predicted that at least as high a concentration of species and sections should also exist in the South Caucasus and nowhere else in the entire area of the genus *Centaurea.* However, the data obtained by us greatly exceed all our expectations. On the relatively tiny territory of the Republic of Armenia, occupying less than one grid square, there occur about 70 species of *Centaurea* s.l. from 25 sections (Fig. 3.15). On the adjacent territory of Nakhichevan Republic, 30 species from 20 sections are known. There are even fewer species in adjacent territories of Georgia and Azerbaijan. Concerning the distribution of narrow endemics in *Centaurea,* according to Wagenitz (1986), the number of such species is the highest in the Cilician Taurus (6 species) and in the adjacent region (5) and also in eastern Anatolia and the Zagros mountains (5 species each). In the Lesser Caucasus, however, there are at least 14 (perhaps even more) narrow endemics. Of these, six are isolated species without close allies.

Wagenitz (1986, pp. 18–19) writes: "Great diversity in the number of species and sections points to eastern Anatolia and adjacent Transcaucasia, Iran and Iraq as the foremost centre of evolution of this genus. The coherent distribution of most groups and the high degree of endemism seem to indicate speciation in relatively recent times and *in situ*." Obviously, the figures of the number of species, sections, and endemic species of the genus *Centaurea* in Armenia distinguish this territory as an explicit primary center of speciation of this genus.

Thirty-two species of the genus *Centaurea* s.l. have been studied karyologically. Of these, 26 are diploids, six are tetraploids, and four species have B-chromosomes.

A large number of endemic species are found in the genus *Cousinia* (*Asteraceae*). The distribution area of *Cousinia* is almost entirely in the Irano-Turan region. The maximum of its species diversity is confined to Central Iran, where about 300 species are concentrated (Rechinger 1972, 1986). It can be confidently asserted that the main, primary center of speciation of this genus is located in the Iranian Highlands. However, as the study of the distribution of all *Cousinia* species across the range of the genus has shown, there are several secondary centers of speciation along the periphery of the range (Cherneva 1974). One of these centers is located in Armenia (Tamanyan and Fayvush 1987). It should be noted that in the Armenian center, speciation takes place most intensively in the mountains in the southern part of the region, in the floristic regions of Meghri and South Zangezur in Armenia, and in the Nakhichevan Republic. Of the 23 species of the genus growing in Armenia, five have been karyologically studied; all of them are diploids, having a very narrow distribution. From a systematic point of view, speciation is found here more intensively in the sections *Cirsioideae* and *Cynaroideae* (Tamanyan and Fayvush 1987). The narrow-endemic *Cousinia* species occupy in most cases small areas on mountain slopes. This is facilitated by the life form of many representatives of this genus – "roll-field." These species practically have no possibility of spreading seeds for long distances from the mother plant.

All 12 species of the genus *Crepis* (Asteraceae) growing in Armenia have been karyologically studied; 10 of them are diploids, one is an octaploid, and four species have B-chromosomes. Particular attention is drawn to the Pannonian-Pontic-Sarmatian species *Crepis pannonica*, in which five diploid cytoraces with different basic numbers, triploid and aneuploid cytoraces, and numerous specimens with different numbers of B-chromosomes are registered in Armenia. In this case, it is important to note that in the main area of distribution of this species (the steppe region from Pannonia to Sarmatia), such karyological diversity was not found; in general, only specimens with $2n = 8$ are known, and no B-chromosomes have been registered (Dimitrova et al. 2003). Most likely, this species formed and spread over the steppe zone of Eurasia as the glaciers retreated, while in Armenia, having fallen into new, extreme conditions, it gave rise to a whole spectrum of cytoraces.

The second largest family of Armenian flora is *Poaceae*. Although this family is characterized by a high level of ploidy (38*x*, 20*x*, 18*x*, etc.), only tetra- and hexaploid cytoraces have been registered among karyologically studied species of cereals in Armenia. Among the polyploids, tetraploid cytoraces predominated, and the ratios of tetra- and hexaploids are 58% and 42%, respectively. *Colpodium versicolor* and *Zingeria biebersteiniana* had the smallest number of chromosomes ($2n = 2x = 4$), whereas *Aeluropus littoralis* ($2n = 6x = 60$) and *Echinochloa crus-galli* ($2n = 9x = 54$) had the highest number of chromosomes (Ghukasyan 2010). Among the karyologically studied annuals, species with diploid cytoraces predominate, and among perennials, there are more polyploid species.

Of the 42 wild species of the genus *Allium* (Alliaceae) growing in Armenia, 30 have been karyologically studied. The vast majority of these are diploids, but three species have diploid and tetraploid cytoraces, and one has di-, tetra-, and hexaploid cytoraces. In addition, one species is tetraploid and one is hexaploid; five species have B-chromosomes. Polyploid species in their distribution are mainly associated with the Caucasus and the Armeno-Iranian province, although there are species that are widespread in the Holarctic and the Ancient Mediterranean.

The genus *Geranium* (Geraniaceae), represented in Armenia by 20 species, is a typical allochthonous component of the flora of Armenia (Fayvush and Adamyan 2015). The vast majority of species are migrants from the north, from the Boreal Subkingdom, because they have the southern border of their distribution in Armenia. Of these, 14 have been karyologically studied (nine diploids, four tetraploids, and one hexaploid) with different basic chromosome numbers. An aneuploid series of basic chromosome numbers $x = 9$, 10, 13, and 14 is present in this genus. The species of the section *Rotundifolia* growing in the territory of Armenia (*G. pyrenaicum, G. pusillum, G. molle, G. rotundifolium, G. divaricatum,* all have $2n = 26$) are characterized by the basic main chromosome number of $x = 13$. The main basic chromosome number of x = 9 is typical for species of the section *Columbinum* growing in the territory of Armenia (*G. columbinum,* 2n = 18). The main basic chromosome number of $x = 10$ is typical for the species of the section *Robertiana* (*G. lucidum,* $2n = 40$). The main basic chromosome number of $x = 14$ is typical for the species of section *Geranium* (*G. ibericum* and G. *sylvaticum,* both have $2n = 28$). An asymmetric karyotype was observed in all the karyologically investigated species of the genus *Geranium* (Ghukasyan et al. 2022).

The distribution of cytoraces across altitudinal belts revealed that the greatest diversity of this region was confined to the middle mountain belt and was likely associated with the greatest richness of the flora of this belt and not with any extrinsic factors.

When considering the confinement of various cytoraces to certain types of vegetation and habitats, tetraploid, hexaploid, and octaploid species were mostly observed in steppe and meadow vegetation (Table 3.8; Fig. 3.16). In addition, species represented by two or more cytoraces, as well as those with B chromosomes, were most represented in steppes and meadows.

When considering the confinement of various cytoraces to certain types of vegetation, tetraploid, hexaploid, and octaploid species were best represented in the

Table 3.8 Cytoraces across the main habitats of Armenia

Habitat		Di-	Tetra-	Hexa-	Octa-	Tri-	Penta-	Aneu-	High ploidy
C3.24	Medium-tall nongraminoid waterside communities	2	4	3				1	
D2	Valley mires, poor fens, and transition mires	2	2		1				
D4.1	Rich fens, including eutrophic tall-herb fens and calcareous flushes and soaks	51	30	12			1	1	1
D6.2	Inland saline or brackish species-poor helophyte beds	5	2	2					
E1.2E	Irano-Anatolian steppes	287	92	29	8	4	1	4	2
E1.33 & E1.45	East Mediterranean xeric grassland & Sub-Mediterranean wormwood steppes	154	39	15	4	1		1	2
E2.32	Ponto-Caucasian hay meadows	179	57	24	4	2	2	3	2
E4.3A	Western Asian acidophilous alpine grassland	54	16	10		1		5	2
E4.4	Ponto-Caucasian alpine grassland	44	17	9		1		4	2
E5.4	Moist or wet tall-herb and fern fringes and meadows	40	16	2	1	3			1
E5.5A	Ponto-Caucasian tall-herb communities	3	5			1			1
F2.33	Subalpine mixed brushes	46	9	6	1	2		1	
F5.13 & G3.93	Juniper matorral & Grecian juniper – *Juniperus excelsa* – woods	8	5						
F5.34	Western Asian pseudomaquis	66	18	9	2	2		3	1
G1.6H	Caucasian beech forests	43	9	7		1		1	1
G1.A1	Oak-ash-hornbeam woodland	78	27	13	2	2		1	1
H2.3	Temperate-montane acid siliceous screes	37	10	7	1			1	
H2.4	Temperate-montane calcareous and ultrabasic screes	17	2	4	1			1	
H2.5	Acid siliceous screes of warm exposures	21	13	6	1			1	
H2.6	Calcareous and ultrabasic screes of warm exposures	34	13	5	1			2	
H3.1	Acid siliceous inland cliffs	31	11	4					1

(continued)

Table 3.8 (continued)

Habitat		Di-	Tetra-	Hexa-	Octa-	Tri-	Penta-	Aneu-	High ploidy
C3.24	Medium-tall nongraminoid waterside communities	2	4	3				1	
H3.2	Basic and ultrabasic inland cliffs	27	3	2					
H5.32	Stable sand with very sparse or no vegetation	9	4	2					
I1.3	Arable land with unmixed crops	16	3	1			1		
I1.53	Fallow un-inundated fields with annual and perennial weed communities	28	10	6					
I2.22	Subsistence garden areas	6							

composition of steppe and meadow vegetation (Fig. 3.16). In addition, in steppes and meadows, species represented by two or more cytoraces, as well as species with B-chromosomes registered in the karyotype, are best represented.

We paid special attention to representativeness of *Poaceae* species in some arid grasslands of Armenia. Their role in the composition of these ecosystems is extremely important; usually they are dominants or codominants the grass cover. We analyzed grasses growing in "steppes with wild wheat species dominance" (E1.2E22), "sandy patches in Ararat valley" (H5.323), "salt steppes and solonchaks grasslands" (E6.25), and "wormwood semi-desert of Armenia" (E1.451).

In the *Poaceae* family of Armenia, the vast majority of species are widely distributed, and they migrated to the territory of Armenia from the side of the Boreal subkingdom or from the side of the Ancient Mediterranean subkingdom. From the ecosystem "steppes with wild wheat species dominance" (E1.2E22), 13 species of grasses were investigated karyologically, 12 of them are associated with the Ancient Mediterranean, and one is distributed mainly in the Boreal subkingdom; nine species are diploids and four polyploids. In the ecosystem "sandy patches in Ararat valley" (H5.323), 19 species of *Poaceae* were investigated. Among them, 14 diploids and species associated with the origin of the Ancient Mediterranean subkingdom (17) predominate. The two diploid species are widely distributed in Eurasia and throughout the world. All five polyploid species originated in the Ancient Mediterranean subkingdom and then migrated to the Caucasus and Armenia as the Tethys dried up. In the ecosystem "salt steppes and solonchaks grasslands" (E6.25), 33 species of grasses were investigated. By their distribution and origin, most of them (23) are associated with the Ancient Mediterranean and migrated to Armenia, most likely in the Tertiary Period, as the Tethys dried up. In the ecosystem "wormwood semi-desert of Armenia" (E1.451), 27 species of grasses were investigated. Here, there are significantly more diploids (19) than polyploids (7). Of these species, 21 are associated with the Ancient Mediterranean. Six species, widely distributed in the northern regions of the Old and New Worlds, most likely appeared on the

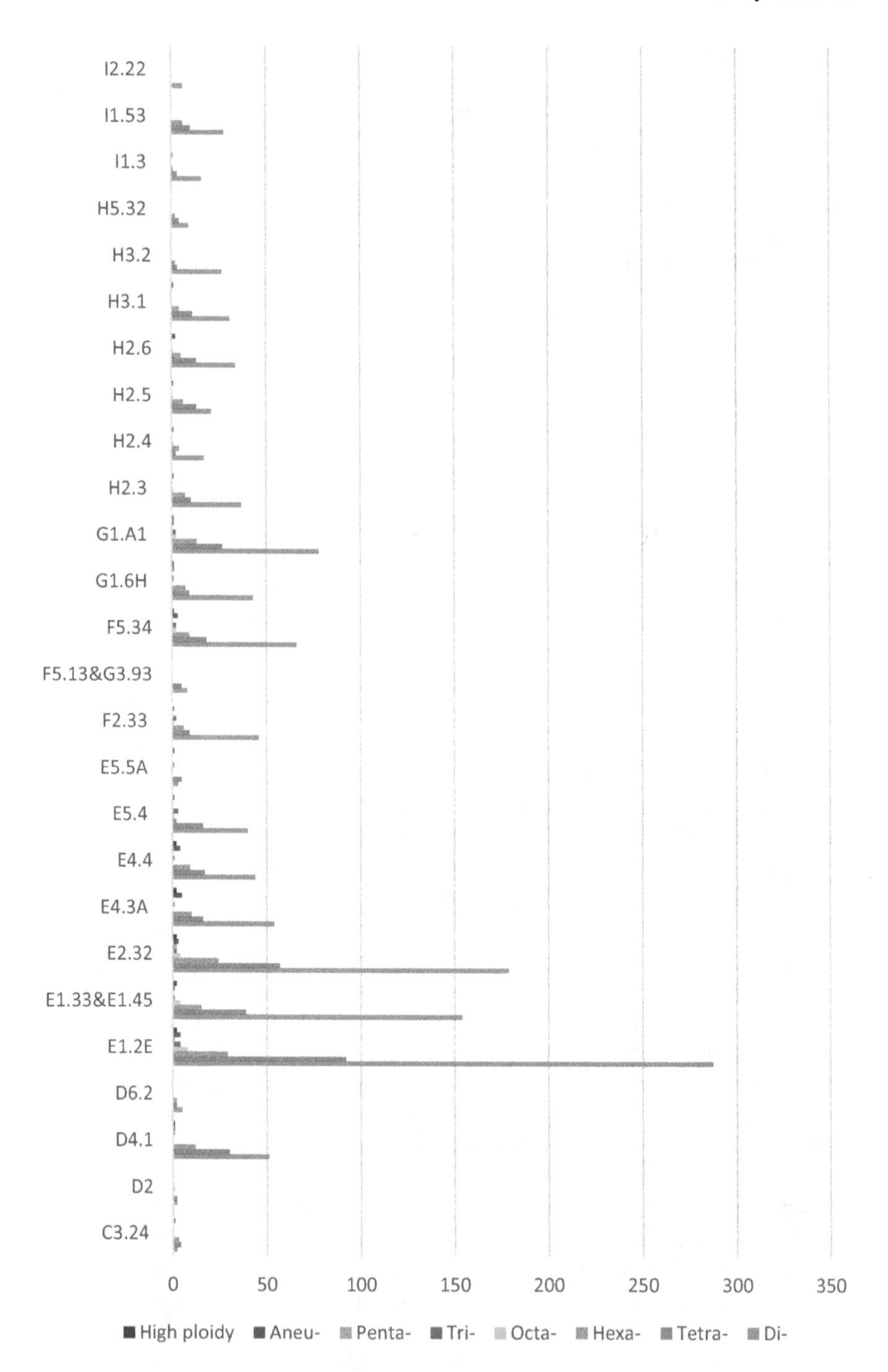

Fig. 3.16 Plant cytoraces across different Armenian habitats

territory of Armenia either during or after the Ice Age. On the basis of our investigation, we can conclude that the gramineous fraction of four ecosystems is predominantly of Ancient Mediterranean origin and in general is an allochthonous component of the Armenian flora.

For the chorological analysis of the studied species and determination of geographical elements, we used the scheme of Portenier (2000), adapted by us for the conditions of the South Caucasus (Fayvush and Adamyan 2015). Adapting Portenier's scheme, we paid more attention to the ecosystems encompassing the Caucasian and Armeno-Iranian provinces, particularly those in the Armenian, Atropatenian, Armeno-Atropatenian, Armeno-Iranian, and Caucaso-Armeno-Iranian areas.

Diploid species predominate in all geographical areas; in particular, only diploids are represented in the steppe area.

Among the species represented only by cytoraces with polyploidy (4n, 6n, 8n, and more), species with a wide Euro-Ancient Mediterranean range prevailed (30), as well as species with Armenian and Caucasian areas (26 each). Species in which only tri-, penta-, and aneuploid cytoraces have been recorded belong to the Armenian (5), Armeno-Iranian (2), Armeno-Atropatenian (1), Ancient Mediterranean (2), and Palearctic (2) geographical elements (Fig. 3.17; Table 3.9).

Most of the species represented by two or more cytoraces with different ploidy (2n–8n and more) have Armenian and Armeno-Iranian (nine each) types of ranges, and another six species each are Atropatenian and Ancient Mediterranean elements.

Of the 22 species with B chromosomes in their karyotype, five were confined to Armenia, and three species each had a Palearctic, East Ancient Mediterranean,

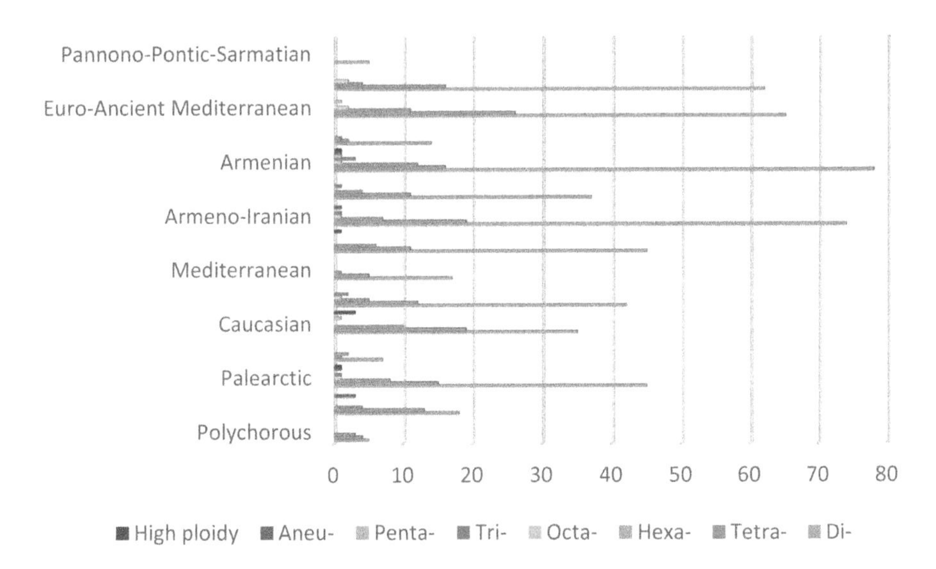

Fig. 3.17 Distribution of different cytoraces across the studied geographical elements (following the scheme of Takhtajan 1986)

Table 3.9 Geographical elements of Armenian flora (karyologically investigated species)

	Geographical element (types of area)	Species
Special type	Polychorous	11
Boreal subkingdom (173 species)	Holarctic	32
	Palearctic	65
	European	14
	Caucasian	62
Ancient Mediterranean subkingdom (408 species)	Ancient Mediterranean	48
	Mediterranean	25
	East Ancient Mediterranean	83
	Armeno-Iranian	89
	Armeno-Atropatenian	41
	Armenian	106 (19 local endemics)
	Atropatenian	16
Transitional type (174 species)	Euro-Ancient Mediterranean	95
	Caucaso-Armeno-Iranian	79
Special (Steppe) type	Pannono-Pontic-Sarmatian	5

Armeno-Atropatenian, Atropatenian, and Euro-Ancient Mediterranean types of ranges.

Based on the position that allo- and autochthonous trends in the evolution of the Armenian flora were approximately equivalent (Fayvush 1990), it should be recognized that approximately half of the species migrated to the territory of Armenia from other regions, while the second half formed on the territory of the republic. Given that nature does not recognize administrative boundaries, the autochthonous trend should be extended to the Armenian Highlands as a whole and partially to the Caucasus. Thus, it should be assumed that the Armenian and Caucasian species, which do not have diploid cytoraces in Armenia but have only polyploid ones, formed in this region, while the diploid cytorace was most likely forced out to be less competitive. Simultaneously, attention is drawn to the fact that polyploid cytoraces are best represented in the composition of steppe and semidesert vegetation and in arid woodlands, which is most likely the result of a morphogenetic "explosion" in large genera in arid ecosystems (Agakhaniants 1981). Caucasian polyploids, on the other hand, are mainly confined to alpine and subalpine vegetation and most likely have a glacial and postglacial age, adapting to and occupying these habitats as glaciers retreat. A large number of polyploid species with the Euro-Mediterranean type range are so widespread that it is possible that diploid cytoraces exist somewhere in this space; only polyploids have reached the territory of Armenia.

The predominance of representatives of the Ancient Mediterranean geographic element among species with two or more cytoraces with different ploidy, namely, species in the Armeno-Iranian, Armenian, Atropatenian, and Ancient Mediterranean areas, is also most likely evidence of intensive speciation processes in the arid regions of the Armenian Highlands and of Armenia in many genera (e.g., *Centaurea, Astragalus, Cousinia*, etc.) (Fayvush and Aleksanyan 2020). This is also evidenced by the predominance of representatives of the Armenian, Armeno-Iranian, and Armeno-Atropatenian elements among the species in the karyotype of which B-chromosomes are registered.

3.4.1 Endemism

When considering the problems of endemism, it is first of all necessary to clearly indicate the geographical region which endemism is being studied. As mentioned above, the territory of Armenia is distributed between two floristic provinces (Caucasian and Armeno-Iranian) belonging to two floristic subkingdoms (Circumboreal and Ancient Mediterranean).

The flora of the entire Caucasian ecoregion (including Armenia, Georgia, Azerbaijan, Russian part of the Great Caucasus, and neighboring parts of Turkey and Iran) is estimated at about 6000–6500 species of vascular plants; 40–45% of them (2791 species) are endemics (Solomon et al. 2013). Taxonomic analysis indicates that there are 21 genera of vascular plants endemic across the Caucasus region. These genera mostly are monotypic or oligotypic, and some have unclear relationships. The vast majority of the genera endemic to the Caucasus grow in the Greater Caucasus, while in the Lesser Caucasus (including Armenia), only four endemic genera exist: *Agasyllis* (Apiaceae; 1 species), *Grossheimia* (Asteraceae; 6 spp.), *Pseudovesicaria* (Brassicaceae; 1 sp.), and *Zuvanda* (Brassicaceae; 1 sp.). Species of these genera grow mainly in the highlands and are confined to stony habitats, screes, rocks, and subalpine meadows.

Although generic endemism is very characteristic of the entire Irano-Turanian region, it is particularly strong in its eastern part (Kamelin 1965; Hedge and Wendelbo 1970, 1978). In the northeastern part of the Armenian Highlands, there are eight genera of Ancient Mediterranean roots: *Peltariopsis, Pseudoanastatica, Takhtajaniella* (all from Brassicaceae), *Smyrniopsis, Stenotaenia, Szovitsia* (all from Apiaceae), *Huynhia* (Boraginaceae), and *Callicephalus*(Asteraceae).

In Armenia, 146 species of the flora (3.8%) are considered as local endemics. *Rosaceae* is the richest in endemic species (c. 30 species), thanks to a great diversity in the genus *Pyrus* (11 species), followed by *Asteraceae* (27 species), *Fabaceae* (17), *Poaceae* (15), *Caryophyllaceae* (10), and *Scrophulariaceae* (8). On the genus level, *Astragalus* has the largest number of species in Armenia and has 12 local endemic species, followed by *Pyrus* (11 species), *Psephellus* (8), and *Centaurea* (7). The number of endemics in different floristic regions is shown on the map of Armenia (Fig. 3.18).

Wagenitz (1986) notes that a large number of endemic *Centaurea* taxa are known only from their locus classicus and some have not been gathered again for more than a hundred years and, quite possibly, have already disappeared. Apparently, they might have been the nonviable mutants. Unfortunately, among the 146 endemic species from different genera described from Armenia, a fairly large number is also known only by type specimens and only from the locus classicus and have not been

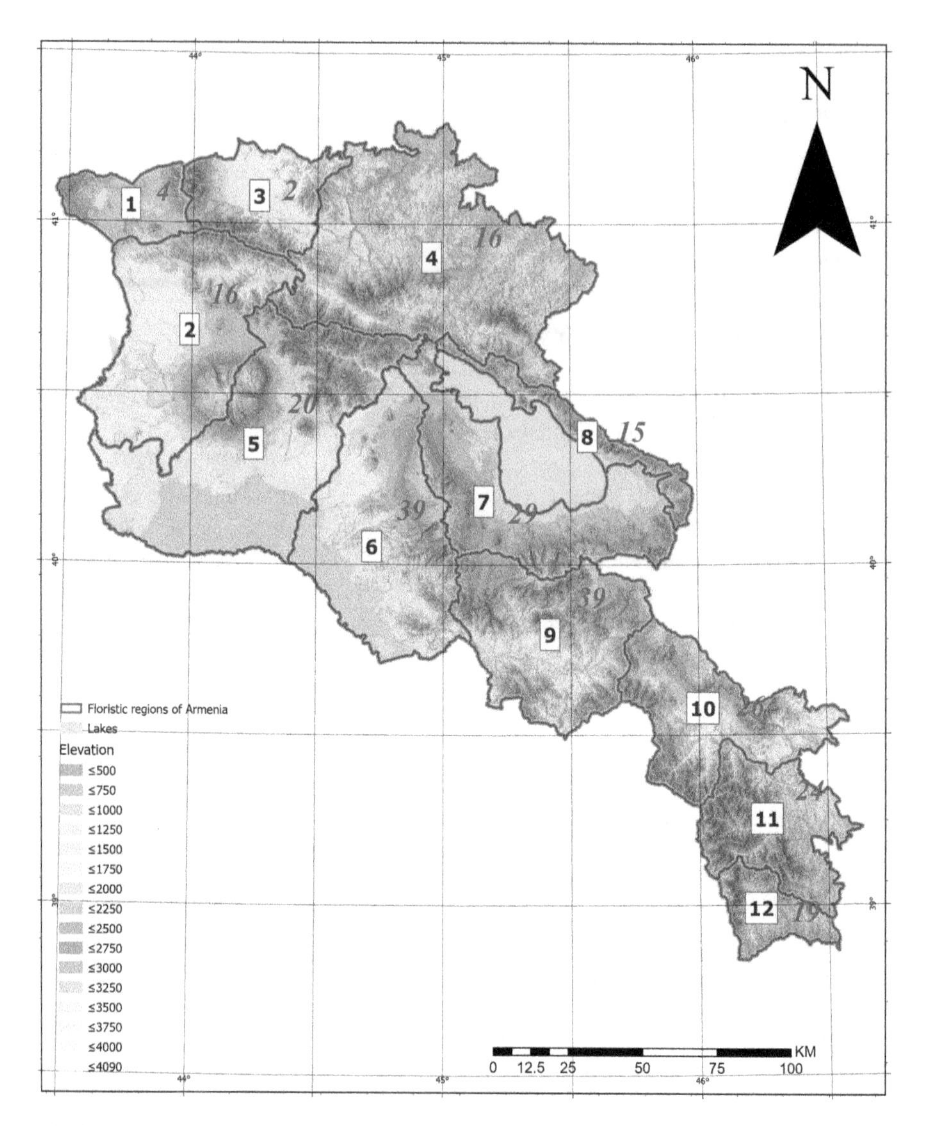

Fig. 3.18 Number of endemics in different floristic regions of Armenia

collected for many years. The current state of their populations is unknown, and it is also quite possible that they have become extinct, being unviable mutants.

Considering the distribution of endemic taxa along high-altitude belts, it is clear that they are more concentrated in the middle and partly in the upper montane belts. Obviously, this montane belt has the greatest variety of habitats, and their conditions are most conducive to speciation. In the lower as well as in the subalpine and especially the alpine belts, the extreme soil and climatic conditions make adaptation of new mutations difficult. Approximately, three-quarter of all endemic taxa grow in arid plant communities – steppes, arid open forests, and shibliak (communities with *Paliurus spina-christi* dominance), steppe shrubs communities, and semideserts, often preferring petrophilous variants of these vegetation types. Apparently, the vast majority of these species are neo-endemic, and their abundance in these types of vegetation confirms the hypothesis of a "morphogenetic explosion" in arid habitats (Agakhaniants 1981). Approximately, a quarter of the endemic species are found in more humid plant communities – forests, meadows, subalpine tall grasses, and wetlands. These taxa are probably paleo-endemics, preserved in the humid refugia of arid mountains or in forests. A good illustration of this situation is the distribution of the number of endemics in the floristic regions of Armenia (Fig. 3.18).

References

Abrahamyan AA (1984) Lichen flora of the Lake Sevan. Dissertation, Yerevan

Abrahamyan AA (1996) The list of lichens of basin of the Lake Sevan. Botanicheskiy J 81:23–29

Abramova AL, Abramov II (1950) Novye mkhi dlja flory Kavkaza (New mosses for the flora of the Caucasus). Bot J 35(5):514–516

Abramova AL, Abramov II (1959) O mkhah Armenii (About mosses of Armenia). Proc Bot Inst AS USSR Spore Plants 12(2):360–366

Abramova AL, Abramov II (1977) Taxonomicheskaja structura bryoflory SSSR (Taxonomic structure of bryofloras of the USSR). News of taxonomy of lower plants, 14, pp 191–200

Agakhaniants O (1981) Aridnye gory SSSR (Arid mountains of the USSR), Moscow

Alverdieva SM (2018) Current state of the lichen flora of Azerbaijan (Unpublished doctoral Dissertation), Baku

Arup U, Søchting U, Frödén P (2013) A new taxonomy of the family Teloschistaceae. Nord J Bot 31:016–083

Asatryan VL, Hambaryan LR, Ghukasyan EK (2022) Gorizontal'noe rasprostranenie biomassy phytoplanktona ozera Sevan: metod rascheta s ispol'zovaniem techologij GIS (Horizontal distribution of the phytoplankton biomass of Lake Sevan: a method of calculation using GIS technologies. Scientific bases for the conservation and restoration of natural resources of Lake Sevan. Rostov-on-Don, pp 323–332

Asplund J, Wardle DA (2017) How lichens impact on terrestrial community and ecosystem properties. Biol Rev 92:1720–1738

Avakyan KG, Dildaryan BI (1989) Stepnaja rastitel'nost' juzhnyh sklonov khrebta Teghenis (The steppe vegetation of southern sides of Techenis mountain range). Proc YSU Nat Sci 3(171):155–157

Avakyan KG, Dildaryan BI (1990) Lesnaja rastitel'njst' severo-vostochnykh sklonov khrebta Teghenis (The forest vegetation of the north-eastern slopes of Techenys ridge in Armenia). Proc YSU Nat Sci 3(175):120–124

Avetisyan MS, Dildaryan BI (1978) Materialy k bryoflore okrestnostey Leninakana (Materials for the bryoflora of surrounding of Leninakan). Young scientific researcher. Natural Sciences, 1, Yerevan, pp 117–120

Avetisyan MS, Dildaryan BI (1983) Materialy k flore mkhov poluostrova Artanish Sevanskogo natsional'nogo parka (Materials of the Flora of Mosses of the Sevan National Park of Artanish Peninsula). Biol J Armenia 36(9):797–799

Avetisyan MS, Dildaryan BI (1986) Uchastie mkhov v formirovanii rastitel'njsti na pochve Tsovinarskogo parka "Sevan" (The participation of mosses in the formation of vegetation on the exposed soils of the Tsovinar Park "Sevan"). Mat. VII Transcaucasian Conf. on Spore Plants, Yerevan, p 8

Avetisyan MS & Dildaryan BI (1987) Nekotorye zakonomernosti v rasprostranenii listostebel'nykh mkhov d lesakh v okrestnostjakh Tsahkadzora (Some regularities in the distribution of leafy mosses in the forests of the surroundings of Tsaghkadzor). Questions of biology, Yerevan, 4, pp 29–36

Badalyan KL (2006) Studies of phytoplankton community of Hrazdan water ecosystem, Yerevan

Barinova SS, Kukhaleishvili L, Nevo E, Janelidze Z (2011) Diversity and ecology of algae in the Algeti National Park as a part of the Georgian system of protected areas. Turk J Bot 35:729–774

Barkhalov SO (1983) Lichen flora of the Caucasus, Baku

Barseghyan AM (1959) Geobotanicheskaja characteristica osnovnyh formacij vodno-bolotnoj rastitel'njsti Araratskoj ravniny (Geobotanical characteristics of the main formations of wetland vegetation of the Ararat Valley). Works Bot Inst AS ArmSSR 12:41–91

Barseghyan AM (1969) Algae. Nat Armenia 11:7–9

Barsegyan AM (1990) Vodno-bolotnaja rastitel'nost' Armjanskoj SSR (Water-marsh vegetation of Armenian SSR), Yerevan

Belanger C (1833–1837) Voyage aux Indes-Orientales par le Nord l'Europe, les provinces du Caucase, la Georgie, l'Armenie et la Perse,... Botanique II. Partie Cryptogamie, Paris

Boissier E, Buhse F (1860) Aufzoellung des auf einer Reise durch Transkaukasien and Persien gesammelten Pflanzen. Nouv Mem Soc Nat Moscau 12:234–237

Carmichael WW (2001) Health effects of toxin-producing cyanobacteria: "The CyanoHABs". Hum Ecol Risk Assess 7:1393–1407

Carta A, Peruzzi L, Ramírez-Barahona S (2022) A global phylogenetic regionalization of vascular plants reveals a deep split between Gondwanan and Laurasian biotas. New Phytol 233(3):1494–1504. https://doi.org/10.1111/nph.17844

Chakraborty S, Ramesh P, Dutta S (2015) Toxic phytoplankton as a keystone species in aquatic ecosystems: stable coexistence to biodiversity. First published. https://doi.org/10.1111/oik.02322

Cherneva O (1974) Kratkiy analiz geographicheskogo rasprostranenija vidov roda *Cousinia* Cass. (Short analysis of geographical distribution of the species of *Cousinia* Cass. genus). Bot J (Moscow-Leningrad) 59:183–191

Danielyan AA (2009) Prirodookhrannaja otsenka i perspectivy ustojchivogo razvitija reki Debed i ee vodosbornogo bassejna (Environmental assessment and prospects for sustainable development of the Debed River and its catchment area), Yerevan

Davis P (1965) Flora of Turkey and the East Aegean Islands. V. 1, Edinburgh

Derzhavin AN (1940) Otchet ob obsledovanii nekotoryh vodokhranilishch Armenii s cel'ju organizatsii prudovyh khozajstv (Report of the survey of some reservoirs of Armenia in order to organize a pond feed farm). Proceedings of the Sevan hydrobiological station, 6

Dildaryan BI (1969) Bryoflora lesov verkhnego gornogo pojasa Severo-vostochnoj Armenii (Bryoflora of forests of upper mountain belt of North-East Armenia). Young Scientific Researcher, Yerevan 2(10):91–96

Dildaryan BI (1971) Bryoflora lesov Severo-vostochnoj Armenii (Bryoflora of forests of North-East Armenia). PhD thesis, Yerevan

Dildaryan BI (1972) Rasprostranenie mkhov po vysotnym pojasam v lesakh Severo-vostochnoj Armenii (The distribution of mosses in altitudinal belts in forests of North-East Armenia). Abstract IV Transcaucasian meeting on Spore Plants, Yerevan, pp 69–70

Dildaryan BI (1979) Razvitie epigejnykh mkhov v dubovo-grabovykh lesah Central'noj Armenii (The development of epigeic mosses in Quercus-Carpinus forests of Central Armenia). Materials of V conference Lower plants of Transcaucasica, Baku, 123–124

Dildaryan BI, Avetisyan MS (1981) Materialy k flore mkhov juzhnykh sklonov Pambakskogo khrebta (Materials for the moss flora of the southern slopes of the Pambak range). Interuniversity proceedings "problems of biology", Yerevan, 2, pp 44–48

Dildaryan BI, Avetisyan MS (1983) Listostebel'nye mkhi na osvobozhdennykh pochvakh bassejna ozera Sevan (Leafy mosses of exposed soils of the Lake Sevan basin). Materials of IV Transcaucasian conference on Spore Plants, Tbilisi, pp 136–137

Dildaryan BI, Avetisyan MS (1987) Dopolnenija k flore mkhov juzhnykh sklonov Pambakskogo khrebta (Contributions to the moss flora of southern slopes of Pambakskij Range). Interuniversity proceedings of problems of biology, Yerevan, 4, pp 44–48

Dildaryan BI, Petrosyan AM (1987) Ecologicheskoe i tsenoticheskoe rasprostranenie mkhov na territorii Dilijanskogo zapovednika (Ecological and coenotic distribution of mosses in the territory of the Dilijan Reserve). Interuniversity proceedings "problems of biology", YSU, 4, pp 74–79

Dildaryan BI, Karapetyan NA, Poghosyan AV (2002) O structurnykh osobennostjakh gametophyta listostebel'nykh mkhov (On the structural peculiarities of gametophyte of leaved mosses). Proc YSU Nat Sci 2(198):105–109

Dimitrova D, Fischer MA, Kaestner A (2003) Crepis pannonica (Asteraceae-Lactuceae): karyology, growth-form, phytogeography, occurrence and habitats in Austria; including subsp. blavii comb, et stat. nov. Neilreichia 2–3:107–130

Dokulil MT, Teubner K (2000) Cyanobacterial dominance in lakes. Kluwer Academic Publishers

Dolukhanov AG (1949) Lesa Zangezura (Forests of Zangezur). Proc BIN AN ArmSSR, Yerevan, 6, pp 65–134

Dylevskaya IV (1949) Materialy k poznaniju vechnozelenykh mkhov v Gruzii (Materials for the knowledge of deciduous mosses in Georgia), J State Museum of Georgia, 14a

Dylevskaya IV (1959) Dopolnenija k flore mkhov Malogo Kavkaza (Contributions to mosses flora of Caucasus Minor). Notes of sistem., geogr. of plants. Bot Inst AS GruzSSR 21:4–6

Dylevskaya IV, Barseghyan AM (1971) Materialy k bryoflore Armenii (Materials for the bryoflora of Armenia). Biol J Armenia 24(8):89–94

Dylevskaya IV, Barseghyan AM (1975) Materials for the wetland flora and moss vegetation of Armenia. Flora, Vegetation and Plant resources of Armenian SSR, 5, pp 27–72

Egorova TV (1999) The sedges (*Carex* L.) of Russia and adjacent states (within the limits of the former USSR). Missouri Botanical Garden Press, Saint-Louis

Ermolaev VI (1981) Phytoplankton reki Pjasiny (Zapadnyj Taimyr) (Phytoplankton of r. Pyasina (Western Taimyr). New data on the phytogeography of Siberia. Nauka Sibiri, pp 16–29

Farra MS (1991) Comparative anatomy and ultrastructure of gametophytes of mosses in Syria. Aftoref. PhD thesis, Moscow

Fayvush GM (1990) Ob autochtonnoj i allochtonnoj tendencijah v razvitii stepey Armjanskoj SSR (On autochthonous and allochthonous tendencies in development of steppes of Armenian SSR). Biol J Armenia 43(3):220–225

Fayvush GM, Adamyan RG (2015) Botanico-geographicheskij analiz semejstva *Geraniaceae* Juss. v Armenii (Phyto-geographical analysis of *Geraniaceae* Juss. family in Armenia). Bot Herald North Caucasus 1:106–123

Fayvush G, Aleksanyan A (2020) The Transcaucasian Highlands. In: Noroozi J (ed) Plant biogeography and vegetation of high mountains of Central and South-West Asia. Springer, pp 287–313

Fayvush G, Ghukasyan A (2023) Karyo-geographical analysis of Armenian flora. Front Biogeogr 15(2):1–8. https://doi.org/10.21425/F5FBG58420

Gabrielian E, Fajvush G (1989) Floristic links and endemism in the Armenian Highlands. In: The Davis & Hedge Festschrift, Edinburgh, pp 191–206

Gabrielian E, Fayvush G (1986) Botaniko-geographicheskij analiz armjanskih vidov roda *Scrophularia* (Phyto-geographical analysis of the Armenian species of the genus *Scrophularia*). Biol J Armenia 39(2):170–173

Gasparyan A, Aptroot A (2016) *Verrucaria juglandis*, a new corticolous lichen species from Armenia. Herzogia 29:103–107

Gasparyan A, Sipman H (2013) New lichen records from Armenia. Mycotaxon 123:491–492

Gasparyan A, Sipman H & Brackel W von (2014) A contribution to the lichen-forming and lichenicolous fungi flora of Armenia. Willdenowia, 44, pp. 263–267

Gasparyan A, Sipman H (2016) The epiphytic lichenized fungi in Armenia: diversity and conservation. Phytotaxa 281(1):1–68

Gasparyan A, Sipman H (2020) The first record of *Lobaria pulmonaria* from Armenia. Herzogia 33:554–558

Gasparyan A, Aptroot A, Burgaz AR, Otte V, Zakeri Z, Rico VJ, Araujo E, Crespo A, Divakar PK, Lumbsch HT (2015) First inventory of lichens and lichenicolous fungi in the Khosrov Forest State Reserve, Armenia. Flora Mediterranea 25:105–114

Gasparyan A, Aptroot A, Burgaz AR, Otte V, Zakeri Z, Rico VJ, Araujo E, Crespo A, Divakar PK, Lumbsch HT (2016) Additions to the lichenized and lichenicolous mycobiota of Armenia. Herzogia 29(2):692–705

Gasparyan A, Sipman H, Lücking R (2017) *Ramalina europaea* and *R. labiosorediata*, two new species of the *R. pollinaria* group (Ascomycota: Ramalinaceae), and new typifications for *Lichen pollinarius* and *L. squarrosus*. The Lichenologist 49(4):301–319

Getsen M V (1985) Algae in ecosystems of the Far North. Nauka, L.: 165 p. (In Russian)

Gevorgyan G, Rinke K, Schultze M et al (2020) First report about toxic cyanobacterial bloom occurrence in Lake Sevan. Int Rev Hydrobiol Armenia:1–12. https://doi.org/10.1002/iroh.20200206012

Ghukasyan AG (2010) Karyological investigations of the family *Poaceae* of the Armenian flora. Abstract international conference "Study of the Flora of the Caucasus". Pyatigorsk, pp 33–34

Ghukasyan AG (2011) Speciation in the genus *Zingeria (Poaceae)* of the flora of Armenia. Takhtajania 1:142–143

Ghukasyan A, Akopian J (2018) Karyological study of the endangered species of Armenian flora *Vavilovia formosa* (Fed.) Steven (Fam. *Fabaceae*) from Geghama mountains population. Electron J Nat Sci 31(2):32–34

Ghukasyan A, Janjughazyan K (2015) Chromosome numbers of some rare flowering plants of Armenian flora. Electron J Nat Sci 24(1):23–26

Ghukasyan A, Adamyan R, Poghosyan A (2022) Karyological study of some representatives of the family *Geraniaceae* of the flora of Armenia. Botanicheskii zhurnal 107(8):794–799. https://doi.org/10.31857/S0006813622080063

Golovenkina NI (1967) Interesnye i redkie diatomovye iz neogenovyh kontinental'nyh otlozhenij Sisianskogo rajona ArmSSR (Interesting and rare diatoms from the Neogene continental deposits of the Sisian region of the Armenian SSR). News of taxonomy of lower plants, pp 38–46

Grubov VI (2010) Schlussbetrachtung zum Florenwerk "Rastenija Central'noj Azii" [Die Pflanzen Zentralasiens] und die Begründung der Eigenständigkeit der mongolischen Flora. Feddes Repertorium 121(1–2):7–13. https://doi.org/10.1002/fedr.201011123

Hambaryan LR (2001) Successija phytoplanktona v period povtorjajushchegosja snizhenija ozera Sevan (Phytoplankton succession during the period of repeated lowering of the level of Lake Sevan), Yerevan

Hambaryan LR, Mamyan AS (2016) Phytoplankton of the tributies of Lake Sevan. Lake Sevan ecological state during the period of water level change. Yaroslavl, pp 61–78

Hambaryan LR, Hovsepyan AA, Hovhannesyan RO (2007) Dinamika kolochestvennogo razvitija phytoplanktonnogo soobshchestva ozera Sevan v 2005 godu (Dynamics of the quantitative development of the phytoplankton community of Lake Sevan in 2005). Agronauka, Yerevan, pp 355–360

Hambaryan LR, Stepanyan LG, Hovhannisyan NA, Yesoyan SS (2019) Environmental risks and consequences of flowering of Cyanobacteria in the summer period of the high mountain Lake Sevan (Armenia), International conference microbes: biology& application, Abstract book, Yerevan, p 72

Hambaryan LR, Stepanyan LG, Mikaelyan MV, Gyurjyan QG (2020a) The bloom and toxicity of Cyanobacteria In Lake Sevan. Proc Yerevan State Univ Chem Biol 54(2):168–176

Hambaryan L, Khachikyan T, Ghukasyan E (2020b) Changes in the horizontal development of phytoplankton of the littoral of Lake Sevan (Armenia) in conditions of water level fluctuations. Limnol Freshw Biol 4:662–664

Harutyunyan S, Mayrhofer H (2009) A contribution to the lichen mycota of Armenia. Bibliotheca Lichenologica 100:137–156

Harutyunyan S, Wiesmair B, Mayrhofer H (2011) Catalogue of the lichenized fungi in Armenia. Herzogia 24:265–296

Hayrapetyan AM, Ghukasyan AG (2015) Palynological and karyological study of endemic species of the genus *Rhaponticoides* Vaill. (*Asteraceae*), included in the Red Book of plants of Armenia. Rep NAN RA 115(2):156–162

Hayrapetyan AM, Ghukasyan AG (2021) Palynological and karyological features of an endemic species *Cousinia fedorovii* Takht. (*Asteraceae*), included in the Red Book of plants of Armenia. Rep NAN RA 121(3):245–250

Hedge IC, Wendelbo P (1970) Some remarks on endemism in Afghanistan. Israel J Bot 19:401–417

Hedge IC, Wendelbo P (1978) Patterns of distribution and endemism in Iran. Notes Roy Bot Gard 36:441–464

Hovsepyan AA (2013) Changes in the phytoplankton community under conditions of lake water level rise in Lake Sevan, Yerevan

Hovsepyan AA (2017) Features of the development of phytoplankton of Lake Sevan in 2015. Biol J Armenia 4(69):94–100

Hovsepyan AA, Khachikyan TG (2016a) Phytoplankton littoral'noj zony i zataplivaemyh territorij ozera Sevan (Phytoplankton of the littoral zone and flooded areas of the coast of Lake Sevan). Lake Sevan. Ecological state during the period of water level change. Yaroslavl, pp 15–34

Hovsepyan AA, Khachikyan TG (2016b) Phytoplankton pelagiali ozera Sevan (Phytoplankton of the pelagial of Lake Sevan). Lake Sevan. Ecological state during the period of water level change. Yaroslavl, pp 35–56

Hovsepyan AA, Hambaryan LR, Hovhanesyan RO, Gusev ES (2010) Planktonnye vodoroali ozera Sevan (Planktonic algae of Lake Sevan. Ecology of Lake Sevan during the period of its level increase. Research results of the Russian-Armenian biological expedition for hydroecological survey of Lake Sevan (Armenia) (2005–2009). Makhachkala, pp 90–104

Hovsepyan AA, Khachikyan TG, Hambaryan LR, Martirosyan AE (2013) Qualitative structural features of phytoplankton community in the Lake Sevan and its catchment basin. Ann Agrarian Sci 11(1):80–85

Hovsepyan AA, Khachikyan TG, Hambaryan LR (2014) Influence of Lake Sevan catchment basin phytoplankton community structure on the same of the lake. AASSA Regional Workshop proceedings "Sustainable management of water resources and conservation of mountain lake ecosystems of Asian countries", Yerevan, pp 102–112

Hovhannisyan RO, Ivanyan MS, Davtyan AA et al. (2010) Hydroecological studies of Lake Sevan and its drainage basin. Biological Journal of Armenia 62(2):81–85

Ignatov MS, Afonina OM, Ignatova EA (2006) Check-list of mosses of East Europe and North Asia. Arctoa 15:1–130

Inashvili T, Kupradze I, Batsatsashvili K (2022) A revised catalog of lichens of Georgia (South Caucasus). Acta Mycol 57:1–46

Kalantaryan NK (2020) The study of the biodiversity of microalgae and cyanobacteria and the assessment of the biotechnological potential of the *Parachlorella kessleri* microalgae, Yerevan

Kamelin R (1965) O rodovom endemisme vo flore Srednej Asii (On the generic endemism in the flora of Middle Asia). Bot J (Moscow-Leningrad) 50:1702–1710

Kara-Murza EN (1931a) Report on geo-botanical activities of Sevan expedition 1927–1928. In: Berg LS (ed) , vol 2. Lake Sevan basin. Leningrad, pp 113–188

Kara-Murza EN (1931b) Otchet o geobotanicheskih issledovanijakh Sevanskoj expedicii v 1927–1928 gg. (Report of geobotanical exploration of Sevan expedition in 1927–1928). Basin of lake Sevan. Leningrad 2(2):113–188

Karimova BK (1972) On the flora of algae in some saz reservoirs of the Alai Valley. Flora and importance of spore plants in Central Asia, Tashkent

Khachikyan TG (2013) Modern characteristics of phytoplankton in the rivers of the catchment basin of Lake Sevan, Yerevan

Khachikyan TG, Stepanyan LG (2018) Species diversity of phytoplankton community in the main rivers of Lake Sevan catchment basin. In: 2nd International Young scientists' conference on biodiversity and wildlife conservation ecological issues Tsaghkadzor, pp 106–108
Khandjian N (2001) *Carex* L. In: Takhtadjan AL (ed) Flora of Armenia, vol 10, pp 456–529
Kharitonov VG (1981) The diatoms from Mayorskoe Lake (Anadir District). Bot Zhurn 66(4):542–549
Khudoyan AA (1994) Phytoplankton osnovnyh pritokov ozera Sevan (Phytoplankton of the main tributaries of Lake Sevan), Moscow
Kiselev LA (1969) Plankton of the seas and continental water bodies. I. Introductory and general questions of planktology, Leningrad
Koopman J, Aleksanyan T, Aleksanyan A, Fayvush G, Oganesian M, Vitek E, Więcław H (2021) The genus *Carex* (Cyperaceae) in Armenia. Phytotaxa 494(1):1–41
Kotseruba V, Pistrick K, Kumke K, Weiss O, Rutten T, Blattner FR, Fuchs J, Endo T, Nasuda S, Ghukasyan A, Houben A (2010) The evolution of the hexaploid grass *Zingeria kochii* (Mez) Tzvel. (2n=12) was accompanied by complex hybridization and uniparental loss of ribosomal DNA. Mol Phylogen Evol Today 56(1):146–155
Kotseruba V, Pistrick K, Gernand D, Meister A, Ghukasyan A, Houben A (2012) Characterisation of the low-chromosome number grass *Colpodium versicolor* (Stev.) Schmalh. (2n=4) by molecular cytogenetics. Caryologia 58(3):241–245
Kulumbaeva AA (1982) Phytoplankton of Lake Issyk-Kul, Frunze: Kyrgyzstan
Legovich NO (1979) "Tsvetenie" vody ozera Sevan ("Blooming" of the water of Lake Sevan). Ecol Hydrobionts Lake Sevan 17:51–74
Levadnaya GD (1992) Benthic diatoms in the river ecosystem. Phytoplankton monitoring, Novosibirsk
Lücking R, Hodkinson BP, Leavitt SD (2017) The 2016 classification of lichenized fungi in the Ascomycota and Basidiomycota – approaching one thousand genera. Bryologist 119:361–416
Lürling M, Mello MM, van Oosterhout F, de Domis L, Marinho MM (2018) Response of natural cyanobacteria and algae assemblages to a nutrient pulse and elevated temperature. Front Microbiol 9:1851
Magakyan A (1941) Rastitel'nost' Armjanskoj SSR (Vegetation of Armenian SSR), Moscow-Leningrad
Mamyan AS (2013) Vozdejstvie antropogennogo factora na phytoplanktonnye soobshchestva rek Pambak i Tandzut (Impact of the anthropogenic factor on the phytoplankton community of the Pambak and Tandzut rivers), Yerevan
Manakyan VA (1969) Dopolnenija k bryoflore Zangezura II (Contributions to the bryoflora of Zangezur. II). Biol J Armenia 22(11):78–86
Manakyan VA (1970) Mkhi juzhnogo Zangezura (Mosses of South Zangezur). PhD thesis, Yerevan
Manakyan VA (1975) Mkhi lesov Meghri (Mosses of forests of Megri). Biol J Armenia 28(9):70–76
Manakyan VA (1983) Novye dannye k bryoflore Zangezura i Armenii (New data on bryoflora of Zangezur and Armenia). Biol J Armenia 36(10):859–862
Manakyan VA (1986) Rod *Sphagnum* L. v Armenii (Genus *sphagnum* L. in Armenia). Materials of VII Transcaucasian conference on Spore Plants, Yerevan, p 54
Manakyan VA (1989a) Obzor bryoflory Armenii (A review of bryoflora of Armenia). Problems of bryology in USSR, Leningrad, pp 157–165
Manakyan VA (1989b) Listostebel'nye mkhi Jugo-vostochnoj Armenii (Leafy mosses of south-eastern Armenia), Yerevan
Manakyan VA (1995) Resultaty bryologicheskikh issledovanij v Armenii (Results of bryological studies in Armenia). Arctoa 5:15–33
Manakyan VA, Vanya J, Duda J, Abramova AL (1999) Materialy k Hepaticae Armenii (Materials to Hepaticae of Armenia). Flora, vegetation and plant resources of Armenia, 12, 17–25
Martín-Bravo S, Jiménez-Mejías P, Villaverde T, Escudero M, Hahn M, Spalink D, Roalson EH, Hipp AL (2019) A tale of worldwide success: behind the scenes of *Carex* (Cyperaceae) biogeography and diversification. J Syst Evol 57:695–718. https://doi.org/10.1111/jse.12549

Melikyan AP, Dildaryan BI (1975) Tipy anatomicheskoj structury steblja ganetofitov v razlichnykh ecologicheskih gruppakh mkhov v Armenii (Gametophyte stem anatomical structure types in various ecological groups of moss in Armenia). Biol J Armenia 28(1):19–24

Meshkova TM (1962) Sovremennoe sostojanie planktona ozera Sevan (The current state of plankton in Lake Sevan). Proceedings of Sevan hydrobiological station, 16, pp 15–88

Mikaelyan AL (1996) Allogennaja successija phytoplanktonnogo soobshchestva ozera Sevan (Allogeneic succession of the phytoplankton community of Lake Sevan), Yerevan

Mnatsakanyan AT (1984) Izmenenija v vidovom sostave i biomasse planktona v ozere Sevan (Changes in the species composition and biomass of phytoplankton in Lake Sevan). Limnology of mountain reservoirs, pp 172–173

Molot LA, Schiff SL, Venkiteswaran JJ, Baulch HM, Higgins SN, Zastepa A, Verschoor MJ, Walters D (2021) Low sediment redox promotes cyanobacteria blooms across a trophic range: implications for management. Lake Reservoir Manag 37:120–142

Morillas L, Roales J, Cruz C, Munzi S (2022) Lichen as multipartner symbiotic relationships. Encyclopedia 2(3):1421–1431

Nanagulyan SG, Poghosyan AV, Shakhazizyan IV, Petrosyan AM, Zakaryan NA (2017) Vozdejstvie medno-molibdenovoj promyshlennosti na sostav vidov mkhov v Armenii (Influence of copper-molybdenum production on the species composition of mosses in Armenia). In: Proceedings of the nature protection and regional development: harmony and conflicts Orenburg, II, pp 72–74

Nazarova EA (2009) Karyosystematicheskoe issledovanie armjanskih predstavitelej roda *Vicia* L. II. Seccii *Faba, Narbonensis, Peregrinae* (Fabaceae) (Karyosystematic investigation of the Armenian representatives of genus *Vicia* L. II. Sections *Faba, Narbonensis, Peregrinae* (Fabaceae). Flora, vegetation and plant resources of Armenia 17, pp 39–42

Nazarova EA (2011) Chechevitsy (*Lens,* Fabaceae) v Armenii (Lentils (*Lens*, Fabaceae) in Armenia). Takhtajania 1:138–141

Nazarova EA, Gukasyan AG (2004) Chromosome numbers of flowering plants of Armenian flora, Yerevan

Nikoghosyan VG (1963) Representatives of lichen flora of Armenia from genus *Ramalina* and *Parmelia*. Biol J Armenia 16(10):69–76

Nikoghosyan VG (1964a) To flora of lichens in Armenia. Biol J Armenia 17(4):89–99

Nikoghosyan VG (1964b) About several lichens of mountain regions in Armenia. Biol J Armenia 17(11):41–48

Nikoghosyan VG (1965a) Representatives of lichen flora of Armenia from genus *Lecanora*, *Xanthoria* and *Physcia*. Biol J Armenia 18(5):72–79

Nikoghosyan VG (1965b) New species of the genus *Dermatocarpon* in Armenia. Novosti Sist Nizsh Rast 2:205–207

Nikoghosyan VG (1966) New data on lichen flora of Armenia. Biol J Armenia 19(3):106–113

Nikulina VN (2005) Structure and long-term dynamics of blue-green algae – an indicator of the ecological state of the internal estuary of the Neva River. Actual problems of modern algology. Materials international conference Kharkov, pp 111–112

Nikulina VN, Mnatsakanyan AT (1984) Phytoplankton ozera Sevan v 1979–1981 gody (Phytoplankton of Lake Sevan in 1979–1981). Experimental and field studies of hydrobionts of the Lake Sevan, pp 18–43

Oganesyan RO (1994) Lake Sevan yesterday, today... NAS RA. Yerevan.

Oganezova GH (2013) Nekotorye osobennosti vidov *Merendera* i *Bulbocodium* (Colchicaceae): geographia, biologia, morphologia i chislo chromosom v svjazi c ih taksonomicheskim polozheniem (Some peculiarities of *Merendera* and *Bulbocodium* species (Colchicaceae): geography, biology, morphology and chromosome number connected with their taxonomic range). Takhtajania 2:60–68

Oganezova GH, Pogosian AV (2004) Issledovanie anatomicheskih i ultrastructurnykh osobennostej mkhov s gory Arailer (Study of the anatomical and ultra-structural peculiarities of some mosses from Arailer (Ara mountain, Armenia). Flora, Vegetation and Plant resources of Armenia, 15, pp 55–60

Oganezova GH, Poghosyan AV, Manakyan VA (2005) Perspectivy sokhranenija mkhov Armjanskoj flory (the perspectives of protection the mosses species from Armenian flora). Proceedings of the actual problems of bryology, St. Petersburg, pp 142–147

Paerl HW, Fulton RS, Moisander PH, Dyble J (2001) Harmful freshwater algal blooms, with an emphasis on cyanobacteria. Sci World J 1:76–113

Paerl HW, Hall NS, Calandrino ES (2011) Controlling harmful cyanobacterial blooms in a world experiencing anthropogenic and climatic-induced change. Sci Total Environ 409(10):1739–1745

Pakhunova VG (1933) About some features in the structure of the representatives of genus *Ramalina* from Armenia. Works Tbilisi Bot Inst 1:349–351

Parparov AS (1979) Pervichnaja produkcija i soderzhanie chlorophylla "a" v phytoplanktone ozera Sevan (Primary production and content of chlorophyll 'a" in the phytoplankton of the Lake Sevan) Proceedings of Sevan hydrobiological station, 17, pp 89–99

Pimm SL, Joppa LN (2015) How many plant species are there, where are they, and at what rate are they going extinct? Ann Missouri Bot Garden 100(3):170–176. https://doi.org/10.3417/2012018

Poghosyan AV, Shakhazizyan IV, Nanagulyan SG (2015) Bryoflora lesnoj zony gory Arailer (Armenia) i osobennosti ultrastructury nekotoryh typov listostebl'nykh mkhov (Bryoflora of the forest zone of Mount Arailer (Armenia) and features of the ultrastructural structure of some types of leaf-stemmed mosses). In: Proceedings of the VIII international scientific and practical conference "science, education, culture and outreach – the basis for the sustainable development of mountain territories" Vladikavkaz, pp 711–714

Pogosyan AV (2003a) Bryoflora vulkanicheskogo massiva Arailer (The bryoflora of the volkanic massive of Arailer). PhD thesis, Yerevan

Pogosyan AV (2003b) Bryoflora vulkanicheskogo massiva Arailer (Bryoflora of volcanic massif Arailer) (Republic of Armenia). Arctoa 12:187–190

Portenier NN (2000) The system of geographical elements of the Caucasian flora. Bot Zhurn 85(9):26–33

Poretsky VS (1953) Diatoms of Lake Teletskoye and related rivers. Diatom collection. Leningrad. 107–172

POWO Plants of the World Online (2022) Facilitated by the Royal Botanic Gardens, Kew. Available at http://www.plantsoftheworldonline.org/. Accessed 23 June 2022

Ranker TA, Sundue MA (2015) Why are there so few species of ferns? Trends Plant Sci 20(7):402–403. https://doi.org/10.1016/j.tplants.2015.05.001

Rechinger KH (1972) Compositae-Cynareae I: *Cousinia*. In: Rechinger KH (ed) Flora Iranica, Lfg. 90, Graz

Rechinger KH (1986) Cousinia: morphology, taxonomy, distribution and phytogeographical implications. Proc R Soc Edinb 89B:45–58

Remigailo PA (2011) Systematic structure of phytoplankton of large rivers central Yakutsk floral region. Flora of Asiatic Russia 2(8):20–27. http://www.izdatgeo.ru

Reynolds CS (2006) The ecology of phytoplankton. Cambridge University Press

Rodrigues LC, Simoes NR, Bovo-Scomparin VM et al (2015) Phytoplankton alphadiversity as an indicator of environmental changes in a neotropical floodplain. Ecol Indic 4:334–341

Rykovsky GF (1989) Epiphytnye mkhi kak ecologicheskaja gruppa extremal'nykh mestoobitanij (Epiphytic mosses as an ecological group of extreme habitats). Problems of Bryology in USSR, Leningrad, pp 190–201

Solomon J, Shulkina T, Schatz G (2013) Red List of the endemic plants of the Caucasus. Missouri Botanical Garden Press.

Salmaso N, Capelli C, Shams S, Cerasino L (2015) Expansion of bloom-forming *Dolichospermum lemmermannii* (Nostocales, Cyanobacteria) to the deep lakes south of the Alps: colonization pat-terns, driving forces and implications for water use. Harmful Algae 50:76–87

Starr JR, Janzen FH, Ford BA (2015) Three new early diverging *Carex* (Cariceae, Cyperaceae) lineages from East and Southeast Asia with important evolutionary and biogeographic implications. Mol Phylogen Evol 88:105–120

Stepanyan LG (2009) Gidrobiologicheskie i gydrokhimicheskie issledovanija Yerevanskogo uchastka gydroecosystema reki Rasdan (Hydrobiological and hydrochemical studies of the Yerevan section of the hydroecosystem of the Hrazdan River), Yerevan
Stepanyan LG (2019) Qualitative, quantitative indices of phytoplankton of the Marmarik river and ecogeographic characteristics. Biol J Armenia 3(71):32–38
Stepanyan LG, Ghukasyan EK (2019) Kolichestvennaja i sezonnaja dinamika fitoplanktonnogo soobshchestva v srednem techenii reki Rasdan (Armenia) v 2018 godu (The quantity and seasonal dynamics of the phytoplankton community in the middle reaches of the Hrazdan River (Armenia) in 2018). Ecosyst Transform 2(2):53–61
Stepanyan LG, Hambaryan LR (2014) Taksonomicheskij sostav fitoplanktona reki Rasdan (Aemwnia) (Taxonomic composition of the phytoplankton of the Hrazdan River (Armenia). Materials international scientific and practical remote conference. Ukraine, pp 21–23
Stepanyan LG, Hambaryan LR (2016) Otsenka skhodstva osnovnyh indikatorod fitoplanktonnogo soobshchestva v razlichnyh tipah vodnyh ecosystem (Assessment of the similarity of the main indicators of the phytoplankton community of different types of aquatic ecosystems). Biol J Armenia 68(4):6–12
Sterner RW, Reinl KL, Lafrancois BM, Brovold S, Miller TR (2020) A first assessment of cyanobacterial blooms in oligotrophic Lake Superior. Limnol Oceanogr 65:2984–2998
Stroykina VG (1952) Fitoplankton pelagiali ozera Sevan (Phytoplankton of the pelagial of the Lake Sevan). Proceedings of Sevan hydrobiological station, 13
Stroykina VG (1958) Materialy po flore vodorosley stojachih ozer i prudov Armenii (Materials for the algae flora of small stagnant water bodies of Armenia). Izv AS ArmSSR (Biol Agric Sci) 11:6
Takhtajan A (1941) Botanico-geographicheskij ocherk Armenii (Phyto-geographical review of Armenia). Proceedings of the Institute of Botany of Armenian branch of USSR Academy of Sciences, 2, pp 3–156
Takhtajan A (1986) The floristic regions of the world. University of California Press
Tamanyan K, Fayvush G (1987) Botaniko-geographicheskij analyz armjanskih vidov roda *Cousinia* (Asteraceae) (Phyto-geographical analysis of the Armenian species of the genus *Cousinia* (Asteraceae)). Biol J Armenia 40(6):464–469
Tambian NN (1962) Materialy po flore vodorosley Araratskoj ravniny (Materials for the algae flora of the Ararat flatland). Izv AS ArmSSR (Biol Sci) 15:8
Tambian NN (1965) Novie taksony diatomovyh dlja algoflory Armjanskoj SSR (New taxa of diatoms for the algoflora of the Armenian SSR). Proc AS ArmSSR 18(3):87–89
Tambian NN (1967) Novye vodorosli dlja Armjanskoj SSR (new algae for the Armenian SSR). news of taxonomy of lower plants, pp 103-106
Tambian NN (1977) O sine-zelenyh vodorosljah Armjanskoj SSR (On blue-green algae of the Armenian SSR). Message Institute of Agrochemical Problems and Hydroponics, 16, pp 16–20
The fifth national report to convention on biological diversity (2014) Yerevan
Tonakanyan GA (1948) O vysokogornoj skal'noj rastitel'nosti juzhnykh otrogov Zangezurskogo khrebta (On the high-mountainous rocky vegetation of the southern tip of the Zangezur Range). Izvest AN ArmSSR 1:21–34
Urbanavichus GP (2015) The lichen flora of the northern Caucasus and its contribution to the diversity of the lichen flora of Russia. Bot Herald North Caucasus 1:93–105
Vardanyan MK (1993) Successija fitoplanktona ozera Sevan v period ego evtrofikacii (The succession of phytoplankton of the Lake Sevan during its eutrophication). International conference CIS countries and the Inter-Parliamentary Assembly "Ecological problems of Lake Sevan" (b), 14, 58, 100
Vladimirova KS (1944) Ob issledovanii microflory rek i ozer Armenii (On the study of the microflora of rivers and lakes in Armenia). News of the AS ArmSSR, 4
Vladimirova KS (1947) Fitoplankton ozera Sevan (Phytoplankton of Lake Sevan). Proceedings of Sevan hydrobiol. station, 9

Von Konrat M, Soederstroem L, Renner MAM, Hagborg A, Briscoe L, Engel JJ (2010) Early Land Plants Today (ELPT): how many liverwort species are there? Phytotaxa 9:22–40

Wagenitz G (1975) Floristic connections between the Balkan Peninsula and the near East as exemplified by the genus *Centaurea*. In: Jordanov D et al (eds) Problems of Balkan flora and vegetation, Sofia

Wagenitz G (1986) *Centaurea* in south-West Asia: patterns of distribution and diversity. Proc R Soc Edinb 89B:11–21

Waterway MJ, Hoshino T, Masaki T (2009) Phylogeny, species richness, and ecological specialization in Cyperaceae tribe Cariceae. Bot Rev 75:138–159

Więcław H, Kalinka A, Koopman J (2020) Chromosome numbers of *Carex* (Cyperaceae) and their taxonomic implications. PLoS One 15(2):e0228353

Winter JG, DeSellas AM, Fletcher R, Heintsch L, Morley A, Nakamoto L, Utsumi K (2011) Algal blooms in Ontario, Canada: increases in reports since 1994. Lake Reservoir Manag 27:107–114

Yaroshenko PD (1946) Obzor rastitel'nosti Gorisskogo tayona (An essay of vegetation of Goris District). Proc Bot Inst Akad Sci ArmSSR 4:157–186

Zakeri Z, Gasparyan A, Aptroot A (2016) A new corticolous *Megaspora* (Megasporaceae) species from Armenia. Willdenowia 46(2):245–251

Zedelmeyer OM (1931) Otchet o geobotanicheskih issledovanijakh na jugo-vostochnom i juzhnom poberezhjakh ozera Sevan letom 1928 g. (Report on the geobotanical study of the southeastern and southern shores of the lake Sevan in the summer of 1928). Basin Lake Sevan 2(2):189–200

Zedelmeyer OM (1933) Geobotanicheskij obzor rastitel'nosti zapadnogo poberezhja ozera Sevan (Gokcha) v 1929 g. (Geobotanical review of the vegetation of the western shore of the lake Sevan (Gokcha) 1929). Basin Lake Sevan 3(3):79–100. http://www.theplantlist.org/1.1/browse/B/

Chapter 4
Mycobiota of Armenia

Siranush Nanagulyan and Lusine Margaryan

Republic of Armenia is a land-locked mountainous country characterized by an exceptional variety of climatic conditions and plant communities, which considerably affected the diversity of systematic groups of fungi, which is the objective of our research. Fungi are included in an independent kingdom of eukaryotic organisms that permeate our environment. They are chlorophyll-free, heterotrophic organisms that feed on readily available organic matter.

Mushrooms take a special place in the organic world system as they populate all over the planet. They are found everywhere in the air, soil, water, animals, people, plants, etc. Fungi are a large group of organisms that resemble both animals and plants. Fungi enter into the composition of all types of natural ecosystems and in artificially created phytocoenosis. Being heterotrophic components or blocks of ecosystems, fungi play an essential role in the decomposition of organic materials and take part in a number of biogeochemical cycles, promoting the formation of humus and participating in soil formation processes.

Currently, the subject of mycological research conducted in Armenia is devoted to the biodiversity of wild fungal species, as well as fungi that contaminate food products, industrial materials, the human environment (residential and office buildings, hospitals, museums, libraries), etc. Research conducted in the field of medical mycology is very relevant today. The goal is not only to identify the composition of the primary pathogens of human diseases but also the opportunistic and conditional pathogenic mycobiota accompanying them. Studies are being carried out on the diversity of the local mycobiota at the species level and the alien invasive species which are associated with the globalization process taking place in the modern world.

S. Nanagulyan (✉) · L. Margaryan
Yerevan State University, Yerevan, Republic of Armenia
e-mail: snanagulyan@ysu.am; lusinemargaryan@ysu.am

G. Fayvush (ed.), *Biodiversity of Armenia*,
https://doi.org/10.1007/978-3-031-34332-2_4

4.1 Short Review of the Classification and Taxonomy of Fungi

The history of the study of fungi of the globe is divided into three periods: (1) initial, from ancient times to the middle of the nineteenth century; (2) cumulative, from the middle of the nineteenth to the middle of the twentieth century; (3) modern, since the middle of the twentieth century till today. The first two periods are characterized by the accumulation of information about the morphology and ecology of fungi, thus the creation of artificial classification systems. Subsequently, researchers searched for phylogenetic relationships between individual groups of fungi and made the first attempts to create natural classification systems.

Based on the same facts, different authors interpreted questions of fungal taxonomy in different ways. Taking into account the complex taxonomic macro- and microscopic features (ontogenesis, cytology, biochemistry, morphogenesis, genetics, etc.) of fungi, researchers subdivide them into a number of divisions, subdivisions, classes, subclasses, orders, and families. On the one hand, this approach partially clarified the taxonomy of fungi, and on the other hand, it made the position of some groups of fungi in a particular class controversial.

There was no consensus among mycologists for a long time as to whether fungi are a separate kingdom of living beings or whether they are a highly specialized or an integral part of the plant kingdom.

Many authors present various classification schemes due to the fact that the taxonomy of fungi is still poorly developed. Despite the complexity and intricacy of the relationship between the individual groups of fungi and a number of parallel evolution lines, fungi can be considered a single group that diverged from prokaryotes in the early stages of evolution.

Fundamental changes have taken place in the taxonomy of fungi. Both the research methods and the structure of classification systems have been constantly modernized in recent decades. For a number of reasons, fungi represent a group of organisms that is difficult to classify. The reasons for this are the convergent similarity of even phylogenetically very distant representatives and their sheer number. Thus, the paleomycological materials for phylogenetic analysis are insignificant (Sidorova 2003).

Various methods were implied to create and improve the classification of fungi. Since the 90s of the last century, molecular methods based on the analysis of proteins, RNA, and DNA have become dominant in the taxonomy of fungi. Using genosystematics and cladistics make it possible to construct phylogenetic trees and understand the phylogenetic relationships of taxa of any given rank. Methods of molecular systematics have an undoubted advantage. They are objective and reproducible, thus allowing the creation of valid data banks (Bridge 2002; Maheshwari 2011; Kropp et al. 2010; Hawksworth and Lücking 2017, etc.).

The result of the accumulation and analysis of that kind of data is the gradual destruction of traditional fungal taxa. Also, their polyphyly or paraphyly was identified. Since the beginning of the 1990s, ideas about the polyphyly of fungi in their traditional sense have finally formed. In all megasystems, regardless of the number of kingdoms accepted by their authors, fungi are distinguished into an independent kingdom (Fungi, Eumycota) which differs in volume depending on authors, namely, the position in the Chytridiomycota system. The kingdom of Fungi (Mycota, Mycetalia) corresponds to the Eumycota group and includes flagellate fungi Chytridiomycota (Cavalier-Smith 1987, 1998; Barr 1992; Hawksworth et al. 1995; Kendrick 2001).

The macrosystem of fungi has not undergone significant changes over the past decade, except for the ranks of taxa, except for the latest version of the system of the organic world of T. Cavalier-Smith (1998), according to which organisms that previously belonged to the kingdom Mycota (fungi in the traditional sense) are subdivided into three kingdoms: Mycota, Chromista, and Protozoa.

On the basis of this classification lies a number of features: composition of the cell wall, lysine synthesis, mobility in the vegetative state, and the structure of zoospores and gametes, which corresponds to their origin in three independent evolutionary lines.

In the classic version, adopted in the 7th and 8th editions of the dictionary by J. Ainsworth and G. R. Bisby (Hawksworth et al. 1983, 1995), the kingdom Fungi includes four divisions – Chytridiomycota, Zygomycota, Ascomycota, and Basidiomycota. Deuteromycota is preserved in a number of systems at the rank of division, while in others, it forms a group of "mitotic fungi," which is a formal taxon. Molecular data generally support these ideas.

Most modern classification systems lack division Deuteromycetes, as molecular phylogenetics suggest the integration of anamorphic species into the teleomorphic fungal taxa.

The introduction of molecular methods into the taxonomy of fungi since the end of the last century has led to a revision of the taxonomic groups. Molecular systematics or genosystematics makes it possible to judge the phylogenetic relationships of taxa of any rank, their monophyly or polyphyly, and create a natural phylogenetic system of fungi using the determination of the nucleotide sequences of individual genes and cladistic methods for constructing phylogenetic trees. As a result, the traditional taxa of fungi are gradually destroyed. For example, such classes as Ascomycetes, Homobasidiomycetes, Gasteromycetes, the Aphyllophorales order, etc., have lost their taxonomic status (Kusakin and Drozdov 1997; Sidorova 2003; Zmitrovich and Wasser 2004; Hibbet et al. 2007; Shnyreva 2011).

However, the main shortcomings of the modern molecular system of fungi are the number of orders and families, and the fact that the taxonomic rank of many groups remains debatable.

In the ninth edition of mentioned Dictionary, a revised classification of fungal phyla reflecting the latest molecular evidence, including a revision of the Ascomycota, Basidiomycota, and a full integration of anamorphic genera in the classification is shown (Kirk et al. 2001). Despite the rapid speed at which changes are made, particularly by molecular sequence data, the 10th edition continues the link between the old and the new hierarchies. According to the Dictionary, fungi are a polyphyletic group and must be divided into at least three kingdoms – Protozoa,

Chromista, and Fungi. This publication clearly differentiates materials on "true" fungi of the kingdom Fungi – Ascomycota, Basidiomycota, Chytridiomycota, Glomeromycota, Microsporidia, and Zygomycota. Deuteromycetes are considered anamorphs of the corresponding taxa of Ascomycetes and Basidiomycetes. In the 9th edition, published in 2001, anamorphic fungi positioned at the subphylum level. In the 10th edition, over 30% of the anamorphs were included in a class (Kirk et al. 2008). Information on the systematic position of anamorphic fungi (the rank of their teleomorphs) is constantly refined. For the vast majority of anamorphic species, there is still no data on the refined taxonomic position of teleomorphs.

A review of molecular phylogenetic studies of Ascomycetes shows that they are not always compatible with the traditional systems of this group. However, a number of mycologists are making more or less successful attempts to rebuild the system, taking into account the latest data. An example is the systems proposed by Eriksson and Winka (1997, 1998), etc. The significant differences between the Kingdoms are summarized in Ainsworth and Bisby's Dictionary (Kirk et al. 2008). The tenth edition presented a fully new classification of the kingdom Fungi based on phylogenetic research, a major revision of the classification of the Basidiomycota and Ascomycota, and enhanced distinctions between the true fungi and unrelated groups traditionally studied by mycologists.

Our understanding of fungal evolution has been significantly influenced by molecular and genome technologies. This technology is quickly becoming routine, representing the starting point for an increasing number of studies. In this way, the number of species rapidly increases which can be incorporated into genome-scale phylogenies (Spatafora et al. 2017).

Thus, there have been significant changes in views on the position of fungi in the system of the organic world, content, and system recently. There is no single and universally recognized classification system of fungi, and the taxonomy is constantly being adjusted. So, for the convenience of presenting the material, we have taken as a basis the system of fungi given in the 10th edition of the Ainsworth and Bisby's Dictionary of the Fungi (Kirk et al. 2008), including some changes and additions given in the multivolume edition of Mycoflora of Armenia (1967–2013), a number of articles and monographs published in recent years and annually reviewed species checklist of fungi (https://www.mycobank.org/).

4.2 Study and Distribution of Fungal Diversity in Armenia

The first mentioned information about the mushrooms of Armenia was as early as the fifteenth century in the "Useless for Ignoramuses" manuscript, the author of which was the great Armenian physician and scientist A. Amasiatsi. The original text was published in Vienna in 1926 and republished in Moscow in 1990 (Amasiatsi 1926, 1990). The medicinal properties of several species of fungi belonging to the macromycetes (Ascomycota and Basidiomycota) were described. The experience of Armenian classical and traditional medicine in the field of pharmacognosy is summarized (Nanagulyan and Taslakhchyan 1998). The work also provides information about the medicinal properties of some macro- and microfungi.

The first publications that mention the fungi of Armenia are by Nevodovsky (1912) and Voronov (1915). The authors indicate several species found on the territory of Armenia among the fungi recorded for the Caucasus.

The work of Voronikhin (1927) "Materials for the flora of the Caucasus" provides data on the flora of the Caucasus, among them about 100 species of micro- and macromycetes, mainly found on trees and shrubs in various regions of Armenia.

The intensive study of parasitic micromycetes of Armenia began in the 1930s. Initially, parasitic fungi as pathogens of plant diseases were the main object of mycological research in Armenia. The first and most complete work on the mycobiota of Armenia was published by Teterevnikova-Babayan and Babayan (1930). The authors provide a list of 193 species of parasitic fungi of cultivated and wild plants, common in lowland and some highland regions of the republic, including several species of macromycetes.

The mycobiota of Armenia in the second half of the twentieth century was investigated with scrupulous attention. The research connected with different fungal ecological and taxonomical groups: coprotrophs (M. G. Taslakhchyan, S. G. Nanagulyan); predatory (A. H. Yesayan) and mycophilous fungi (L. L. Osipyan, J. H. Abrahamyan, S. G. Nanagulyan); fungi in the air of domestic, industrial, and hospital premises; urban air (L. L. Osipyan, H. H. Batikyan, J. H. Abrahamyan, S. G. Nanagulyan, I. M. Eloyan, Y. Kh. Hovhannisyan); etc., made a significant contribution (Osipyan and Nanagulyan 2008).

Biotechnological investigations are conducted in recent years. Macromycetes are studied as a source of food and animal feed biomass for agriculture. The antioxidant activity of some species with pharmacological significance was also studied (S. G. Nanagulyan, M. G. Taslakhchyan, L. V. Margaryan, V. S. Gevorgyan).

Research connected with the biodegradation of industrial materials and products made from them deserves attention. Measures were recommended to improve the fungus resistance of polymer glues and component materials (J. H. Abrahamyan, I. V. Shahazizyan, S. G. Nanagulyan).

Medicinal fungi pathogenic to humans are an independent ecological group. Pathogens of candidiasis, onychomycosis (L. L. Osipyan, E. Y. Sarkisyan), and mycological diseases of ENT-organs (J. H. Abrahamyan, S. G. Nanagulyan, I. M. Eloyan, Y. Kh. Hovhannisyan) are studied jointly with doctors of clinical hospitals.

A systematic study of the composition of fungi in different regions of the republic and various taxonomic groups of fungi allowed researchers to publish a number of manuscripts reflecting information about new species of fungi for the republic or for science, partially presented in a multivolume edition (Mycoflora/Mycobiota of Armenia, 1967–2013) and numerous articles. Initially, the parasitic fungi as pathogens of plants were the main object of mycological research in Armenia. Eight volumes of Mycoflora of Armenia were mainly devoted to them. One of the widespread diseases of cultivated and wild plants in Armenia is Septoria, the study of which is described in the monograph by Teterevnikova-Babayan (1962). She noted 155 species, most of which parasitize wild herbaceous plants.

The order Peronosporales is studied monographically, published in volume I "Mycoflora of the Armenian SSR" (Osipyan 1967). The author lists 125 types of downy mildew fungi, which parasitize mainly herbaceous plants from all floristic regions of Armenia. The foundation was laid for multidisciplinary research and saprotrophic fungi from different trophic and ecological groups, which play an important role in medicine and industry as causative agents of human diseases, contaminants of food products, toxin producers, and also destroyers of industrial materials.

A significant contribution to the parasitic hyphal fungi research in the republic was conducted by Osipyan (1975) and summarized in volume III "Mycoflora of the Armenian SSR." The monograph describes 368 species of fungi, most of which infect wild and cultivated plants.

The results of the study of the rust fungi in Armenia were summarized by D. N. Teterevnikova-Babayan in the IV volume of "Mycoflora of the Armenian SSR" (1977). The book provides information on the species of rust fungi, their distribution by host plant families, the types of their development cycles, and the allocation of rust fungi in the ecological and climatic zones of Armenia. Additional information about rust fungi found in different floristic regions of the republic was published later (Simonyan 1978, 1990; Simonyan et al. 1993; Osipyan 2009, 2013).

The sixth volume of "Mycoflora of the Armenian SSR" is devoted to the sphaeropsidal fungi of Armenia, where Teterevnikova-Babayan et al. (1983) calculated all information about fungi with unicellular colorless conidia. Some are obligate or facultative parasites of valuable wild and cultivated plants such as forage grasses.

Simonyan contributed to the research of powdery mildew in Armenia significantly. The complete summary of these micromycetes is published in the volume VII of "Mycoflora of Armenia" (Simonyan 1994). The book provides a systematic treatment of the order Erysiphales of Armenia, represented by 106 species parasitizing 786 host plant species. Along with floristic investigations, the author made the ecological and systematic analysis of the order with the most harmful species.

The eighth volume of "Mycobiota of Armenia" (Osipyan 2013) is a continuation of the previously published "Mycoflora of the Armenian SSR" which is devoted to smut fungi (Ustilaginomycetes) and supplements the previously published volumes with new for Armenia species of fungi from the orders Peronosporales, Hyphales, Uredinales, Sphaeropsidales.

The second volume of "Mycoflora of the Armenian SSR" described macromycetes (Melik-Khachatryan and Martirosyan 1971). The first part is devoted to the results of studying the Gasteromycetes of the Republic. The publication summarizes all the data on Gasteromycetes of Armenia (56 species) and their ecology. And the second part is devoted to the species composition of Aphyllophorales s.l., which made it possible to identify 70 species, 20 forms, and varieties.

Melik-Khachatryan significantly contributed to the research of agaricoid fungi in Armenia. The complete summary of these macromycetes of the republic is presented in volume V of "Mycoflora of the Armenian SSR" (Melik-Khachatryan

1980). The book summarizes all information available at that time (392 species). The author, along with floristic studies, carried out works on ecology and phytocoenology and studied the physiological activity and biochemical composition of some species of agaric fungi.

Rare works include paleomycological studies that have revealed some fossil fungi (Teterevnikova-Babayan and Taslakhchyan 1969).

The study of the mycobiota of South Armenia is devoted to the works of Simonyan (1968, 1969) in the Meghri region, distinguished by unique ecological and climatic conditions. The mycobiota of botanical gardens and arboretums of Armenia has been studied in detail and comprehensively. The results of many years of research were summarized by Simonyan (1981) in the monograph "Mycoflora of botanical gardens and arboretums of the Armenian SSR" where data on 1036 species of mushrooms are published.

The study of the pathogenic mycobiota of forage grasses in Armenia, identifying the taxonomic composition, conducting a comparative analysis, and clarifying their ecological characteristics started in 2002. About 160 different species, varieties, and forms of fungi found on legumes and cereal forage grasses were registered in Armenia (Nanagulyan and Soghoyan 2009, 2012).

Information regarding Gasteromycetes of Armenia was published in 2000. It cataloged 83 species (Nanagulyan and Osipyan 2000). A book on the cap mushrooms of Armenia presents the results of studies on 565 species of Agaricales in the Armenian republic (Nanagulyan 2008).

Specially protected nature areas of Armenia, such as the Dilijan and Khosrov reserves, were studied (Nanagulyan and Taslakhchyan 1991). Studies identify the taxonomic composition of 636 species varieties and forms of macromycetes, conduct a comparative analysis of fungi, and establish their ecological characteristics.

As a result of studies of macromycetes in the Sevan National Park, 66 species of fungi belonging to 42 genera and 20 families from the classes Ascomycetes and Basidiomycetes were found. Order Agaricales included the highest number of species (51) in the study area (Nanagulyan and Charchoghlyan 1986).

Since 2009, we have started special studies of macromycetes in the Shikahogh State Reserve (including Plane Grove State Sanctuary), Armenia, to reveal the complete taxonomic composition. According to our materials (Margaryan et al. 2015) and the literature data, 436 species of macroscopic fungi are currently registered in the study area. Of the total number of macromycetes found, 12 species and 2 genera are recorded in Armenia for the first time, and 417 species are new for the Shikahogh Reserve (Nanagulyan et al. 2019).

In the last years, the bibliography of scientific work on the mycology of the Armenian Republic, which is an index of the research articles on fungi in Armenia from the end of the nineteenth century up to 2014, was published (Osipyan et al. 2017). It also includes notes by foreign authors on the discovery of specific fungal species in the territory of the country.

Armenia involves the opulence and originality of fungal biota. Fungi and fungus-like organisms are widely distributed in nature. They are included in terrestrial and

Table 4.1 Systematic diversity of Armenian mycobiota

Kingdom	Phylum	Division	Number of species
Protista	Myxomycota		44
Chromista	Oomycota		143
Fungi	Eumycota	Zygomycota	140
		Ascomycota	2758
		Basidiomycota	1492
Total: 3	3	3	4577

aquatic ecosystems as heterotrophic components. Fungi are found everywhere and have certain requirements for the specific environments in which they live.

In the Republic of Armenia, it has been revealed 4577 species of fungi, referring to the phylum Myxomycota (Kingdom Protista), Oomycota (Kingdom Chromista), and Eumycota (Kingdom Fungi) (Table 4.1).

A total of 44 species of Myxomycota have been recorded from various floristic regions of Armenia. Previously, the Oomycota was included in the true fungi group, now they are part of the group of fungus-like organisms. The biochemistry, ultrastructure, and molecular sequences of these organisms indicate that they belong to the Chromista. In Armenia, there are 143 species, varieties, and forms of Peronosporales.

Kingdom Fungi included phylum Eumycota with traditional divisions Zygomycota (140 species), Ascomycota (2758), and Basidiomycota (1492). The multivolume edition of "Mycobiota of Armenia" (Osipyan 1975, 2013) presented information about representatives of Deuteromycotina (2153 species), which now are considered anamorphs of the corresponding taxa of Ascomycota and Basidiomycota. Forty species of macromycetes are included in the Red Book of Higher Plants and Fungi of Armenia (Tamanyan et al. 2010).

4.3 Ecology of Macromycetes

Macromycetes are the most significant heterotrophic component of ecosystems and perform such important functions as the decomposition of organic matter and mycorrhiza formation. They are of great theoretical and applied importance. Many macroscopic fungi have been widely used for food for a long time. Some species are used as animal feed or in medicine, which also makes their intensive study a necessity (Wasser 2010).

The solution to the main problems of mycoecology currently reduced to three main areas: identification of the species composition of fungi in various habitat conditions, study of the influence of biotic and abiotic environmental factors on the vital activity and distribution of fungal taxa, and identification of the trophic structure and determination of the relationship of fungi with the substrate (saprotrophism, symbiotrophism, parasitism). Ecological studies of macroscopic fungi are mainly aimed at identifying the species composition by fruiting bodies within certain communities and studying their phenology and the influence of physical environmental factors on the development of macromycetes. As well as developing methods for mycoecological and mycocoenological studies and establishing the trophic structure of fungal groups.

4.4 Trophic Structure of Macromycetes

In the process of long-term evolution based on trophic and topical relationships under various environmental conditions, fungi have developed countless adaptations. The so-called ecological groups emerged as a result. There is still no unified and universally recognized nomenclature in the name of the ecological and trophic groups of fungi. Each researcher uses his terminology and identifies different groups of symbiotrophs, saprotrophs, and parasites. Fungi affect humans in many ways in biotechnology and everyday life. They are important members of ecosystems, acting as saprobes, parasites (plant and animal), mutualists (mycorrhizae, endophytes, lichens), and commensals.

When analyzing the trophic groups of macromycetes in Armenia, we adopted the following differentiation of groups (Nanagulyan 2008): I. parasites (on tree species, mosses, ferns, fungi), II. Symbiotrophs, III. saprotrophs (xylotrophs, humus saprotrophs, litter saprotrophs, coprotrophs, herbotrophs, psammotrophs, phyllotrophs, carbotrophs, carpotrophs, calcetrophs, technotrophs).

Parasitic macromycetes in Armenia are represented with 54 species and forms of Pezizomycotina (18 species) and Agaricomycotina (36). Parasitic Agaricomycotina cause various types of rots of trunks and roots of different tree species. The most commonly distributed and harmful parasites in Armenia are *Fomes fomentarius*, *Polyporus squamosus*, *Laetiporus sulphureus*, some species of the genera Inonotus, Phellinus, etc.

According to the feeding substrates, parasitic macromycetes of Armenia are distributed as follows:

I. Tree species parasites – 39 species:
   (a) On both coniferous and deciduous trees – eight species (*Armillaria mellea*, *Laetiporus sulphureus*, etc.)
   (b) On deciduous trees – 22 species (*Colpoma quercina*, *Fomes fomentarius*, *Polyporus squamosus*, species from genera *Inonotus*, *Phellinus*, etc.).
   (c) On coniferous trees – nine species (*Pyrofomes demidoffii*, *Porodaedalea pini*, *Lachnellula calyciformis*, species of *Hypoderma*, *Lophodermium*, etc.).

II. Moss parasites – three species: *Hemimycena rickenii*, *Crepidotus epibryus*, *Entoloma conferendum*

III. Fern parasites – one species: *Cryptomycina pteridis*

IV. Fungal parasites – 11 species: *Asterophora lycoperdoides*, *Collybia cookei*, *C. cirrhata*, *Orbilia epipora*, *Rosellinia clavariae*, etc.

In addition to obligate parasites, facultative saprotrophs are also found in the republic, settling on both living and dead wood. For example, oyster mushroom (*Pleurotus ostreatus*) is one of the most common wood-destroying fungi in the forests of Armenia. Its abundant fruiting is observed both on healthy and half-decayed trees, stumps, fallen trunks of beech, oak, hornbeam, and sometimes pine. The same can be said about the velvet shank (*Flammulina velutipes*), species of the genera *Daedaleopsis*, *Fomitopsis*, *Pholiota*, and some others (Figs. 4.1 and 4.2)

Symbiotrophs Macromycetes, which form mycorrhiza (root fungus) mainly on the roots of trees and shrubs, make up a significant part of the macroscopic fungi known at present.

The mycotrophy of tree species is widespread in nature. It is the basis for the existence of forests in almost all temperate zones. Mycorrhizal fungi form a unique ecological group of macromycetes characterized by symbiotic relationships with higher plants, the absence of cellulose and lignin-degrading enzymes, and the energy dependence of the fungus on the symbiont organism.

Based on many years of personal observations and research, we specified the main species composition of mycorrhizal woody and shrubby plants in Armenia. About 217 species of symbiotrophs are registered on the territory of Armenia. The vast majority of symbiotrophs grow in Northeastern Armenia where the main forest

a - ***Boletus edulis***, b - ***Pleurotus ostreatus***, c - ***Macrolepiota procera***, d - ***Lactifluus piperatus***, e - ***Suillus granulatus***, f - ***Agaricus bernardii***

Fig. 4.1 Macromycetes in Armenia

a - ***Boletus edulis***, b - ***Mutinus caninus***, c - ***Laetiporus sulphureus***, d - ***Lycoperdon perlatum***, e - ***Ganoderma lucidum***, *f* - ***Polyporus squamosus***

Fig. 4.2 Macromycetes of Armenia

tracts of the republic are concentrated. These forests are consisting of highly mycotrophic tree species where mild, moderately humid climatic conditions prevail.

Most species of mycorrhizal organisms under our conditions belong to Agaricomycotina (211 species). All the species of the families Amanitaceae, Boletaceae, Gomphidiaceae, Russulaceae, Strobilomycetaceae, and Xerocomaceae; all species of the genera *Amanita, Amanitopsis, Boletus, Cantharellus, Cortinarius, Hygrophorus, Lactarius, Russula, Suillus*, and *Xerocomus*; some representatives of the genera *Entoloma, Hebeloma*, and *Inocybe*; etc are symbiotrophs. There are no mycorrhiza-forming species in the families of Pleurotaceae, Coprinaceae, Strophariaceae, etc.

Six species of mycorrhizal Pezizomycotina are also found in Armenia: *Gyromitra esculenta*, *Elaphomyces granulatus*, *Peziza badia*, *Pseudocraterellus undulatus*, *Sarcosoma globosum* (Fig. 4.3), and *Tuber aestivum.*

The distribution and species diversity of symbiotrophs are indicators of habitat conditions for trees. In particular, they indicate deterioration of their vital condition as a result of recreational impacts.

The species composition of mycorrhizal organisms varies depending on the forest type and tree species. Symbiotrophs found in Armenia are more common in broad-leaved forests (116 species). There are 76 species in mixed forests, and 94 species inhabit pure coniferous forests. Mushrooms of this group have different

a - *Pleurotus eryngii*, b - *Boletus satanas*, c - *Hericium erinaceum*, d - *Agaricus xanthodermus*, e - *Sarcosoma globosum*, f - *Amanita gemmata*

Fig. 4.3 Macromycetes of Armenia

obligateness in relation to tree species. Most of the registered mycorrhizal species belong to ecologically broadly tolerant species. These include *Amanita pantherina*, *Inocybe fastigiata*, *Cantharellus cibarius*, *Lactarius vellereus*, *Paxillus involutus*, *Russula delica*, *R. foetens*, *R. lilacea*, *R. xerampelina*, *R. adusta*, *Xerocomus subtomentosus*, and *Laccaria laccata*. Some species of the genera *Boletus*, *Cortinarius*, *Inocybe*, etc., form mycorrhiza with more than one species of woody and shrubby plants. All these species appear in forest communities of fungi and plants at their specific phenological periods of development (Fig. 4.4).

Narrowly tolerant species are confined to single tree species. *Chroogomphus rutilus*, *Lactarius helvus*, *L. deliciosus var. pini*, *Russula queletii*, *Cortinarius sanguineus*, *Entoloma cuneatum*, *E. venosum*, *Gomphidius roseus*, *Suillus granulatus*, *S. luteus*, and others form mycorrhiza only with pine; the following species enter into symbiosis only with oak *Lactarius piperatus* и *L. quetus*; and the following species cohabitate only with deciduous trees: *Boletus rubellus*, *Hygrophorus eburneus*, *Cortinarius armeniacus*, *Entoloma rhodopolium*, *Inocybe asterospora*, *Lactarius pallidus*, *L. blennius*, *Russula alutacea*, etc.

Saprotrophic macromycetes use dead organic matter as a food source. They have formed a specific set of enzymes that allow them to decompose especially resistant lignocellulosic compounds in the process of evolution. Saprotrophs have conquered extremely diverse ecological niches due to their high activity and metabolic rate, biochemical adaptation, and the ability to quickly transition to anabiosis under adverse environmental factors. This allowed researchers to identify several ecological and trophic groups of saprotrophic fungi. The main of these are xylotrophs,

a - *Amanita phalloides*, b - *Cortinarius caerulescens*, c - *Cortinarius armillatus*, d - *Amanita pantherina*, e – *Amanita muscaria*, f – *Coprinopsis picacea*

Fig. 4.4 Macromycetes of Armenia

humus and litter saprotrophs, coprotrophs, herbotrophs, psammotrophs, phyllotrophs, carbotrophs, carpotrophs, calcetrophs, and technotrophs.

Xylotrophs The trophic group of xylotrophic macromycetes is the most studied due to its great practical importance. They are the main causative agents of wood rot. In the mycobiota of Armenia, xylotrophs are the most abundant macromycetes species (502). Representatives of this group, being wood destroyers, develop in forests and gardens on dead wood, fallen trunks, buried wood, rotten wood, and stumps. Xylotrophs are found in all taxonomic groups of macromycetes. Most of them belongs to Basidiomycota. Almost all of its representatives are found on trees (361 species). Wood-destroying species were also found among Ascomycota (141).

The results of the research showed that the majority of xylotrophs are able to decompose the wood of various deciduous and coniferous species. Certain species of wood destroyers are confined to certain tree species (*Fomitopsis betulina* on birch, *Pyrofomes demidoffii* on juniper). Many species live on wood only such as deciduous species (*Neolentinus lepideus, Tricholomopsis rutilans, Hypholoma capnoides*, etc.) or coniferous species (*Polyporus varius, Flammula alnicola, Kuehneromyces mutabilis*, etc.). Most xylotrophs are polytrophic as noted above. Thus, they are able to decompose wood of both coniferous and deciduous species (*Schizophyllum commune, Hypholoma fasciculare, Flammulina velutipes, Coltricia perennis, Pleurotus cornucopiae, Lycoperdon pyriforme*, etc.).

Humus saprotrophs have a specific enzymatic apparatus which allows them to break down complex organic compounds into simpler ones in the course of their vital activity. They supply humic acids with different amount and activity to the soil which accumulate in it.

Representatives of humus saprotrophs are found in various ecosystems – forests, meadows, steppes, and on soils that are very different in composition and structure. In terms of diversity of species composition, humus saprotrophs in Armenia rank second among the saprotrophs of the studied mycobiota (279 species). Most representatives of Pezizomycetes belong to this group (63 species). Among them are quite common *Aleuria aurantia, Discina ancilis,* and *Helvella lacunosa* species; some species of *Peziza*, *Geopyxis*, and *Humaria* genera; and all the species of *Morchella, Octospora, Otidea, Verpa,* and other genera. About 215 species are humus saprotrophs among Agaricomycetes – *Boletopsis leucomelas, Bovista nigrescens, B. plumbea, Coltricia perennis, Bankera fuligineoalba, Macrotyphula juncea, Lycoperdon pratense, and Phallus impudicus* – all the species of *Agaricus, Hydnellum, Leucoagaricus, Macrolepiota, Melanoleuca, Hygrocybe*, and *Phellodon* genera; and some representatives of the genera *Agrocybe*, *Calvatia*, *Entoloma*, *Geastrum*, *Inocybe*, *Lepista*, *Lycoperdon*, *Mutinus*, and others.

There are facultative mycorrhiza formers among humus saprotrophs. Such species in our studies are *Inocybe fastigiata*, *I. praetervisa*, *Laccaria laccata*, and *Melanoleuca melaleuca*.

Litter Saprotrophs Forest litter plays an extremely important role in the forest ecosystem. It decomposes and mineralizes plant residues. The litter is a visual and complete expression of the results of ecological processes. At present, litter is considered an integral dynamic part of the soil profile and, at the same time, an independent component of forest ecosystems.

The forest floor can be arranged into layers of different degradation states. These layers differ biochemically and ecologically since it decomposes over a long period. The stable nature of the environment in terms of the main habitat factors leads to the species and numerical stability of destructor groups, in particular saprotrophic macromycetes, the mycelium of which is located in this layer.

About 105 species of litter saprotrophs are found in Armenia, of which the maximum species diversity was noted in agaricoid fungi (100 species). Agaricomycetes are distinguished by abundance, developing on the litter in broad-leaved (68 species) and mixed (44 species) forests. About 35 species were found in conifer forests.

The species of this trophic group have a relatively narrow selectivity for tree species. For example, in pine forests on the litter, species *Auriscalpium vulgare*, *Clitocybe ericetorum*, *C. angustissima*, *Leucopaxillus paradoxus*, and *Strobilurus esculentus* are frequent. Of the fungi that mineralize the litter in broad-leaved forests can be noted *Pseudoclitocybe expallens*, *Clitocybe metachroa*, *Marasmius dryophilus*, *Lepista flaccida*, etc.

Species *Leucocybe candicans* and *Marasmius dryophilus* are found in the oak forests, and species *Mycena pura*, *Clitocybe gibba*, *C. nebularis*, *Collybia dryophila*, *Lepista nuda*, etc., grow in a variety of forest formations on the litter of deciduous and mixed forests.

Five species of the Pezizomycotina litter saprotrophs have been recorded. These are *Cudonia circinans* and *Pseudoplectania nigrella* in mixed forests and *Lachnum virgineum* and *Exarmidium hemisphaericum* in deciduous forests.

Coprotrophs Coprotrophic fungi utilize organic substances found in animal excrement. This substrate is the only source of their nutrition and, therefore, determines the distribution of coprotrophic fungi in nature. Representatives of this trophic group are characterized by several adaptive traits that have been developed in them in the course of evolution and have allowed them to conquer this particular ecological niche. Coprotrophs are found near settlements, on pastures, in greenhouses on moist soil, in the vicinity of farms, and inside and outside of forests.

Studies of the species composition of coprotrophic macromycetes by incubation of samples of herbivore droppings in a humid chamber as well as collections in natural conditions made it possible to identify 73 species in Armenia. Pezizomycotina (34) and Agaricomycotina (25) host the greatest number of species.

Among discomycetes, the Ascobolaceae family includes 17 species belonging to the genera *Ascobolus* (6), *Saccobolus* (7), *Ascophanus* (3), *Iodophanus*, and *Thecotheus* (1 species each). Of the sac fungi, 16 species belonging to the genera *Coprotus* (4), *Lasiobolus* and *Rhyparobius* (3 each), *Thelebolus* (2), and *Ascodesmis*, *Cheilymenia*, *Coprobia*, and *Trichobolus* (one species each) belong to coprotrophs. Only the species *Peziza fimeti* found on the animal excrements belonging to the Pezizomycetes. There are 14 species of coprotrophic Sordariomycetes from the genera *Podospora* (8), *Sporormia* (5), and *Sordaria* (1) of the Sordariaceae family.

The family Coprinaceae is the most diverse in its species composition among agaricoid fungi with 12 species from the genera *Coprinus* (8), *Anellaria* (2), *Panaeolus*, and *Panaeolina* (1 species each). The family Bolbitiaceae is represented by two species of the genus *Conocybe* and 1 species from each the genera *Agrocybe*, *Bolbitius*, and *Pholiotina*. From the families Strophariaceae and Pluteaceae, one species of the genera *Psilocybe*, *Stropharia*, and *Volvariella* was found on the excrements of herbivores.

The maximum number of species was noted on the excrements of cows and horses. And the smallest on the excrements of sheep, goats, donkeys, rabbits, and birds. Pezizomycetes representatives such as *Peziza fimeti*, *Ascobolus glaber*, *A. immersus*, *Saccobolus violascens*, and *Ascophanus testaceus*; several species of *Podospora* and *Sporormia*; and the following Agaricomycetes species – *Coprinus ephemerus*, *C. comatus*, *C. niveus*, *Conocybe pubescens*, and *Stropharia semiglobata* – are mostly found both in nature and in laboratory conditions (Nanagulyan and Taslakhchyan 1998).

Herbotrophs inhabiting herbaceous plants is represented in Armenia by 39 species. Almost all representatives belong to Pezizomycotina (36 sp.) – herbotroph species belong to genera Phyllachora and Clypeoporthe as well as species *Nectria dacrymycella*, *Diaporthe pardalota*, etc. Discomycetes living on the remains of herbaceous plants include 21 species. These are mainly representatives of the families

Hyaloscyphaceae (8), Leotiaceae (7), and Dermateaceae (6). Herbotrophic discomycetes are unevenly distributed among families of host plants. Most often they settle on the remains of large herbaceous plants from the families Asteraceae, Dipsacaceae, and Boraginaceae.

Of the basidial macromycetes, the steppe oyster mushroom (*Pleurotus eryngii*) develops on the dead roots of large umbellifer plants. *Cyathus olla* and *Crucibulum laeve*, in addition to dung, often settle on the remains of herbaceous plants.

Psammotrophs Mushrooms that are living on sandy soils constitute the trophic group of psammotrophs. This is a very interesting ecological group. Almost all of its representatives are found in treeless habitats in areas with a dry climate and sandy soils. Only 22 species are found in Armenia, of which six belong to Pezizomycotina (e.g., *Helvella atra*, *Rhizina undulata*, *Geopora arenosa)*; 16, to Agaricomycotina (e.g., *Psathyrella fulvescens*, *Coprinopsis cinerea*, *Gyrophragmium dunalii)*; almost all species of the genus *Tulostoma* (6 sp.); and one each of the genera *Astraeus*, *Battarrea*, *Bovista*, *Disciseda*, *Geastrum*, *Myriostoma*, and *Podaxis*.

Phyllotrophs The trophic group of phyllotrophs can be found on alive and fallen leaves of plants. This group includes only 14 species of macromycetes. Fallen leaves of tree species are inhabited mainly by representatives of Pezizomycotina. In Armenia, these are species of *Calycellina punctata* and *Mollisia tumidula*, which settle on fallen leaves of oak, maple, and poplar.

Carbotrophs Carbotrophs are a specific trophic group associated with a narrow ecological niche and live in places of fires, conflagrations and on burnt wood. Of the macromycetes identified in Armenia, this includes 11 species. From the sac fungi, this trophic group includes *Geopyxis carbonaria*, *Plicaria trachycarpa*, *Peziza sepiatra*, *Geoscypha violacea*, *Pyronema omphalodes*, *Trichophaea gregaria*, etc. *Peziza sepiatra* is also found on wood submerged in the soil which makes it a xylotroph as well. The remaining species are obligate carbotrophs. Of the Agaricomycotina, species of *Hebeloma anthracophilum*, *Lyophyllum atratum*, and *Pholiota carbonaria* live on burnt wood.

Carpotrophs On the fruits of higher plants in the conditions of the republic, eight species of macroscopic fungi develop. Among them are four species of Pezizomycotina: *Hymenoscyphus epiphyllus* on fallen beech fruits, *H. fructigenus* on fallen beech and oak fruits, *Stromatinia pseudotuberosa* on acorn pericarp, and *Monilinia fructigena* on apple and quince fruits. The substrates for the pyrenomycetes *Hypocrea fungicola* and *Xylaria filiformis*, in addition to the fallen fruits of hardwoods, are, first, for rotting basidiophores of cap mushrooms, and second, for forest litter. Of the agaric fungi on beech fruits, decaying wood, and leaf litter, one species, *Crepidotus variabilis*, has been recorded.

Technotrophs The trophic group of technotrophs in Armenia is represented by the only species of *Peziza ostracoderma* found on a rubber water hose in a mine. The discomycete *Peziza muralis* lives on cement substrates and wet plaster. It was found repeatedly on the cement floor in garages and on damp walls in living quarters. Under the conditions of Armenia, on damp walls in buildings, the fungus *Coprinus sp*. was also often noted, which could not be identified to the species level.

4.5 Useful and Harmful Species of Macrofungi in Armenia

Medicinal and nutritional properties have been attributed to mushrooms for thousands of years. Many species of fungi are well known in traditional medicine and scientific literature as producers of biologically active substances (Nanagulyan et al. 2002; Nanagulyan et al. 2013; Rogers 2006; Wasser 2010).

Despite the medicinal value of macrofungi and their widespread use in various countries, the species composition of the medicinal macrofungi has not been approved (Hawksworth 2001). Based on modern taxonomy in China, a critical analysis of medicinal macromycetes was carried out, and a list of 692 species of basidial medicinal macrofungi was obtained (Wu et al. 2019).

The main value of macroscopic mushrooms is not only low-calorie content or high-protein content but also in the presence of biologically active substances with healthful properties. Unlike microscopic fungi, macromycetes in this aspect have been poorly studied, and only in recent decades have begun to attract the attention of official medicine. Data on the edibility and toxicity of a particular species are extremely contradictory, which is associated with both environmental conditions and the traditional knowledge of the population of various countries and nationalities.

For instance, in most Muslim countries, harvesting mushrooms and eating them is considered a sin. Each country in Western Europe eats different species of mushrooms. For example, in Italy the porcini or king bolete (*Boletus edulis*) is of great value; the Germans and Swiss prefer the golden chanterelle (*Cantharellus cibarius*); and among the Spanish, the deliciosus milkcap (*Lactarius deliciosus*) takes the first place (Nanagulyan et al. 2021).

Bioactive compounds of fungi are mainly their secondary metabolites, which accumulate both in the mycelium of the fungi and in the fruiting body. It has been shown that several fungal metabolites, particularly those isolated from macromycetes *Cantharellus cibarius* and species of Pleurotus genus, have high potency against trematodes yet cause no major side effects in humans (Tsyganov et al. 2018; Rodríguez-Barrera et al. 2021). On the market, fungi biological products are presented mainly in the form of capsules, powders, tinctures, teas, coffee, and cocoa. The pharmacies of our republic also sell tinctures and tablets from single species of mushrooms with medicinal properties (*Inonotus obliquus*, *Ganoderma lucidum*, *Tricholoma matsutake*) which received from foreign countries.

In recent years, the consumption of mushrooms for food has increased all over the world, including Armenia. In 1970–1980 in the markets of the republic, champignon (*Agaricus bisporus*) and steppe oyster mushroom (*Pleurotus eryngii*) were mainly sold.

Currently, Armenian markets are represented by such species as *Pleurotus ostreatus*, *Calocybe gambosa*, *Cantharellus cibarius*, *Lactarius deliciosus*, *Suillus granulatus*, and *Lepista nuda* and some species from the genera *Russula*, *Tricholoma*, etc. (Nanagulyan et al. 2020).

As a result of the study of the taxonomic composition of the useful and harmful macromycetes of Armenia, 484 species of macroscopic fungi were identified, from which 300 species are edible, 64 species are poisonous, and 126 species with medicinal properties (Table 4.2). The detected macromycetes belong to the divisions Ascomycota, which includes the subdivision Pezizomycotina (class Pezizomycetes) and Basidiomycota with the subdivision Agaricomycotina (class Agaricomycetes and Tremellomycetes).

From the division of Ascomycota the class Pezizomycetes is represented by one order Pezizales with 13 species of edible fungi and three species of fungi with medicinal properties.

Table 4.2 shows that the main part of edible species belongs to the division Basidiomycota (class Agaricomycetes). The lead order is Agaricales with 206 species, and the second place with the number of edible species is the order Russulales with 29 species. With species richness, the third place belongs to the order Boletales (27 species). The order Polyporales includes seven species of edible mushrooms. Other orders of mentioned class are represented by a smaller number of edible species. The class Tremellomycetes with order Tremellales includes one genus and two species of edible macromycetes and one genus and two species of fungi with medicinal properties.

Poisonous mushrooms, which contain various toxins that lead to poisoning and even death, play a dangerous role in human life. Currently, 64 species of poisonous macromycetes have been identified in Armenia, of which 59 species belong to the order Agaricales and five species – order Boletales. Poisoning with poisonous mushrooms is associated not only with the presence of toxins in them but also if the macrofungi are exposed to bacterial, fungal, or chemical contamination. In the first case, the fungi are exposed to mechanical damage or penetration of insects into the damaged parts, which leads to their infection with bacteria or pathogenic fungi. The second case is associated with the ability of many species of macroscopic fungi to accumulate and adsorb heavy metals, pesticides, and fungicides that are toxic to humans. The toxins of poisonous mushrooms are divided into three groups. Toxins that cause digestive disorders belong to the first group of toxins. These toxins are contained in the species of the genera *Agaricus*, *Tricholoma*, etc. The second group of toxins is contained in some fungi from the genera *Amanita*, *Cortinarius*, *Entoloma*, *Hebeloma*, etc., and leads to disorders of the central nervous system. The most dangerous group of toxins is the third group, which is mainly fatal; poisoning occurs due to the use of *Amanita phalloides* and *Amanita virosa*, etc. Poisoning mainly occurs in the summer-autumn period, when the collection and consumption of mushrooms increases.

Currently, medicinal cap fungi are used as dietary food, nutritional supplements, a new class of medicines, cosmetics, natural biocontrol agents in plant protection, etc. In Armenia, 300 species of macrofungi have nutritional value, and 126 species

Table 4.2 Quantitative distribution of important species of macromycetes in Armenia

Division/subdivision	Class	Order	Edible fungi		Poisonous fungi		Medicinal fungi	
			Number of genera	Number of species	Number of genera	Number of species	Number of genera	Number of species
Ascomycota/ Pezizomycotina	Pezizomycetes	Pezizales	8	13	–	–	3	3
Basidiomycota/ Agaricomycotina	Agaricomycetes	Agaricales	68	206	16	59	34	67
		Auriculariales	1	2	–	–	1	3
		Boletales	14	27	4	5	5	11
		Cantharellales	4	5	–	–	3	3
		Gloeophyllales	1	1	–	–	1	1
		Gomphales	2	5	–	–	1	1
		Hymenochaetales	–	–	–	–	2	3
		Phallales	1	1	–	–	1	1
		Polyporales	4	7	–	–	16	21
		Russulales	2	29	–	–	3	10
		Thelephorales	2	2	–	–	–	–
	Tremellomycetes	Tremellales	1	2	–	–	1	2
Total: 2/2	3	13	108	300	20	64	71	126

are considered medicinal fungi. Among the species of medicinal mushrooms of Armenia, the species *Agaricus xanthodermus*, *Amanita muscaria*, *Calvatia craniiformis*, *Coprinus comatus*, *Fistulina hepatica*, *Lepista nuda*, *Schizophyllum commune*, etc., have antitumor and anticancer activity. Several species of macromycetes (*Agaricus arvensis*, *Amanita citrina*, *A. phalloides*, *A. rubescens*, *Calocybe gambosa*), which are used for gastrointestinal diseases and also promote metabolism and stimulate digestion, have been identified in Armenia. It should also be noted that, since macrofungi are considered difficult to digest food, in case of gastrointestinal diseases, they should be consumed in small quantities. Most of the species found in Armenia have antibacterial, antiviral, and antifungal properties, which are associated with chemicals such as terpenoids, purines, phenolic derivatives isolated from fruit bodies, and fungal mycelium. These are species *Agaricus campestris*, *Agrocybe dura*, *Clitocybe geotropa*, *Flammulina velutipes*, *Hypholoma fasciculare*, *Kühneromyes mutabilis*, *Pholiota destruens*, *Lepista nebularis*, *Lepista nuda*, *Oudemansiella mucida*, *Laccaria laccata*, etc. (Nanagulyan et al. 2021).

Thus, the above indicates great potential for the use of macromycetes, both edible and poisonous species in medicine and pharmacology.

References

Amasiatsi A (1926) Useless to the ignorant. Vienna

Amasiatsi A (1990) Nenuzhnoe dlja neuchej (Useless to the ignorant), vol 13. Moscow

Barr DJS (1992) Evolution and kingdoms of organisms from the perspective of a mycologist. Mycologia 84(1):1–11

Bridge PD (2002) The history and application of molecular mycology. Mycologist 16:90–99

Cavalier-Smith T (1987) The origin fungi and pseudofungi. Evolutionary biology of the fungi. Cambridge University Press, Cambridge, pp 339–353

Cavalier-Smith T (1998) A revised six-kingdom system of life. Biol Rev 73(3):203–266

Eriksson OE, Winka K (1997) Supraordinal taxa of Ascomycota. Myconet 1(1):1–16

Eriksson OE, Winka K (1998) Families and higher taxa of Ascomycota. Myconet 1(2):17–24

Hawksworth DL (2001) Mushrooms: the extend of the unexplored potential. Int J Med Mushr 3:333–337

Hawksworth DL, Lücking R (2017) Fungal diversity revisited: 2.2 to 3.8 million species. Microbiol Spectr 5(4):5–4

Hawksworth DL, Sutton BC, Ainsworth GC (1983) Ainsworth and Bisby's dictionary of the fungi, including the Lichens, 7th edn. Kew

Hawksworth DL, Kirk PM, Sutton BC, Pegler DN (1995) Ainsworth and Bisby's dictionary of the fungi, 8th edn. CAB International, Wallingford

Hibbet DS, Binder M, Bischoff JF et al (2007) A higher level phylogenetic classification of the fungi. Mycol Res 3:509–547

Kendrick B (2001) The fifth kingdom, 3rd edn. Mycologue Publications

Kirk PM, Cannon PF, David JC, Stalpers JA (eds) (2001) Ainsworth & Bisby's dictionary of the fungi, 9th edn. CABI Publishing, Wallingford

Kirk PM, Cannon PF, Minter DW, Stalpers JA (eds) (2008) Ainsworth & Bisby's dictionary of the fungi, 10th edn. CABI Publishing, Wallingford

Kropp BR, Matheny PB, Nanagyulyan SG (2010) Phylogenetic taxonomy of the *Inocybe splendens* group and evolution of supersection "Marginatae". Mycologia 102(3):560–573

Kusakin OG, Drozdov AL (1997) Phylema of the living beings. St Petersburg
Maheshwari R (2011) Fungi: experimental methods in biology, 2nd edn, vol 28. CRC Press, p 358
Margaryan LV, Hovhannisyan YK, Nanagulyan SG (2015) Biota of macroscopic fungi in Shikahogh State Reserve (Armenia): taxonomical analyses. In: Materials of the VIII international scientific and practical conference, Vladikavkaz, pp 702–706
Melik-Khachatryan JH (1980) Mycoflora of the Armenian SSR, Agaricales. Yerevan
Melik-Khachatryan JH, Martirosyan SN (1971) Mycoflora of the Armenian SSR, Gasteromycetes and Aphyllophoroid fungi. Yerevan
Nanagulyan SG (2008) Shljapochnye griby Armenii (Agaricoid Basidiomycetes) (Cap Fungi of Armenia (Agaricoid Basiodiomycetes). Yerevan
Nanagulyan SG, Charchoghlyan AA (1986) K issledovaniju macromycetov Natsional'nogo parka "Sevan" (To the study of macromycetes of the National Park "Sevan"). Materials of the VII Transcaucasian conference on spore plants, Yerevan, p 66
Nanagulyan SG, Osipyan LL (2000) Konspect makroskopicheskih gribov Armenii. Gasteromycetes (Conspectus of macroscopic fungi of Armenia. Gasteromycetes). Yerevan
Nanagulyan SG, Soghoyan YY (2009) Ecological features of the most important representatives of forage grass micromycetes in the desert-semidesert zone. Immunopathol Allergol Infectol 1:95–96
Nanagulyan SG, Soghoyan YY (2012) Distribution of parasitic forage grass fungi in floristic regions of Armenia. Modern mycology in Russia. In: Abstracts of the third congress of mycologists of Russia, Moscow, vol 3, p 298
Nanagulyan SG, Taslakhchyan MG (1991) Makromicety Dilijanskogo i khosrovskogo gosudarstvennyh zapovednikov Armjanskoj SSR (Macromycetes of Dilijan and Khosrov State Reserves of the Armenian SSR). Yerevan
Nanagulyan SG, Taslakhchyan MG (1998) History of investigation of Armenian macromycetes. Biol J Armenia 4(51):302–307
Nanagulyan SG, Sirunyan AL, Hovhanisyan EK (2002) Biodiversity and ecology of the medicinal mushrooms of Armenia. Int J Med Mushr 4(1):71–76
Nanagulyan SG, Hovhannisyan YK, Margaryan LV (2013) Conservation of medicinal mushrooms in Armenia. In: The 7th international medicinal mushroom conference, China, p C2-0-13
Nanagulyan SG, Margaryan LV, Hovhannisyan YK, Boyajyan ES (2019) New for Armenia species and genera of Basidiomycetes from Shikahogh State Reserve. Proc YSU Chem Biol 1(53):29–32
Nanagulyan S, Zakaryan N, Kartashyan N et al (2020) Wild plants and fungi sold in the markets of Yerevan (Armenia). J Ethnobiol Ethnomed 16(1):1–27
Nanagulyan SG, Margaryan LV, Hovhannisyan YK, Shahazizyan IV (2021) Useful and harmful properties of agaricoid basidiomycetes of the Shikahogh State Reserve (Republic of Armenia). Proc YSU B Chem Biol Sci 55(256):266–275
Nevodovsky GS (1912) Fungal pests of cultivated and wild useful plants of the Caucasus in 1911. Appendix to the proceedings of the Tiflis Botanical Garden, Tiflis, vol 2, no 2
Osipyan LL (1967) Mycoflora of the Armenian SSR, Peronosporales. Yerevan
Osipyan LL (1975) Mycoflora of the Armenian SSR, Hyphal fungi. Yerevan
Osipyan LL (2009) Addition to the 4th volume of "Mycoflora of Armenian SSR. Rust fungi". Fl. Veg. Pl. Resources Armenia 17:102–103
Osipyan LL (2013) Mycobiota of the Armenian SSR. Yerevan
Osipyan LL, Nanagulyan SG (2008) Development of mycological research in Armenia. Actual problems of botany in Armenia. Institute of Botany, Yerevan, pp 31–37
Osipyan LL, Nanagulyan SG, Soghoyan YY (2017) The bibliography of scientific work on mycology of Armenian Republic. Yerevan
Rodríguez-Barrera TM, Téllez-Téllez M, Sánchez JE, Castañeda-Ramirez GS, Acosta-Urdapilleta M, Bautista-Garfias CR, Aguilar-Marcelino L (2021) Edible mushrooms of the genus *Pleurotus* as biocontrol agents of parasites of importance for livestock. Scientia fungorum, p 52
Rogers R (2006) The fungal pharmacy: medicinal mushrooms of Western Canada. Edmonton

Shnyreva AV (2011) Molecular systematics and species concept in fungi: approaches and resolutions. Mikol Fitopatol 45(3):209–220
Sidorova II (2003) Macrosystems and methodology of fungi: Methodology and changes of the last decade. News in the systematics and nomenclature of fungi, Moscow, pp 7–71
Simonyan SA (1968) Review of mycoflora of Meghri region. Biol J Armenia 21(5):79–85
Simonyan SA (1969) Materials for the mycoflora of the Meghri region of the Armenian SSR. Biol J Armenia 22(12):60–65
Simonyan SA (1978) New materials on the flora of rust fungi in Armenia. Proc YSU 1:156–159
Simonyan SA (1981) Mycoflora of botanical gardens and arboretums of the Armenian SSR. Yerevan
Simonyan SA (1990) Materials for the mycobiota of Shirak. Rust fungi (order Uredinales). Biol J Armenia 7(43):587–591
Simonyan SA (1994) Mycoflora of the Armenian SSR, Powdery mildew of Armenia. YSU Press, Yerevan, p 7
Simonyan SA, Mamikonyan TO, Barseghyan AK (1993) New materials on the microbiota of the Ararat basin. Phototrophic micromycetes of the Ararat basin and Araler mount, pp 41–50
Spatafora JW, Aime MC, Grigoriev IV, Martin F, Stajich JE, Blackwell M (2017) The fungal tree of life: from molecular systematics to genome-scale phylogenies
Tamanyan K, Fayvush G M, Nanagulyan SG, Danielyan TS (2010) The Red Book of plants of Armenia. Higher plants and fungi. Ministry of Nature protection of Republic of Armenia, Yerevan
Teterevnikova-Babayan DN (1962) Overview of fungi from the genus *Septoria*. Yerevan
Teterevnikova-Babayan DN (1977) Mycoflora of the Armenian SSR, Rust fungi. Yerevan
Teterevnikova-Babayan DN, Babayan AA (1930) Materials on the study of mycoflora of Armenian SSR
Teterevnikova-Babayan DN, Taslakhchyan MG (1969) The development of paleomycological research in the Soviet Union and the results works carried out in the Armenian SSR. In: Abstracts of reports 3rd Transcaucasian conference on the history of science. Tbilisi, pp 198–199
Teterevnikova-Babayan DN, Taslakhchyan MG, Martirosyan IA (1983) Mycoflora of the Armenian SSR, Sphaeropsidales with colorless unicellular conidia. Yerevan
Tsyganov M, Vishnivetskaya G, Kovner A, Sorokina I, Dushkin A, Mordvinov V, Avgustinovich D (2018) A search for potential anthelminthic drugs using the model of Opisthorchis felineus – induced Opisthorchiasis. Syst Biol Biomed:856–863
Voronikhin NN (1927) Materials for the flora of the fungi of the Caucasus. In: Proceedings of botonical museum. Academy of Sciences of the USSR Press, Leningrad, pp 87–252
Voronov YN (1915) List of fungi still known for the flora of the Caucasus. In: Proceedings of the Tiflis Botanical Garden, Tiflis, vol 1, pp 3–200
Wasser SP (2010) Medicinal mushroom science: history, current status, future trend, and unsolved problems. Int J Med Mushr 12(1):1–16
Wu F, Zhou LW, Yang ZL et al (2019) Resource diversity of Chinese macrofungi: edible, medicinal and poisonous species. Fungal Divers 98:1–76
www.mycobank.org
Zmitrovich IV, Wasser SP (2004) Modern view on the origin and phylogenetics reconstruction of Homobasidiomycetes fungi. Evolutionary theory and processes. Modern Horizons, Dordrecht/Boston, pp 230–263

Chapter 5
Fauna of Armenia

**Mark Kalashian, Karen Aghababyan, Noushig Zarikian,
Bardukh Gabrielyan, Marine Arakelyan, and Astghik Ghazaryan**

5.1 Terrestrial Invertebrates

Mark Kalashian, Karen Aghababyan, and Noushig Zarikian

Armenia is situated on the border of large biogeographical divisions – provinces and sub-provinces and its fauna was historically formed under the influence of different surrounding faunas – European, Mediterranean and Irano-Turanian. Due to this biogeographical position as well as due to diversity of landscapes, variations in altitude and mountainous nature, the invertebrate fauna of Armenia is characterized by rich species diversity and high level of endemism. According to the First National Report to the Convention on Biological Diversity (Republic 1998), the number of species is very approximately estimated as 17,000, including 12,000 species of insects, and more than 300 species are considered in this publication as endemic of the country.

Like in majority of countries, invertebrates are studied much less than vertebrates and very unevenly meaning in particular coverage level of large taxonomic groups.

M. Kalashian (✉) · N. Zarikian · B. Gabrielyan
Scientific Center of Zoology and Hydroecology, National Academy of Sciences of Armenia, Yerevan, Armenia
e-mail: mark.kalashian@sczhe.sci.am; noushig.zarikian@sczhe.sci.am; gabrielb@sci.am

K. Aghababyan
BirdLinks Armenia NGO, Yerevan, Armenia
e-mail: karen.aghababyan@env.am

M. Arakelyan · A. Ghazaryan
Yerevan State University, Yerevan, Armenia
e-mail: arakelyanmarine@ysu.am; astghik.ghazaryan@ysu.am

G. Fayvush (ed.), *Biodiversity of Armenia*,
https://doi.org/10.1007/978-3-031-34332-2_5

There are not so many comprehensive works specially dedicated to the Invertebrate fauna of Armenia so far. First to be mentioned are the volumes of the series Fauna of Armenian SSR, which cover some groups of insects, and special volume is dedicated to molluscs. Some identification guides were published as well. Besides, some data can be extracted from works of applied content dedicated to agricultural or forest pests.

The level of study of different groups of Armenian terrestrial invertebrates is briefly discussed below.

Volume on Mollusca (Akramowski 1976) includes 155 species; more detailed review of this group is presented below.

From class Arachnida, mites and ticks are rather well studied. Special work is dedicated to the mite family *Phytoseiidae* (Harutyunian 1977), and review of all Acarina: Parasitiformes Reutyer, 1909 including 571 species was recently published (Harutyunian and Dilbarian 2006). Also family *Tenuipalpidae* was revised (Dilbarian and Kocharyan 2014), 16 species of it are reported for Armenia. There are data of presence of 108 species of oribatid mites (Oribatei) belonging to 35 families in Ararat Plane (Khanbekian and Kalashian 1992). Scorpion fauna includes three species reviewed by Richter (1945). The state of the researches on spiders (Arachnida) is presented separately.

Armenian Spiders' fauna is one of the poorly studied groups, along with other invertebrates. Spiders from Armenia were first studied by Koch (1878) and Kulczyński (1895). Few papers on Arachnofauna of recent Armenian area have been published by Russian arachnologists (Charitonov 1956; Eskov 1987; Tanasevitch 1987, 1990; Dunin 1992; Dunin and Zacharjan 1991; Marusik 1989; Mikhailov 1986, 2013, 2016; Mikhailov et al. 2017). Important papers for the Armenian spider fauna were published by Logunov (Logunov 2015; Logunov and Guseinov 2008; Rakov and Logunov 1997) and N. Zarikian (Zarikian 2020, 2021, 2022a, b; Zarikian and Kalashian 2021; Zarikian et al., 2022; Zarikian et al. 2023). A large number of spider species from the Armenia fauna are still waiting to be recorded and, moreover, much new species to be described, which will increase our understanding of the spider's biodiversity of our country.

From insect orders, dragonflies (Odonata) are rather well studied. Akramowski (1948) reported for Armenia 53 taxa of dragonflies; this order is further studied by V. Ananian who together with foreign specialists published an updated list which includes 56 species (Tailly et al. 2004).

One of the best studied groups are Orthopteran insects on which were published two volumes from the series Fauna of Armenian SSR, dedicated to locusts – Acridoidea (Avagian 1968) and grasshoppers – Tettigonoidea (Avagian 1981). For the Armenian fauna, the author mentioned 105 species of locusts and 42 species of grasshoppers.

Among hemipterans (order Hemiptera) scale, insects (Coccoidea) are rather well studied. Borkhsenius (1949) revising the group reported for Armneia 119 species from seven families, and later Ter-Grigorian (1973) enlarged this list by 41 species of Mealybugs (family *Pseudococcidae*). True bugs (Heteroptera) are much less studied. Akramowskaya (1983) proposed identification guide for the family

Coreidae, including 31 species, and Mirzoyan (1977) in his book of dendrophilous insects of Armenia listed 16 species of the group. Besides, he listed 20 species of dendrophilous cicadas (Cicadoidea) and 105 species of aphids (Aphidoidea).

Beetles (Coleoptera) were continuously in the focus of researches of many entomologists. Ter-Minassian (1947) and Plavilstshikov (1948) provided identification guides for weevils (Curculionidae) and longhorn beetles (Cermabycidae), respectively. First work includes about 560 species, and 214 species of longhorns are elaborated in the second one. Beetle fauna of Armenia during decades was studied by Iablokoff-Khnzorian and his students. Due to their activity, six volumes of the series Fauna of Armenian SSR dedicated to some families were published, including majority of ground-beetles (Carabidae) (Iablokoff-Khnzorian 1967), lamellicorn beetles (Scarabaeoidea) (Iablokoff-Khnzorian 1967), blister beetles (Meloidae) and comb-clowed beetles (Alleculidae, currently considered subfamily of darkling beetles – Tenebrionidae) (Iablokoff-Khnzorian 1983), as well as click beetles (Elateridae) (Mardjanian 1986) and seed beetles Bruchidae (now subfamily of leaf beetles – Chrysomelidae) (Karapetian 1985). Recently, fauna of darkling beetles (without Alleculinae) was reviewed (Nabozhenko et al. 2021). Currently, the work on beetle fauna of Armenia is continued in the Scientific Center of Zoology and Hydroecology NAS RA. Data on Armenian beetles are generalised and discussed separately.

Among lepidopteran insects (Lepidoptera), best studied are butterflies which fauna includes 236 species and is discussed separately below. Comprehensive work on geometer moths (Geometridae) was carried out by Wardikjan (1980) who provided identification guide on harmful geometer moths including 73 species and reported for the fauna of Armenia as a whole 306 species (Wardikjan 1985). Data on 253 species of dendrophilous moths belonging to 26 families (except Geometridae) are provided in the book of Mirzoyan (1977).

For such an important group as Hymenopterans, only two revisions were published. One, dedicated to ants (family Formicidae) includes comprehensive study of 116 species (Arakelian 1994). Another well-reviewed group is family Encyrtidae which includes 170 species from Armenian fauna. Besides, there are data on about 100 dendrophilous hymenopterans from 11 families in the book of Mirzoyan (1977).

From dipteran insects (Diptera), Terterian (1968) published review of black flies (Simuliidae) which includes 45 species. This author also conducted research on horse-flies (Tabanidae), but, unfortunately, his data on the fauna composition were not published. Besides, Mirumyan is studying gall midges (Cecidomyiidae) and provided data on about 130 species of Armenian fauna (Mirumyan 2011; Mirumyan and Skuhravá 2022).

It must be emphasized that some important and diverse groups of invertebrates have not yet been studied specifically. These are, for instance, terrestrial crustaceans – woodlices (Isopoda), millipedes (Diplopoda) and centipedes (Chilopoda), earthworms (Annelida), Collembola, several insect orders (mayflies – Ephemeroptera, cockroaches – Blattodea, mantises – Mantodea, lace wings – Neuroptera, etc.). Understudied are Hymenopterans except of few families, majority of Dipteran families, most of true bugs (Heteroptera), very few data exists on majority of Armenian moths.

5.1.1 *Molluscs*

Mollusc fauna of Armenia, according to Akramowski (1976), includes 155 species; majority of them belongs to Gastropods (class Gastropoda) – 141 species belonging to 5 orders, 20 families and 85 genera. Bivalves (class Bivalvia) are less diverse and includes 14 species from two orders, 6 families and 6 genera.

5.1.1.1 Geographical Analysis of Armenian Molluscs

Detailed geographical analysis of Armenian mollusks was conducted by Akramowski (1976). Simplified version of his analysis is presented in Table 5.1. The following faunistic elements are considered:

Table 5.1 Division of Armenian molluscs by faunistic elements and vegetation types (according to Akramowski, 1976, simplified)

	Faunistic elements	Vegetation types					
		Total	Semidesert	Phryganoids and arid open forest	Steppe	Forest	Subalpine and alpine
1.	Widely distributed	33	17	8	28	21	16
2.	European	47	19	18	31	39	14
3.	Mediterranean	4	1	3	1	2	–
4.	Eastern Mediterranean (Euxinian)	6	1	1	1	2	1
5.	*Anteroasiatic mesophilous*	10	4	3	4	10	2
6.	Pan-Caucasian	4	1	–	1	2	1
7.	Transcaucasian	24	4	2	5	18	12
8.	Hyrcanian	2	–	–	–	2	–
9.	*Anteroasiatic xerophilous*	2	–	–	1	–	–
10.	Asia Minor	1	–	–	1	–	–
11.	Armenian Highland	3	–	–	2	2	1
12.	Armenian-atropatenian	3	–	3	2	–	–
13.	Northern Iranian	12	7	8	6	4	–
14.	Araratian	4	2	2	2	–	–
	Total	155	56	50	85	92	47

1. Widely distributed species inhabit nearly whole Palearctic, sometimes penetrating to neighbouring provinces.
2. European element – the species included occupy major part of Europe, sometimes penetrating to Caucasus and Western Siberia.
3. Species of Mediterranean element occupy Mediterranean countries sometimes reaching East to Transcaucasia.
4. Eastern Mediterranean species are distributed from Balkans to Minor Asia and Caucasus.

Following elements are considered by Akramowski as Anteroasian, they are as follows:

5. Anteroasiatic mesophilic species inhabit mesophilic habitats of Western Asia, including mountain systems of the Caucasus, Armenian Taurus, Pontic and Zagros Mountains.
6. Pan-Caucasian species range cover whole Caucasus and mountainous territories of Transcaucasia, including Lesser Caucasus.
7. Transcaucasian species occupy Transcaucasia, usually its certain parts.
8. Hyrcanian element range mainly lays in forest belt south of Caspian Sea, two species-rich in Armenia.
9. Anteroasiatic xerophilous species.
10. Asia Minor.
11. Armenian Highland – main range in Armenian Highland.
12. Armenian-atropatenian species live from Northwestern Iran to Armenia.
13. Northern Iranian species range continues from Armenia (Ararat Plane) through North Iran, some species-rich in Kopetdagh Mounts in Turkmenistan.
14. Araratian species are endemics of Ararat Valley in Armenia.

From 155 species known from Armenia, about 60 are considered endemics of Caucasian Ecoregion. There are six Armenian endemics with very narrow distribution, two of them (*Euxina valentini* (Loosjes, 1964) and *E. akramovskii* Likharev, 1962 are endemics of forests of Zangezur Range (including its spur, Bargushat range), *Shadinia akramovskii* (Shadin, 1952) and *Gabbiella araxena* Akramowki, 1970 inhabit water bodies of Ararat valley only, *Orculella ruderals* Akramowki, 1947 is known only from type locality, vicinities of Gnishik village on northern slopes of Daralagjaz Range, and *Pupilla bipupillata* Akramowki, 1947 distributed in Central and Southern Armenia, from Azat river valley to Meghri.

5.1.1.2 Distribution of Armenian Molluscs by Altitude Belts and Habitats

Due to the specific requirements of the habitat, especially with regard to humidity, the presence of terrestrial molluscs usually depends on the availability of suitable shelters of various types.

From semidesert belt, about 55 species are reported. Molluscs distributed here use as the shelters stones, sagebrush (*Artemisia*) and other semi-shrubs, rock cracks, etc. Most common species here is *Xerosecta crenimargo* (Pfeiffer, 1848). Here occur also *Truncatellina callicratis* (Scacci, 1833), *Pupilla triplicata* (Studer, 1820), *P. interrupta* (Reinhardt, 1876), *P. signata* (Mousson, 1873), *Imparietula sieversi* (Mousson, 1873), *Eopolita derbentina* (Boettger, 1886), *Phecolimax annularis* (Studer, 1820), etc. This belt in Armenia is strongly transformed by agriculture use and comprises large territories of orchards, vineyards, arable lands, as well as some wetlands. In such intrazonal and usually mesophilic territories penetrate several mollusc species such as slag *Deroceras caucasicum* (Simroth, 1901), in arable lands can be observed *Xeropicta derbentina* (Krynicki, 1836).

Belt of phryganoids and arid open forest has similar conditions for molluscs and thus is described as a whole. Here are registered 50 species of molluscs, including two endemics of Armenia, *Orculella ruderalis* Akramowski, 1947 (under the stones) and *Pupilla bipapullata* Akramowski, 1947 (in the plant litter), as well as *Euomphalia pisiformis* (Pfeiffer, 1848), *Jaminia isseliana* (Bourguignat, 1865), and other, including several steppe and semidesert species.

Steppe belt is characterized by diverse mollusc fauna comprising of 85 species. In the steppe, the main shelter for molluscs is turf, then soil and stones. Here are presented several rather xerophilous species like *Vallonia costata* (Müller, 1774); under the stones and gravel can be found *Imparietula seductilis* (Rossmaesler, 1837) and *Phenacolimax annularis* (Studer, 1820), in loose soil occur *Imparietula pupoides* (Krynicki, 1833) and *I. tetrodon* (Mortilet, 1854). In the meadow-steppe are common *Truncatellina cylindrica* (Ferrusac, 1807), *Pupilla inops* (Reinhardt, 1876), in the turf besides *Vallonia costata* mentioned above and *V. pulchella* (Müller, 1774), under the stones can be found *Limax flavus* (Linnaeus, 1758), *Vitrinoides monticola* (Boettger, 1881), on grasses *Vertigo substriata* (Jeffreys, 1830), *V. pygmaea* (Draparnaud, 1801), *V. angustior* (Jeffreys, 1830), *Euomphalia selecta* (Klika, 1894).

Most diverse is a fauna of forests of different types which counts 92 species. Mainly in the soil occur *Cecilioides acicula* (Müller, 1774), *Hyrcanolestes armeniacus* (Simroth, 1910), *H. orientalis* (Simroth, 1912), *Chondrula tridens* (Müller, 1774). Forest litter inhabits *Pomatius rivulare* (Eichvald, 1829), *Carychium tridentatum* (Risso, 1826), *Cionella lubricella* (Porro, 1838), *Sucinella oblonga* (Draparnaud, 1801), *Vertigo pusilla* (Müller, 1774), *V. substriata* (Jeffreys, 1830), *V. pygmaea* (Draparnaud, 1801), *Truncatellina costulata* (Nillson, 1822), *Lauria cylindracea* (Da Costa, 1778), *Orcula doliorum* (Brugutere, 1792), *Acanthinula aculeate* (Müller, 1774), *Punctum pygmaeum* (Draparnaud, 1801), and many others. Some species can be found in fallen trunks, for example, *Serrulina serratula* (Pfeiffer, 1847) and *Caspiophaedusa perlucens* (Boettger, 1877). Under the stones occurs *Fruticocampylea narzanensis* (Ktynocki, 1836). Numerous are slags, namely, *Deroceras melanocephalum* (Kaleniczenko, 1851), *D. reticulatum* (Müller, 1774), *D. caucasicum* (Simroth, 1901), *Parmacella ibera* (Eichvald, 1841). In light forests in the shelters (litter, under the stones and fallen trunks, in hollows) live *Limax flavus* (Linnaeus, 1758), *Vitrinoides monticola* (Boettger, 1881), *Monochroma brunneum* (Simroth, 1901). Several species can be observed on trunks and leaves, namely, *Circassina circassica* (Mousson, 1863), *Caucasotachela calligera* (Dubois de Montpereux, 1840), as well as *Ena obscura* (Müller, 1774), *Mentissoidea litotes* (Schidt, 1868), *Idyla foveicollis* (Charpentier, 1852), *Quadriplicata quadriplicata* (Schidt, 1862) and others. Several species occur in forest edge and lawns, some of them penetrate here from steppe and arid open forests, but some are typical for such habitats, for example, *Chondrula tridens* (Müller, 1774), *Pupilla interrupta* (Reinhardt, 1876), as well as species of the genus *Helix: H. lucorum* Linnaeus, 1758, in Northern Armenia also *H. buchi* Dubois de Montpereux, 1839. Fauna of subalpine forest is composed partly by species penetrating here from forest and high mountain ecosystems, but some species such as *Columella edentula* (Draparnaud, 1805), *Euxina somchetica* (Pfeiffer, 1846), *E. tschetschenika* (Pfeiffer, 1866) are typical for this kind of habitats.

Forty-seven species are registered in subalpine and alpine meadows. Some of them, like *Vallonia costata, V. pulchella, Cionella lubrica* (Müller, 1774), etc. are widely distributed in other belts, but there are several species typical for these habitats, for example, *Oxychilus retowskii* (Lindholm, 1914), *Karabaghia bituberosa* (Lindholm, 1927), *Trichia armeniaca* (Pfeiffer, 1846), *Columella columella* (Martens, 1830), *Hesseola solidior* (Mousson, 1873), etc. Most likely, this belt inhabits also red-listed snail *Bithynia troscheli* (Paasch, 1842) which is known so far only by empty shells in the litter on banks of the River Masrik where they were brought from the overlying territories.

Many species of molluscs occur in intrazonal ecosystems presented in several altitudinal belts.

Rocks are characterized by the presence of good shelters, namely, deep cracks. Nearly everywhere except of high mountains can be found *Pyramidula rupestris* (Draparnaud, 1801) and *Armenica brunnea* (Rossmaesler, 1839); phrygana and arid forests inhabit red-listed *Turanena scalaris* (Naegele, 1902), as well as *Levantina djulfensis* (Dubois de Montpereux, 1840), *L. escheriana* (Borguignat, 1864), in steppe lives *Granopupa granum* (Draparnaud, 1801); rocks in forest belt inhabit *Chondrula clienta* (Westerlund, 1863), *Armenica griseofusca* (Mousson, 1876), as well as Armenian endemics *Euxina akramowskii* Likharev, 1962 and *E. valentini* (Loosjes, 1964).

Several species inhabit shores of water bodies with suitable humidity conditions. These are *Succinea putris* (Linnaeus, 1758), *Oxyloma elegans* (Risso, 1826), *O. sarsi* (Esmark, 1886), *Vertigo antivertigo* (Draparnaud, 1801), *V. moulinsiana* (Dupuy, 1849), *Zonitoides nitidus* (Müller, 1774), *Deroceras transcaucasicus* (Simroth, 1901), as well as some meadow species – *Carychium minimum* (Müller, 1774), *Hesseola solidior*, *Cionella lubrica*, etc.

Rich is the fauna of the water bodies of different types.

Typical for springs but sometimes penetrating to streams flowing from them are red-listed endemic species *Shadinia akramovskii* as well as *Sh. terpoghassiani* (Shadin, 1952), *Galba truncatula* (Müller, 1774), *Radix peregra* (Müller, 1774), *Euglesa casertana* (Poli, 1791), *Theodoxus pallasi* Lindholm, 1924, *Physella acuta* (Draparnaud, 1805), red-listed bivalve *Odhneripisidium annandalei* (Prashad, 1925), etc. In the rivers can be observed *Valvata piscinalis* (Müller, 1774), *Anodonta piscinalis* Nillson, 1823, *Ancylus fluviatilis* (Müller, 1774), *Crassiana crassa* (Philipsson, 1788).

Lakes and ponds inhabit *Radix auricularia* (Linnaeus, 1758), *Lymnaea stagnalis* (Linnaeus, 1758), *Euglesa subtruncata* (Malm, 1855), *E. obtusalis* (Lamarck, 1818), *Anodonta piscinalis* above-mentioned, *Bathyomphalus contortus* (Linnaeus, 1758), *Gyraulus acronichus* (Férussac, 1807), *Planorbis planorbis* (Linnaeus, 1758), *Physella acuta* (Draparnaud, 1805), *Musculium hungaricum* (Clessin, 1887), etc. Only in Lake Janlich near Goris in Armenia live red-listed *Musculium strictum* (Normand, 1844). Some species associated with ponds and lakes can live in swamps; in peat bogs occurs *Aplexa hypnorum* (Linnaeus, 1758).

In Lake Sevan and associated small reservoirs were found about 20 species of molluscs, among which *Euglesa personata* (Malm, 1855) in Armenia is known only from the lake; some of these species, for example red-listed *Acroloxus lacustris* (Linnaeus, 1758), *Gyraulus albus* (Müller, 1774), *G. laevis* (Alder, 1838) and *Planorbis carinatus* (Müller, 1774) most likely are extinct here due to the lowering of the lake level and drying of small reservoirs, including Lake Gilli.

5.1.1.3 Rare and Threatened Species

Like other components of Armenian biodiversity, molluscs suffer from the threats described in Chap. 6. Many of them can be considered rare and threatened.

So far 16 species of molluscs are listed in the Red Book of Animals of the Republic of Armenia, including 14 species of gastropods (*Bithynia troscheli* (Paasch, 1842) – CR, *Shadinia akramovskii* (Shadin, 1952) – CR, *Acroloxus lacustris* (Linnaeus, 1758) – CR, *Planorbis carinatus* (O.F. Muller, 1774) – CR, *Anisus leucostomus* (Millet, 1813) – CR, *Gyraulus albus* (O.F. Muller, 1774) – EN, *Gyraulus laevis* (Alder, 1838) – EN, *Gyraulus regularis* (Hartmann, 1841) – CR, *Columella* (Martens, 1853) – CR, *Vertigo angustior* Jeffreys, 1830 – CR; IUCN (ver 2.3) – LR/CD; Annex II, *Orculella ruderalis* (Akramowski, 1947) – CR, *Orculella bulgarica* (P.Hesse, 1915) – CR, *Turanena scalaris* (Naegele, 1902) – CR, *Euxina akramowskii* (Likharev, 1962) – CR, and two species of bivalves (*Musculium strictum* (Normand, 1844) – CR, *Odhneripisidium annandalefi* (Prashad, 1925) – CR).

Besides, in Armenia are presented several other threatened species which deserve special protection measures. These are Armenian endemics not included to the Red Book yet, namely, *Euxina valentini, Pupilla bipupillata, Gabbiella araxena* as well as several species known from Armenia from single and partly threatened habitats and partly have narrow distribution (sub-endemics of Armenia, endemics of Caucasian Ecoregion) such as *Valvata pulchella* (Studer, 1820), *Hippeutis complanatus* (Linnaeus, 1758), *Ena schuschaensis* (Kobelt, 1902), *Armenica disjuncta* (Mortillet, 1854), *A. griseofusca* (Mousson, 1876), *Cecilioides raddei* (Boettger, 1879), *Hyrcanolestes velitaris* (Martens, 1880), *H. armeniacus* (Simroth, 1910), *H. orientalis* (Simroth, 1912), *Karabaghia bituberosa* (Lindholm, 1927) and some others. These species have to be estimated using IUCN Criteria and, if necessary, must be included in the next edition of the country's Red Book.

5.1.2 Arachnida: Araneae

Among the Arthropoda, the spiders, scorpions, pseudoscorpions, harvestmen, mites, and certain other less popular forms constitute a group known to zoologists as the class Arachnida.

5.1.2.1 Order Araneae (Spiders)

Spiders, though they are ubiquitous, have remained one of the neglected groups of Armenian fauna. Nevertheless, the Arachnofauna of Armenia includes 234 species belonging to 34 family (Nentwig et al. 2023; World Spider Catalog 2023), the most diverse family is Salticidae (the jumping spiders), which has 46 species.

Thirteen species of spiders are endemic for Armenia, and 34 species are endemics for the Caucasus region: *Araeoncus caucasicus* Tanasevitch, 1987, *Asianellus potanini* (Schenkel, 1963), *Centromerus minor* Tanasevitch, 1990, *Chalcoscirtus tanasevichi* Marusik, 1991, *Chinattus caucasicus* Logunov, 1999, *Clubiona caucasica* Mikhailov & Otto, 2017, *C. golovatchi* Mikhailov, 1990, *Dysdera armenica* Charitonov, 1956, *Filistata lehtineni* Marusik & Zonstein, 2014, *Geolycosa dunini* Zyuzin & Logunov, 2000, *Harpactea eskovi* Dunin, 1989, *H. nachitschevanica* Dunin, 1991, *Heliophanus dunini* Rakov & Logunov, 1997, *Lycosa piochardi* Simon, 1876, *Mesiotelus caucasicus* Zamani & Marusik, 2021, *Micaria kopetdaghensis* Mikhailov, 1986, *Nomisia conigera* (Spassky, 1941), *Oecobius nadiae* (Spassky, 1936), *Oedothorax meridionalis* Tanasevitch, 1987, *Olios sericeus* (Kroneberg, 1875), *Orthobula charitonovi* (Mikhailov, 1986), *Panamomops fedotovi* (Charitonov, 1936), *Pardosa caucasica* Ovtsharenko, 1979, *P. colchica* Mcheidze, 1964, *Pelecopsis crassipes* Tanasevitch, 1987, *Persiscape gideoni* (Levy, 1996), *Phrurolithus azarkinae* Zamani & Marusik, 2020, *Pseudomogrus albocinctus* (Kroneberg, 1875), *Tenuiphantes contortus* (Tanasevitch, 1986), *Raveniola dunini* Zonstein, Kunt & Yağmur, 2018, *Rhysodromus rikhteri* (Logunov & Huseynov, 2008), *Synageles persianus* Logunov, 2004, *Walckenaeria bifasciculata* Tanasevitch, 1987. Five species seems to be endemics for the Armenian Highland: *Araeoncus clavatus* Tanasevitch, 1987, *Evarcha armeniaca* Logunov, 1999, *Heliophanus curvidens* (O. Pickard-Cambridge, 1872), *Metellina orientalis* (Spassky, 1932), *Proislandiana beroni* Dimitrov, 2020.

Spiders occur in all terrestrial habitats. Forests and open forests have high species richness of Aranea in Armenia and in these acosystems 98 species have been recorded. It is worth to mention that the forests of Armenia introduce as a special habitat of 11 Caucasus endemic species, and three Armenian endemics: *Harpactea secunda* Dunin, 1989 (Dunin, 1989), *Dysdera collucata* Dunin, 1991, *D. mazini* Dunin, 1991 (Dunin, 1991). It has to be noticed that many widespread species like *Evarcha arcuata* (Clerck, 1757), *Heliophanus auratus* C. L. Koch, 1835, *Heliophanus cupreus* (Walckenaer, 1802), *Micaria pulicaria* (Sundevall, 1831), *Pellenes brevis* (Simon, 1868) are presented in the Armenian forests.

Another diverse Arachnofauna presents the desert and semidesert habitats with more than 60 species of which 9 are Armenian endemics: *Harpactea armenica* Dunin, 1989, *H. deelemanae* Dunin, 1989, *H. golovatchi* Dunin, 1989, *H. nenilini* Dunin, 1989, *Heliophanus forcipifer* Kulczyński, 1895, *Pseudicius pseudocourtauldi* Logunov, 1999, *Mesiotelus patricki* Zamani & Marusik, 2021, *Philodromus juvencus* (Kulczyński, 1895) and *Raveniola ambardzumyani* Marusik & Zonstein, 2021 (Kulczyński, 1895, Dunin, 1989, Rakov & Logunov, 1997, Logunov, 1999, Marusik & Zonstein, 2021, Zamani & Marusik, 2021, Zarikian et al. 2022), as well as many widespread species like *Cheiracanthium erraticum* (Walckenaer, 1802) (Fig. 5.1), *Geolycosa dunini* Zyuzin & Logunov, 2000, *Lycosa praegrandis* C. L. Koch, 1836, *Philodromus aureolus* (Clerck, 1757) and *Steatoda paykulliana* (Walckenaer, 1806) (Fig. 5.2).

Grasslands (steppes, meadows, meadow-steppes) are rather interesting habitats for most of Araneidae, Linyphiidae, Philodromidae, some Salticidae and Thomisidae families' representatives. While as an endemic species we could mention only one species *Pardosa condolens* (O. Pickard-Cambridge, 1885) (Schmidt, 1895), many widespread species are represented in these ecosystems, for example *Misumena vatia* (Clerck, 1757), *Oxyopes lineatus* Latreille, 1806 (Fig. 5.3), *Pisaura mirabilis* (Clerck, 1757) (Fig. 5.4) and *Thomisus onustus* Walckenaer, 1805.

Beside all the above-mentioned species, the urban species occupies a special place among the terrestrial spiders. These spiders share our homes and gardens. Those are the fear spreaders causing people to suffer of Arachnophobia. We could record many widespread species like *Pholcus phalangioides* (Fuesslin, 1775), *Scytodes thoracica* (Latreille, 1802), *Steatoda bipunctata* (Linnaeus, 1758), *Salticus*

Fig. 5.1 *Cheiracanthium erraticum*

Fig. 5.2 *Steatoda paykulliana*

Fig. 5.3 *Oxyopes lineatus*

scenicus (Clerck, 1757), *Oecobius nadiae* (Spassky, 1936) and *Philaeus chrysops* (Poda, 1761) (Fig. 5.5) (Zarikian and Kalashian 2021; Otto 2022; Zarikian 2022a, b).

There have been recorded also water spiders *Argyroneta aquatica* (Clerck, 1757) which found in freshwater (Otto 2022) or on bushes near waters, *Argiope*

Fig. 5.4 *Pisaura mirabilis*

Fig. 5.5 *Philaeus chrysops*

bruennichi (Scopoli, 1772) (Fig. 5.6), *Tetragnatha dearmata* (Thorell, 1873), *Tetragnatha extensa* (Linnaeus, 1758) (Fig. 5.7), *Tetragnatha montana* (Simon, 1874), *Tetragnatha nigrita* (Lendl, 1886) (Zarikian & Kalashian, 2021, Otto, 2022).

Fig. 5.6 *Argiope bruennichi*

Fig. 5.7 *Tetragnatha extensa*

Spiders flourish at surprisingly utmost elevations on all the high altitudes of Armenia. The dominant groups here are Eresidae, Salticidae, Gnaphosidae, Thomisidae and Lycosidae families' species, that can be found over the elevation 2500 m.a.s.l. These species are also widely distributed even on the lowlands, for

Fig. 5.8 *Eresus lavrosiae*

example *Xysticus audax* (Schrank, 1803), *X. kochi Thorell*, 1872 and *Eresus lavrosiae* Mcheidze, 1997 (Fig. 5.8).

5.1.3 Beetles (Coleoptera)

Like in nearly all terrestrial ecosystems, beetles in Armenia are the most diverse group of animals (and plants as well). Collection of beetles in the country began in nineteenth century, and some studies were carried out using this material, for instance, in comprehensive works of Reitter. And due to continuous interest to this group, studied by several well-known scientists, of which we can mention Richter, Ter-Minassian, Plavilstshikov and some others, and decades of researches conducted by famous coleopterologist, Dr. Iablokoff-Khnzorian and his students, beetles are one of the most studied in the country.

Back in 1961, Iablokoff-Khnzorian (1961) published the comprehensive work on Armenian beetle fauna. Relying on his own catalogue, he mentioned from Armenia 3319 species and conducted zoogeographic analysis of the fauna. Though the catalogue continued and continues to be updated with new species and now includes about 3900 species newly found or described from Armenia; here we provided analysis based on the 1961 book as general image seems to be similar.

Iablokoff-Khnzorian divided the fauna to several faunistic elements and types of biotopes. These faunstic elements, somewhat simplified are presented below (Table 5.2).

1. Widely distributed species are those which range goes beyond Holarctic.
2. Palearctic species are widely distributed in the province, some penetrated to Nearctic or have range covering nearly whole Eurasia.

Table 5.2 Division of Armenian beetles by faunistic elements and biotopes (according to Iablokoff-Khnzorian 1961, simplified)

NN	Faunistic elements	Biotope types								Total
		Desert and semidesert	Phrygana[a]	Light forest	Broad-leaved forest	Pine Forest	Steppe	Meadows	Wetlands	
1.	Widely distributed	27	–	5	19	–	8	4	14	77
2.	Palearctic	75	1	53	128	12	80	17	95	461
3.	South Palearctic	51	1	17	3		11	1	22	105
4.	Euro-Siberian	18	–	22	107	20	24	13	49	253
5.	European	39	2	72	198	14	44	13	54	436
6.	Mid-European Caucasian	26	2	24	63	–	15	6	18	154
7.	Mediterranean	3	2	3	14	–	–	–	5	28
8.	Eastern Mediterranean	88	14	61	40	4	17	11	32	192
9.	Steppe	143	7	71	49	1	41	15	69	396
10.	Irano-Transcaucasian	33	4	10	2	–	1	2	5	57
11.	Araratian	29	7	17	1	–	9	8	6	77
12.	Ancient Mediterranean	49	5	39	19	3	5	3	32	155
13.	Egeido-Turanian	98	16	20	7	–	4	3	40	188
14.	Hyrcano-Armenian	1	–	6	8	–	2	2	1	20
15.	Pan-Caucasian	28	5	28	74	–	9	23	13	180
16.	Transcaucasian	35	7	17	47	–	5	24	15	150
17.	Armenian endemics	80	34	34	27	–	3	44	32	254
	Total	823	107	499	806	54	279	188	502	3258

[a]Here and further the term Phrygana is used for phryganoid vegetation

3. South Palearctic distributed in Mediterranean, Western and Middle Asia sometimes reaching India and/or Palaearctic.
4. Euro-Siberian are distributed in Europe (including Caucasus), Siberia, sometimes also in Northern Palaearctic and in the mountains of Central Asia.
5. European species are widely distributed in Europe (including Caucasus), sometimes reaching Western Siberia and Mediterranean mountains.
6. Mid-European Caucasian – this peculiar group includes species known from Western Europe and Caucasus and is characterized by big range disjunction.
7. Mediterranean species are widely distributed in Mediterranean reaching Caucasus.
8. Eastern Mediterranean distributed to the East of Italy reaching Iran, Crimea and Caucasus.

9. So-called steppe species are characteristic for steppe zone of Palearctic, to the west sometimes penetrated to Western Europe, to the East reach Northern China.
10. Irano-Transcaucasian species occurs in Iran and Southern Transcaucasia.
11. The group of Araratian species includes those distributed in Anterior Asia and to the Northeast reach Armenia not penetrating to other Caucasian countries.
12. Ancient Mediterranean species distributed from Western Mediterranean to Central Asia – along territories surrounding ancient Tethys.
13. Egeido-Turanian species living in Central Asia and the eastern Mediterranean, including Caucasus.
14. Hyrcano-Armenian distributed besides Armenia also in Talysh Mountains, Northern Iran and Kopetdagh Mountains.
15. Pan-Caucasian are widely distributed in Caucasian Ecoregion, from Ciscaucasia to most Northeastern Turkey and Northwestern Iran.
16. Transcaucasian species occupy Transcaucasia, usually its certain parts.
17. Armenian endemics so far are known only from Armenia, sometimes also in Nakhichevan AR of Azerbaijan which forms a zoogeographic unity with Armenia, and sometimes in opposite bank of Arax River.

In total, in Armenian fauna beetles from 106 families (from 148 mentioned for Palearctic region as a whole) are represented. Some diverse and rather well-studied groups are considered below. The analysis is based on data provided in monographs specially dedicated to the concrete group; total number of species for each group is taken from respective parts of the Catalogue of Palearctic Coleoptera (below mentioned as Catalogue). Species names are provided according to modern nomenclature as it is accepted in the Catalogue. In distribution of the species by habitats are provided data on most preferable ones, if known.

5.1.3.1 Family Ground beetles – Carabidae (Iablokoff-Khnzorian 1976)

The family is one of the most diverse in Armenia being second only to weevils (Curculionidae), possibly also to rove beetles (Staphylinidae). According to Iablokoff-Khnzorian (1967) in Armenia are presented 409 species of the ground beetles; in the work he revised 267 species; taking into account new findings and descriptions in Catalogue (Huber et al. 2017) listed 508 species and subspecies for the country's fauna. There are about 85 species and subspecies of ground beetles endemic for the Caucasus Ecoregion; among these 31 taxa are endemics of Armenia: *Carabus caucasicus antonkozlovi* Cavazzuti, 2014, *C. c. tatianagorokhovae* Cavazzuti & Kozlov, 2014, *C. c. myskai* Cavazzuti, 2004, *C. c. nakagomei* Cavazzuti & Kozlov, 2016, *C. varians alagoensoides* Mandl, 1975, *Nebria sevanensis* Shilenkov, 1983, *Cicindela georgiensis prunieri* Deuve, 2012, *Elaphrus hypocrita araxellus* Iablokoff-Khnzorian, 1963, *Dyschirius sevanensis* Iablokoff-Khnzorian, 1962, *Ocys trechoides* Reitter, 1895, *Duvalius stepanavanensis* Iablokoff-Khnzorian, 1963, *D. suvodolensis* S.B. Ćurčić, Brajkovic & B.P.M. Ćurčić, 2003, *D. yatsenkokhmelevskyi* (Iablokoff-Khnzorian, 1960), *Trechus armenus*

Iablokoff-Khnzorian, 1963, *T. dilizhanicus* Belousov, 1989, *T. infuscatus* Chaudoir, 1850, *T. khalabicus* Belousov, 1990, *T. khnzoriani* Pawłowski, 1976, *Deltomerus khnzoriani* Kurnakov, 1960, *Chilotomus alexandri* Kalashyan, 1999, *Pterostichus capitatus* (Chaudoir, 1850), *P. cucujinus* Reitter, 1892, *P. eriwanicus* (Tschitschérine, 1897), *P. kadjaranze* (Jedlička, 1947), *P. latiusculus* (Chaudoir, 1868), *P. arator perlidaghensis* Reitter, 1902, *Laemosthenus lederi arenicus* (Kalashian, 1979), *L. l. khnzoriani* (Kalashian, 1983), *Zabrus aurichalceus punctipennis* (Chaudoir, 1846), *Z. corpulentus armeniacus* Ganglbauer, 1915, *Z. trinii araxidis* Reitter, 1889.

Ground beetles have very diverse ecology and biology. Majority of them are predators feeding on different kinds of invertebrates (molluscs, different arthropods, earthworms, etc.). On the other hand, there are numerous herbivorous species, feeding mainly on seeds, roots, and leaves of grassy plants. Some species are saprophagous and can consume organic residues. Ground beetles are presented in all types of habitats of Armenia, and in high mountains can be considered the most diverse group among all beetles.

Fauna of **deserts and semideserts** in Armenia includes approximately 100 species. Mainly they live in stony and sometimes clay semideserts, these are *Calosoma maderae dsungaricum* (Gebler, 1833), *Distichus planus* (Bonelli, 1813), *Broscus cephalotes* (Linnaeus, 1758), *Carterus rufipes* (Chaudoir, 1843), *Graniger cordicollis* (Audinet-Serville, 1821), *Penthus tenebrioides* (Waltl, 1838), *Dixus eremita* (Dejean, 1825), *D. semicylindricus* (Piochard de la Brûlerie, 1872), *D. obscurus* (Dejean, 1825), *Ditomus calydonius oriens* (Dvořák, 1993), *Acinopus striolatus* Zoubkoff, 1833, *A. megacephalus* (Rossi, 1794), *A. ammophilus* Dejean, 1829, *Parophonus laeviceps* (Ménétriés, 1832), *Platyderus umbratus* (Ménétriés, 1832), *Gynandromorphus peyroni* Carret, 1905, *Amblystomus metallescens* (Dejean, 1829), *Zabrus rotundicollis* (Ménétriés, 1836), *Z. morio* (Ménétriés, 1832), *Platytarus faminii faminii* (Dejean, 1826), red-listed *Poecilus festivus* (Chaudoir, 1868), *Chlaenius festivus festivus* (Panzer, 1796), *Cymindis lineata* (Quensel, 1806), *C. andreae* Ménétriés, 1832, *Tachys bistriatus bistriatus* (Duftschmid, 1812) *T. turkestanicus* Csiki, 1928, *Microderes brachypus* (Steven, 1809), *Parazuphium chevrolatii schelkownikovi* (Carret, 1898), etc. Only here lives myrmecophilous *Paussus turcicus* Frivaldszky, 1835. Few species prefer sands, for instance, *Scarites salinus* Dejean, 1825, *S. procerus eurytus* Fischer von Waldheim, 1828.

The habitats of phryganoid vegetation inhabit few ground beetles, usually common with steppe and light forest fauna. Rather typical for these habitats are *Carabus convexus rhinopterus* Hampe, 1852, *Cicindela asiatica asiatica* Audouin & Brullé, 1839 (Fig. 5.9), *Calathus libanensis pluriseriatus* Putzeys, 1873, *Lebia trimaculata* (Villers, 1789), as well as some species feeding on seeds of different Asteracea, red-listed endemic *Chilotomus alexandri* Kalashyan, 1999 (Fig. 5.10) and *Tschitscherinellus oxygonus oxygonus* (Chaudoir, 1850).

Not so numerous are ground beetles in **light forests** which fauna includes about ten species more typical for them, for example, *Carabus convexus rhinopterus* Hampe, 1852, above-mentioned, well-known predator of caterpillars *C. sycophanta sycophanta* (Linnaeus, 1758), some *Lebia* species (*L. cruxminor cruxminor* (Linnaeus, 1758), *L. trimaculata* (Villers, 1789), *L. scapularis scapularis* (Geoffroy,

Fig. 5.9 Tiger beetle *Cicindela asiatica asiatica*

Fig. 5.10 *Chilotomus alexandri* (Red Book: CR)

1785)], *Mastax thermarum thermarum* (Steven, 1806), *Brachinus atripensis* Ballion, 1871, etc.

About 65 species and subspecies are typical for **broad-leaved forests**. These are several species of the genus *Carabus* (*Carabus septemcarinatus* Motschulsky, 1840, *C. varians varians* Fischer von Waldheim, 1823, Armenian endemic *C. v. alagoensoides* Mandl, 1975, *C. victor* Fischer von Waldheim, 1836, *C. puschkini puschkini* Adams, 1817, *C. calleyi calleyi* Fischer von Waldheim, 1823, etc., as well as endemics of Armenia or South Transcaucasia subspecies of snail-eater *C. (Procerus) caucasicus* Adams, 1817 – *C. c. antonkozlovi* Cavazzuti, 2014, *C. c. nakagomei* Cavazzuti & Kozlov, 2016 and red-listed *C. c. fallettianus* (Cavazzuti, 1997: Fig. 5.11). Here occur also *Calosoma inquisitor inquisitor* (Linnaeus, 1758), *Leistus fulvus* Chaudoir, 1846, *L. lenkoranus* Reitter, 1885, *Nebria picicornis luteipes* Chaudoir, 1850, two red-listed Armenian endemics of the genus *Duvalius* (*D. stepanavanensis* Iablokoff-Khnzorian, 1963, *D. yatsenkokhmelevskyi* (Iablokoff-Khnzorian, 1960)), *Tachyta nana nana* (Gyllenhal, 1810), *Trechus liopleurus liopleurus* Chaudoir, 1850, *T. quadrimaculatus* Motschulsky, 1850, *Bembidion incommodum* Netolitzky, 1926, *B. articulatum* (Panzer, 1796), *Anisodactylus binotatus* (Fabricius, 1787), *Laemostenus gratus* (Faldermann, 1836), etc.

Steppe habitats occupy about 50 species. These are some *Carabus* (*C. maurus maurus* Adams, 1817 *C. cribratus cribratus* Quensel, 1806) and *Chlaenius* species [*Ch. vestitus* (Paykull, 1790), *Ch. cruralis* Fischer von Waldheim, 1829, *Ch. decipiens* (Dufour, 1820)], *Clivina fossor fossor* (Linnaeus, 1758), *Bembidion lampros* (Herbst, 1784), as well as numerous Harpalinae (*H. tenebrosus tenebrosus* Dejean, 1829, *H. affinis* (Schrank, 1781), *H. caspius* (Steven, 1806), *H. rubripes* (Duftschmid, 1812), *H. serripes serripes* (Quensel, 1806), etc., *Ophonus azureus* (Fabricius, 1775), *Amara aenea* (De Geer, 1774), *A. apricaria* (Paykull, 1790), endemic of

Fig. 5.11 Snail-eater ground beetles *Carabus (Procerus) scabrosus fallettianus* (Red Book: VU)

Armenia *Zabrus trinii araxidis* Reitter, 1889). Common here are *Brachinus crepitans* (Linnaeus, 1758), *B. explodens* Duftschmid, 1812, etc.

Typical for **meadows** are about 50 species of ground beetles. These are representatives of the genus *Carabus* (*C. stjernvalli stjernvalli* Mannerheim, 1830, *C. pumilio* Küster, 1846, *C. wiedemanni gotschi* Chaudoir, 1846, *C. scabripennis* Chaudoir, 1850, *C. chevrolati korbi* Breuning, 1928, *C. tamsi tamsi* Ménétriés, 1832, etc.), *Calosoma breviusculum breviusculum* (Mannerheim, 1830), *Cicindela desertorum desertorum* Dejean, 1825, *Nebria gotschii* Chaudoir, 1846, *Deltomerus khnzoriani* Kurnakov, 1960, *Trechus dzermukensis* Iablokoff-Khnzorian, 1963, red-listed endemic of alpine meadows of Aragats Mount *T. infuscatus* Chaudoir, 1850, *T. magniceps* Reitter, 1898, *T. angelicae* Reitter, 1892, *Bembidion bipunctatum bipunctatum* (Linnaeus, 1760), endemic of Meghri range *Ocys trechoides* Reitter, 1895, *Zabrus aurichalceus punctipennis* (Chaudoir, 1846), *Pterostichus armenus armenus* (Faldermann, 1836), *P. chydaeus chydaeus* (Tschitschérine, 1897), as well as Armenian endemics *P. eriwanicus* (Tschitschérine, 1897) and *P.capitatus* (Chaudoir, 1850), *Laemostenus mannerheimi mannerheimi* (Kolenati, 1845), *Cymindis miliaris* (Fabricius, 1801), etc. Some species can reach altitudes 3000–3600 m (e.g. *Trechus* spp., *Deltomerus, P. eriwanicus* , *P.capitatus*), being sometimes, together with rove beetles (Staphylinidae) the only representatives of the order. Many species can be found near melting snow, sometimes consuming organic residues.

Very diverse is the fauna of **wetlands** which includes more than 100 species inhabiting different types of these habitats. Swamps inhabit *Carabus clathratus auraniensis* Müller, 1903, Armenian endemic *Elaphrus hypocrita araxellus* Iablokoff-Khnzorian, 1963, *E. uliginosus* (Fabricius, 1792), *Pogonus luridipennis* (Germar, 1822), *P. punctulatus* Dejean, 1828, *Chlaenius tristis tristis* (Schaller, 1783), *Agonum rugicolle* Chaudoir, 1846, *Bembidion obliquum* Sturm, 1825, *B. varium* (Olivier, 1795), *Diachromus germanus* (Linnaeus, 1758), *Idiomelas morio* (Ménétriés, 1832), *Stenolophus marginatus* Dejean, 1829, *S. mixtus* (Herbst, 1784), *S. abdominalis persicus* Mannerheim, 1844, *Acupalpus elegans* (Dejean, 1829), *A. maculatus* (Schaum, 1860), etc. In salt marshes live *Siagona europaea europaea* Dejean, 1826, *Cymbionotum semelederi* (Chaudoir, 1861), *Apotomus rufithorax* (Pecchioli, 1837), *Daptus vittatus* Fischer von Waldheim, 1823, *Parophonus hirsutulus* (Dejean, 1829), *Dicheirotrichus ustulatus* (Dejean, 1829), *Harpalus breviusculus* Chaudoir, 1846, *Tachys bistriatus bistriatus* (Duftschmid, 1812) *T. fulvicollis* (Dejean, 1831) *T. scutellaris scutellaris* (Stephens, 1828), *Anisodactylus poeciloides pseudoaeneus* Dejean, 1829, *A. signatus* (Panzer, 1796), etc. Typical for shores of streams and other water bodies are tiger beetles (Cicindelinae), *Calomera fischeri fischeri* (Adams, 1817), *C. caucasica* (Adams, 1817), *C. littoralis winkleri* (Mandl, 1934), *Cylindera sublacerata levithoracica* (Horn, 1891), etc. Very typical for such habitats are *Omophron limbatum* (Fabricius, 1777) and representatives of the tribe Bembidiini – *Asaphidion flavipes* (Linnaeus, 1760), *Sinechostictus lederi lederi* (Reitter, 1888), *Perileptus areolatus areolatus* (Creutzer, 1799), *Elaphropus caraboides* Motschulsky, 1839, species of the genus *Tachyura* [*T. hoemorroidalis* (Ponza, 1805), *T. anomala* (Kolenati, 1845), *T. diabrachys* (Kolenati, 1845)], members of huge genus *Bembidion* (*B. cyaneum*

Fig. 5.12 *Dyschirius sevanensis* (Red Book: EN)

Chaudoir, 1846, *B. kartalinicum* Lutshnik, 1938, *B. tibiale* Duftschmid, 1812, *B. transcaucasicum* Lutshnik, 1938, *B. caucasicum* Motschulsky, 1844, *B. menetriesii* (Kolenati, 1845), *B. persicum* Ménétriés, 1832, *B. ellipticocurtum* Netolitzky, 1935 etc., etc.). Immediate shoreline occupy *Dyschirius* species, including endemic of Lake Sevan basin, red-listed *D. sevanensis* Iablokoff-Khnzorian, 1962 (Fig. 5.12). Shorelines inhabit also *Chalaenius coeruleus* (Steven, 1809), *C. circumscriptus* (Duftschmid, 1812), etc.

5.1.3.2 Superfamily Lamellicorn Beetles – Scarabaeoidea

This rather well-defined group of beetles is currently divided into several families, here it is presented as a whole. Armenian fauna of the superfamily was revised by Iablokoff-Khnzorian (1967) who mentioned for Armenian fauna 217 species and accepted for the group in Armenia three families: Lucanidae, Trogidae and Scarabaeidae; taking into account new findings and descriptions in Catalogue (Nikolajev et al. 2016) listed 289 species and subspecies for the country's fauna; some groups mentioned as subfamilies by Iablokoff-Khnzorian in the Catalogue are given according to a modern system as a separate families.

Family **Earth-boring dung beetles – Geotrupidae** presented in Armenia by three species of the genus *Geotrupes*: *G. mutator* (Marsham, 1802), *G. olgae* (Olsoufieff, 1918), *G. spiniger* (Marsham, 1802), which are rather common in open landscapes from steppe to alpine meadows on dung.

Family **Hide beetles – Trogidae**. From Armenia so far are known three species of the genus *Trox*: *T. eversmannii* Krynicki, 1832, *T. hispidus* (Pontoppidan, 1763) and *T. niger* Rossi, 1792. The species are saprophagous, in nature can be found in some organic residues, for instance, in bird nests in different types of habitats.

Family **Enigmatic scarab beetles – Glaresidae** in Armenia is presented by one endemic subspecies *Glaresis oxiana armena* Iablokoff-Khnzorian, 1967, so far known from semideserts of Ararat Plane.

Family **Stag beetles – Lucanidae**. In Armenia seven species, three of them are endemics of the Caucasus Ecoregion. All the species develop in the dead and already rotten wood, majority of them, namely, *Aesalus ulanowskii* Ganglbauer, 1886, *Dorcus parallelipipedus* (Linnaeus, 1758), *Lucanus ibericus* Motschulsky, 1845 (with two subspecies), *Platycerus caucasicus* (Parry, 1864), *P. primigenius* Weise, 1960 and *P. vicinus* Gusakov, 2003 are living in broad-leaved forests. One species, very rare *D. peyronis* Reiche & Saulcy, 1856 is known by single findings in light forests of Central and Southern Armenia.

Family **Sand-loving scarab beetles – Ochodaeidae**. In Armenia, four species, *Codocera ferruginea ferruginea* (Eschscholtz, 1818) which is recorded from Armenia without exact data, and *Ochodaeus* species: *O. chrysomeloides* (Schrank, 1781) in dry steppes, *O. cornifrons* Solsky, 1876 and *O. integriceps* Semenov, 1891, both typical for semidesert habitats.

Family **Scavenger scarab beetles – Hybosoridae** – in Armenia one widely distributed species, *Hybosorus illigeri* Reiche, 1853 occurring in semideserts of Central and Southern Armenia.

Family **Bumble bee scarab beetles – Glaphyridae**. According to Catalogue in Armenia presented by 20 species and subspecies, nine of which are endemics of the Caucasus Ecoregion, including three Armenian endemics: *Eulasia eichleri* (Zaitzev, 1923), and two red-listed *Glaphyrus* species (*G. calvaster* Zaitzev, 1923, *G. "caucasicus* Kraatz, 1887"; Fig. 5.13), according to our recent study,

Fig. 5.13 *Glaphyrus "caucasicus"*

the latter is a separate species under description). Species of *Glaphyrus* (*G. festivus* Ménétriés, 1836, *G. micans micans* Faldermann, 1835 and two endemics above-mentioned) occupy semidesert and low part of steppes, sometimes penetreting to phryganoids and light forests. According to our current studies the species composition of the genera *Eulasia* and *Pygopleurus* needs revision, the species of both genera occupy open landscapes from semidesert (*Eulasia eichleri* (Zaitzev, 1923)) to subalpine meadows (*Pygopleurus psilotrichius psilotrichius* sensu Iablokoff-Khnzorian), some species can be found in forest edges and glades (e.g. *Eulasia chrysopyga* (Faldermann, 1835), Fig. 5.14).

Family **Scarabs – Scarabaeidae**. In Armenia 256 species, about 55 species and subspecies are endemics of the Caucasus Ecoregion, of these 15 are Armenian endemics, namely, *Chilothorax distinctus sevanicus* (Rakovič, 1991), *Ch. dobrovljanskyi* (Koshantschikov, 1913), *Neagolius aragatsi* (Shokhin & Kalashian, 2015), *Pharaonus caucasicus* (Reitter, 1888), *Adoretus rubenyani* Kalashyan, 2002, *Anisoplia venusta* Baraud, 1991, *Anoxia maljuzhenkoi* (Zaitzev, 1928), *Pseudopachydema medvedevi* Iablokoff-Khnzorian, 1971, *Tanyproctus araxis* Reitter, 1902, *T. rubicundus* Reitter, 1902, *T. vedicus* Kalashian, 1999, *T. antennatus* Iablokoff-Khnzorian, 1953, *Amphimallon helenae* (Iablokoff-Khnzorian, 1983), *Protaetia hajastanica* Ghrejyan and Kalashian, 2017, *P. gayaneae* Ghrejyan and Kalashian, 2022.

The family includes phytophagous species developing in the soil and feeding on roots (Melolonthinae, Rutelinae, *Pentodon*, etc.), as well as saprophagous species, including huge number of dung beetles (Scarabeinae, Aphodiinae) and species developing in rotten wood and other plant residues (Cetoniinae, *Oryctes*, etc.).

Fig. 5.14 *Eulasia chrysopyga*

Taking into account peculiarities of distribution of dung beetles which depends on movement of cattle to a significant extent, they will be reviewed separately.

Non-coprophagous scarabs can be distributed as follows.

Most diverse is fauna of **deserts and semideserts**, which includes about 50 species. Here live *Pleurophorus anatolicus* Petrovitz, 1961, *Rhyssemus germanus* (Linnaeus, 1767), species of the genus *Pentodon* (*P. bidens sulcifrons* Küster, 1848, *P. quadridens distantidens* Reitter, 1899, peculiar wingless *P. reitteri* Jakovlev, 1904), *Blitopertha nigripennis* (Reitter, 1888), *Brancoplia leucaspis* (Laporte, 1840), *Anisoplia reitteriana* Semenov, 1903, *A. austriaca kurdistana* Reitter, 1889, *Adoretus discolor* Faldermann, 1835, red-listed endemic of Ararat Plane *Adoretus rubenyani* Kalashyan, 2002, *Polyphylla adspersa* Motschulsky, 1854, *Microphylla paupera* Hampe, 1852, *Cyphonotus testaceus* (Pallas, 1781), endemic of Arax valley *Anoxia maljuzhenkoi* (Zaitzev, 1928), *Dasytrogus jubatus* (Reitter, 1890), majority of Armenian species of the genus *Tanyproctus* (*T. confinis* Motschulsky, 1860, *T. ovatus* Motschulsky, 1860, *T. satanas* Reitter, 1902 and red-listed Armenian endemics *Tanyproctus vedicus* Kalashian, 1999 (Fig. 5.15) and *Tanyproctus araxis* Reitter, 1902). These habitats prefer *Stalagmosoma albellum* (Pallas, 1781), *Aethiessa rugipennis* (Burmeister, 1842), *Protaetia funebris funebris* (Gory & Percheron, 1833), etc. Typical for sandy desert is red-listed *Pharaonus caucasicus* Reitter, 1888 (Fig. 5.16).

The fauna of **phryganoid vegetation** and **light forests** is rather similar, including about 25 species. These habitats prefer *Tropinota senicula* (Ménétriés, 1832), *Tropinota hirta suturalis* Reitter, 1913, *Oxythyrea albopicta* Motschulsky, 1845, *Protaetia trojana godeti* (Gory & Percheron, 1833), *Amphimallon vernale vernale*

Fig. 5.15 *Tanyproctus vedicus* (Red Book: CR)

Fig. 5.16 *Pharaonus caucasicus* (Red Book: CR)

Fig. 5.17 *Protaetia caucasica*

(Brullé, 1832), *Anisoplia farraria farraria* Erichson, 1847; more typical for light forests themselves are *Oryctes nasicornis latipennis* Motschulsky, 1845, *Omaloplia spireae spireae* (Pallas, 1773) *Melolontha aceris* Faldermann, 1835, *Holochelus fallax fallax* (Marseul, 1879), *Tanyproctus rufidens* (Marseul, 1879), *Cetonia aeratula* Reitter, 1891, *C. aurata aurata* (Linnaeus, 1758), *Protaetia speciosa speciosa* (Adams, 1817), *P. affinis affinis* (Andersch, 1797), *P. caucasica* Kolenati, 1846 (Fig. 5.17), etc.

Some species from light forests penetrate also to **broad-leaved forests,** mainly their edges and glades. In general, there are about 20 species known nearly exclusively from these habitats, for example, *Melolontha pectoralis* Megerle von Mühlfeld, 1812, *Holochelus brenskei* (Reitter, 1888), *Hoplia pollinosa* Krynicki, 1832, *Valgus hemipterus* (Linnaeus, 1758), *Trichius fasciatus* (Linnaeus, 1758), *Oxythyrea funesta* (Poda von Neuhaus, 1761) (Fig. 5.18), etc.

Fauna of non-coprophagous scarabs of the **steppes** includes about 25 typical species. These are *Holochelus rusticus* (Faldermann, 1835), *H. tataricus* (Faldermann, 1835), *Rhizotrogus aestivus* (Olivier, 1789), *Amphimallon solstitiale setosum* Reitter, 1902, *Pentodon idiota idiota* (Herbst, 1789), *Chaetopteroplia segetum velutina* (Erichson, 1847), red-listed endemic *Pseudopachydema medvedevi* Iablokoff-Khnzorian, 1971, *Protaetia ungarica armeniaca* (Ménétriés, 1832), *P. araratica* (Reitter, 1891), recently described Armenian endemics *P. gayaneae* Ghrejyan and Kalashian, 2022 and *P. hajastanica* Ghrejyan and Kalashian, 2017, etc. Here, especially to the low part of steppe belt, penetrate several species more typical for semi-deserts, for example, *Rhyssemus germanus* (Linnaeus, 1767), *Tanyproctus ovatus* Motschulsky, 1860, *Blitopertha nigripennis* (Reitter, 1888), *Brancoplia leucaspis leucaspis* (Laporte, 1840), *Anisoplia farraria farraria* Erichson, 1847, *A. austriaca kurdistana* Reitter, 1889, etc.

Very pure is the fauna of the **meadows**; there are no species typical for them, very few species penetrate to low part of subalpine belt, namely, *Amphimallon solstitiale setosum* Reitter, 1902, *Rhizotrogus aestivus* (Olivier, 1789), somewhere also *Cetonia aurata aurata* (Linnaeus, 1758).

Fig. 5.18 *Oxythyrea funesta*

Wetlands occupy *Psammodius generosus* Reitter, 1892, *Psammodius asper* (Fabricius, 1775), *Rhyssemus interruptus* Reitter, 1892, *Ataenius horticola* Harold, 1869, which live in the sand near shorelines, as well as *Oryctes nasicornis latipennis* Motschulsky, 1845, *Polyphylla olivieri* Laporte, 1840, *Maladera punctatissima* (Faldermann, 1835), *Anomala dubia abchasica* Motschulsky, 1854 developing in plant residues.

Dung beetles consist more than half of the whole Armenian fauna of scarabs and includes more than 160 species. Like non-coprophagous scarabs, dung beetles are most diverse in **deserts and semideserts,** with more than 60 species found here. Only in this belt occur *Mendidius diffidens* (Reitter, 1892), *Plagiogonus syriacus* (Harold, 1863), *Plagiogonus praeustus* (Ballion, 1871), *Erytus cognatus* (Fairmaire, 1860), *E. aequalis* (Schmidt, 1907), *Esymus suturinigra* (Schmidt, 1916), *Phalacronothus fumigatulus* (Reitter, 1892), *Mecynodes kisilkumi* (Solsky, 1876), *Alocoderus hydrochaeris* (Fabricius, 1798), as well as *Scarabaeus pius* (Illiger, 1803), *Copris hispanus cavolinii* (Petagna, 1792), *Euoniticellus pallipes* (Fabricius, 1781), *Euonthophagus atramentarius* (Ménétriés, 1832), *Onthophagus cruciatus* Ménétriés, 1832, *O. dorsosignatus* d'Orbigny, 1898, *O. suturellus* Brullé, 1832, *Onitis damoetas* (Steven, 1806), *O. humerosus* (Pallas, 1771), *Cheironitis pamphilus* (Ménétriés, 1849), *Ch. haroldi* (Ballion, 1871) and some others.

Many species common in this belt, can be found also in **steppes**, for example, *Chilothorax clathratus* (Reitter, 1892), *Acrossus depressus* (Kugelann, 1792), *Nialus varians* (Duftschmid, 1805), *Labarrus lividus* (Olivier, 1789), *Gymnopleurus mopsus mopsus* (Pallas, 1781) (Fig. 5.19), *G. flagellatus flagellatus* (Fabricius, 1787), *Sisyphus schaefferi schaefferi* (Linnaeus, 1758), *Euonthophagus amyntas subviolaceus* (Ménétriés, 1832), *O. lucidus* (Illiger, 1800), *O. marginalis marginalis*

Fig. 5.19 *Gymnopleurus mopsus mopsus*

(Gebler, 1817), *O. truchmenus truchmenus* Kolenati, 1846, *Caccobius mundus* (Ménétriés, 1839), *C. histeroides* (Ménétriés, 1832) etc. More typical for steppes are *Colobopterus erraticus* (Linnaeus, 1758), *Planolinus fasciatus* (Olivier, 1789), *Phaeaphodius dauricus* (Harold, 1863), *Biralus satellitius* (Herbst, 1789), *Acrossus luridus* (Fabricius, 1775), *Amidorus thermicola* (Sturm, 1800), *A. cribrarius* (Brullé, 1832), *Melinopterus prodromus* (Brahm, 1790), *Esymus merdarius* (Fabricius, 1775), *Scarabaeus typhon* Fischer von Waldheim, 1823) (Fig. 5.20), *Ateuchetus armeniacus* (Ménétriés, 1832), *Gymnopleurus flagellatus flagellatus* (Fabricius, 1787), *Paroniticellus festivus* (Steven, 1809), *Onthophagus ruficapillus* Brullé, 1832, *O. gibbulus rostrifer* Reitter, 1892, *O. vacca* (Linnaeus, 1767), *O. fissicornis* (Steven, 1809), etc.

Some species from steppe can penetrate to the low part of subalpine belt, for example, *Euheptaulacus carinatus carinatus* (Germar, 1824), *Amidorus obscurus obscurus* (Fabricius, 1792), *Colobopterus erraticus* (Linnaeus, 1758), *Planolinus fasciatus* (Olivier, 1789), *Onthophagus sacharovskii* Olsoufieff, 1918, etc. Nearly exclusively in subalpine and alpine meadows occur *Copris armeniacus* Faldermann, 1835, *Onthophagus formaneki* Reitter, 1897, *Amidorus alagoezi* (Olsoufieff, 1918) and recently described endemic of alpine meadows of Aragats Mount *Neagolius aragatsi* Shokhin & Kalashian, 2015.

Typical for **broad-leaved forests** are about 20 species, including *Coprimorphus scrutator* (Herbst, 1789), *Teuchestes fossor* (Linnaeus, 1758), *Acrossus planicollis* (Reitter, 1890), *Acrossus rufipes* (Linnaeus, 1758), *Bodilopsis rufa* (Moll, 1782), *Limarus maculatus* (Sturm, 1800), *Melinopterus sphacelatus* (Panzer, 1798), *Melinopterus edithae* (Reitter, 1906), *Oxyomus sylvestris* (Scopoli, 1763), as well as some *Onthophagus* species – *O. taurus* (Schreber, 1759) *O. coenobita* (Herbst, 1783), *O. lemur* (Fabricius, 1781).

Fig. 5.20 *Scarabaeus typhon*

Nearly everywhere, except of driest portions of deserts, dense forest and high mountainous habitats can be found *Eupleurus subterraneus subterraneus* (Linnaeus, 1758), *Otophorus haemorrhoidalis* (Linnaeus, 1758), *Trichonotulus scrofa* (Fabricius, 1787), *Aphodius fimetarius* (Linnaeus, 1758), *Chilothorax distinctus distinctus* (Müller, 1776), *Acanthobodilus immundus* (Creutzer, 1799), *Bodilus lugens* (Creutzer, 1799), *Calamosternus granarius* (Linnaeus, 1767), *Eudolus quadriguttatus* (Herbst, 1783), *Copris lunaris* (Linnaeus, 1758), *Euoniticellus fulvus* (Goeze, 1777), *Gymnopleurus geoffroyi* (Fuessly, 1775), *Onthophagus furcatus* (Fabricius, 1781), *O. fracticornis* (Preyssler, 1790), *Caccobius schreberi* (Linnaeus, 1767) and some others.

There are no species specifically connected with **phryganoids, light forests, pine forests, as well as wetlands**. Fauna of dung beetles here is formed from species penetrating here with cattle from surrounding biotopes.

5.1.3.3 Family Jewel Beetles – Buprestidae

Family Jewel beetles (Burestidae) is another group of the beetles richly presented in Armenia; the country's fauna includes about 190 species (in the Catalogue (Kubáň et al. 2016), there are 197 species listed, but some of these records are based on old and sometimes unreliable references). There are about 35 species of jewel beetles endemic for the Caucasus Ecoregion; among these 13 species are endemics of Armenia: *Acmaeoderella pellitula* (Reitter, 1890), *Sphenoptera geghardica* Kalashian & Zykov, 1994, *S. vediensis* Kalashian, 1994, *S. antoniae* Reitter, 1891, *S. bellamyi* Kalashian, 2014, *S. jakowlewi* Reitter, 1895, *S. khosrovica* Kalashian, 1990, *S. diluta* Jakovlev, 1900, *Anthaxia breviformis* Kalashian, 1988, *A. superba* Abeille de Perrin, 1900, *Sphaerobothris aghababiani* Volkovitsh & Kalashian, 1998, *Agrilus bucephalus* Daniel, 1903, *Cylindromorphus vedicus* Kalashian, 2002; four more species, namely, *Sphenoptera araxidis* Reitter, 1890, *S. infantula* Reitter, 1895, *S. tenax* Jakovlev, 1902 and *Trachys araxicola* Obenberger, 1918 are described from Arax River Valley ("Araxesthal") and known only by type specimens; their exact origin – Armenia or Nakhichevan is not clear.

The representatives of the family are exclusively phytophagous with larvae developing in the wood of trees and shrubs, stems and roots of grassy plants, some species are leaf miners and some are developing freely in the soil. Jewel beetles are thermophilous and mostly xerophilous which determines their distribution and habitat preferences to a significant extent. Another factor determining the distribution of jewel beetles is the presence of host plants; thus, many species can be found in different habitats together with their hosts.

Very diverse is the fauna of **deserts and semideserts**. From here are known more than 50 species. Among them typical are *Julodis faldermanni* Mannerheim, 1837 (Fig. 5.21), *Julodella globithorax* (Steven, 1830), *Acmaeodera ghilarovi* Volkovitsh, 1988, several species from the genus *Acmaeoderella* (*A. elbursi* (Obenberger, 1924), *A. fulvinaeva* (Reitter, 1890), *A.serricornis* (Abeille de Perrin, 1900), *A. obscura* (Reitter, 1889, red-listed *A. pellitula* (Reitter, 1890) and others), numerous

Fig. 5.21 *Julodis faldermanni*

Fig. 5.22 *Sphenoptera artemisiae*

Sphenoptera (*S. hispidula* Reitter, 1890, *S. artemisiae* Reitter, 1889 (Fig. 5.22), *S. khnzoriani* Kalashian, 1996, *S. sancta* Reitter, 1890, *S. micans* Jakovlev, 1886, Armenian endemics *S. bellamyi* Kalashian, 2014 and *S. vediensis* Kalashian, 1994, etc.), *Anthaxia flavicomes* Abeille de Perrin, 1900, *Agrilus albogularis albogularis*

Gory, 1841, *A. araxenus araxenus* Iablokoff-Khnzorian, 1960, *A. vaginalis* Abeille de Perrin, 1897, endemic *Cylindromorphus vedicus* Kalashian, 2002. Due to severe transformation of habitats, two species inhabited this belt, *Capnodis excisa excisa* Ménétriés, 1848 and *Lampetis argentata* Mannerheim, 1837 should be considered regionally extinct as they were not found in Armenia since 1930s.

Biotopes of **phryganoid vegetation** inhabit about 25 species of jewel beetles. Several species live on spiny *Astragalus* (sect. *Tragacantha*), *Acmaeodera chalcithorax* Volkovitsh, 1986, *A. planidorsis* Semenov, 1896, *Acmaeoderella flavofasciata albifrons* (Abeille de Perrin, 1891), *Sphenoptera anthracina* Jakovlev, 1887, *Anthaxia truncata* Abeille de Perrin, 1900, etc. On *Vitis* lives *Agrilus derasofasciatus* Lacordaire, 1835. With different host plants are connected *Sphenoptera sculpticollis* Heyden, 1886 (on *Echinops*), *S. demissa* Marseul, 1865, *Anthaxia anatolica anatolica* Chevrolat, 1838 (on *Ferula* and *Cachrys*), etc. On *Ephedra* lives red-listed Armenian endemic *Sphaerobothris aghababiani* Volkovitsh & Kalashian, 1998 (Fig. 5.23).

Some of these species together with their host plants can be found in different types of **light forests** which fauna consists of about 40 species. Typical for latter biotopes are numerous species associated with tree and shrub vegetation. These are *Capnodis* species – *C. cariosa cariosa* (Pallas 1776) on *Pistacia*, *C. porosa porosa* (Klug, 1829) on *Ziziphus*, *C. henningii* Faldermann, 1835, *C. tenebrionis* (Linnaeus, 1760) on Rosacea (mainly *Amygdalus*), also on Rosacea live *Perotis lugubris longicollis* Kraatz, 1880, *Lamprodila balcanica* (Kirchsberg, 1876), *Sphenoptera zarudnyi schatinensis* Alexeev & Zykov, 1991, *S. anthaxoides* Reitter, 1895, *Anthaxia*

Fig. 5.23 *Sphaerobothris aghababiani* (Red Book: CR)

Fig. 5.24 *Anthaxia superba* (Red Book: CR)

ephippiata Redtenbacher, 1850, red-listed endemic *A. superba* Abeille de Perrin, 1900 (Fig. 5.24), *A. tractata* Abeille de Perrin, 1901, *Agrilus cuprescens cuprescens* Ménétriés, 1832. On *Ulmus* species develops *Lamprodila mirifica nadezhdae* (Semenov, 1909). Here live also several multiphagous species such as *Anthaxia bicolor bicolor* Faldermann, 1835, *A. cichorii* (Olivier, 1790), *A. olympica* Kiesenwetter, 1880, *Melanophila cuspidata* (Klug, 1829), *Agrilus roscidus* Kiesenwetter, 1857 and species which host plants are not known yet, for example, *Anthaxia holoptera* Obenberger, 1914, *A. mirabilis* Zhicharev, 1918, *A. krueperi* Ganglbauer, 1885, *Trachys splendidulus* Reitter, 1890, etc. Juniper forest inhabit *Anthaxia passerinii* Pecchioli, 1837, *A. caucasica* Abeille de Perrin, 1900, *A. discicollis kanaanita* Obenberger, 1912.

Fauna of **broad-leaved forest** includes about 20 species. First of all characteristic are species developing on oak (*Quercus* spp.): *Eurythyrea quercus* (Herbst, 1780), numerous *Agrilus* species (*A. angustulus angustulus* (Illiger, 1803), *A. buresi* Obenberger, 1935, *A. convexicollis* Redtenbacher, 1847, *A. graminis* Kiesenwetter, 185, *A. hastulifer* Ratzeburg, 1837, *A. obscuricollis* Kiesenwetter, 1857, etc.); mainly forests inhabit *Acmaeodera octodecimguttata octodecimguttata* (Piller & Mitterpacher, 1783), *Anthaxia podolica lucniki* Richter, 1949 and *A. manca* (Linnaeus, 1767), *Chrysobothris affinis tetragramma* (Ménétriés, 1832), *Meliboeus fulgidicollis* (Lucas, 1846) and some others. Several species more common for light forests inhabit forest edges and clearings, for example, above-mentioned *Anthaxia bicolor bicolor* Faldermann, 1835, *A. cichorii* (Olivier, 1790), *A. olympica* Kiesenwetter, 1880, *Lamprodila mirifica nadezhdae* (Semenov, 1909), etc. Mainly here lives *Coraebus rubi* (Linnaeus, 1767) developing on *Rubus*.

Fauna of **pine forests** in Armenia includes six species connected with pine. Three species, *Buprestis haemorrhoidalis araratica* (Marseul, 1865), *Chalcophora mariana* (Linnaeus, 1758) and, probably, *Phaenops cyanea* (Fabricius, 1775) need for larval development rather big trees and can be found in relatively old natural forests which cover in Armenia small territories in Tavush and Lori provinces only. Vice versa, species of the genus *Anthaxia* (*A. godeti* Gory & Laporte, 1839 *A. nigrojubata nigrojubata* Roubal, 1913, *A. quadripunctata quadripunctata* (Linnaeus, 1758) (the last one is probably invasive)) which larvae live in branches inhabit also artificial pine forests widely distributed in the country.

To the low part of **steppe** belt penetrate several species more typical for semidesert, for example, *Julodis andreae andreae* (Olivier, 1790), *Acmaeoderella vetusta* (Ménétriés, 1832), *A. villosula* (Steven, 1830), *Agrilus albogularis albogularis* Gory, 1841, *Trachys phlyctaenoides* Kolenati, 1846, etc. In the traganth steppes, rather common are species of *Sphenoptera* and *Anthaxia* namely, *S. tragacanthae* (Klug, 1829), *S. simplex* Jakovlev, 1893, *S. hypocrita* Mannerheim, 1837, *S. anthracina* Jakovlev, 1887, *A. amasina amasina* Daniel, 1903, *A. truncata* Abeille de Perrin, 1900, as well as *Acmaeoderella mimonti decorata* (Marseul, 1865) and some others. Besides, in the steppe biotopes, especially those used for pasture can be found *Capnodis tenebricosa tenebricosa* (Olivier, 1790) on *Rumex*, some *Meliboeus* species on different thistles: *M. parvulus* (Küster, 1852), *M. robustus* (Küster, 1852), as well as *M. graminis* (Panzer, 1799) on *Arthemisia*. On cereals live *Paracylindromorphus subuliformis subuliformis* (Mannerheim, 1837) and *Cylindromorphus filum* (Gyllenhal, 1817). In total there are about 25 species living in this belt.

Due to thermophily of jewel beetles, the high mountain habitats are not suitable for them. Very few species exceed the altitudes 2000 m a. s. l. In meadow-steppe can be found *Meliboeus robustus* (Küster, 1852) on *Onopordum*, some species of *Sphenoptera* more typical for steppe: *S. simplex* Jakovlev, 1893, *S. anthracina* and *S. hypocrita* Mannerheim, 1837 and very variable and plastic *S. tragacanthae* (Klug, 1829) developing in different species of *Astragalus* both soft and spiny and inhabiting different landscapes from desert to low parts of subalpine belt; only one species, *S. fallatrix* Obenberger, 1927, occupies subalpine belt reaching altitudes 2500–2600 m.

Wetlands inhabit several species living on Salicaceae (*Populus, Salix*) characteristic for riversides and tugai forests such as *Dicerca aenea validiuscula* Semenov, 1896, *Poecilonota variolosa variolosa* (Paykull, 1799), *Capnodis miliaris miliaris* (Klug, 1829) 1865, *Anthaxia diadema shelkovnikovi* Obenberger, 1940, *Buprestis salomonii* Thomson, 1878, *Eurythyrea aurata* (Pallas, 1776), *Trachypteris picta decostigma* (Fabricius, 1787), *Trachys minutus minutus* (Linnaeus, 1758), *Agrilus lineola hermineus* Abeille de Perrin, 1907 and some others; several species are connected with *Tamarix*, namely, *Sphenoptera antoniae* Reitter, 1891, *S. balassogloi balassogloi* Jakovlev, 1885, *S. mesopotamica* Marseul, 1866. Some species live on grassy plants, such as *Aphanisticus emarginatus* (Olivier, 1790) on *Juncus*, *Cyphosoma tataricum* (Pallas, 1771) on *Bolboschoenus*, etc.

5.1.3.4 Family Darkling Beetles – Tenebrionidae (Iwan et al. 2020)

The fauna of darkling beetles (Tenebrionidae) of Armenia is recently reviewed by Nabozhenko et al. (2021), who listed for Armenia 123 species with the exception of the subfamily Alleculinae and excluding 29 species listed in Catalogue (Iwan et al. 2020) due to doubtfulness of data; Alleculinae was revised (as a separate family) by Iablokoff-Khnzorian (1983) who mentioned 36 species for the country's fauna. There are about 40 species endemic for the Caucasus ecoregion, among them 18 species are endemics of Armenia: *Armenohelops armeniacus* Nabozhenko, 2002, *Blaps araxicola* Seidlitz, 1893, *B. kovali* Abdurakhmanov & Nabozhenko 2011, *Boromorphus armeniacus* Reitter, 1889**,** *Ceratanisus khnzoriani* Nabozhenko, Ferrer, Kalashian, Abdurakhmanov, 2016, *C. transcaucasicus* Nabozhenko, Ferrer, Kalashian, Abdurakhmanov, 2016**,** *Cylindrinotus erivanus* (Reitter, 1902), *Ectromopsis bogatschevi* (Iablokoff-Khnzorian, 1957), *Entomogonus clavimanus* Reitter, 1903 and red-listed *E. amandanus* (Reitter, 1902), *Microdera urartu* Nabozhenko & Kalashian, 2022**,** *Phtora plagiocnema* (Iablokoff-Khnzorian, 1956), *Scaurus araxinus* Richter, 1945**,** *Zophohelops humeridens* (Reitter, 1902), *Isomira stricta* Khnzorian, 1976, *I. armena* Khnzorian, 1976, *Omophlus emmae* (Khnzorian, 1959), *O. armeniacus* Znojko, 1950.

Darkling beetles are partly saprophagous, some species are phytophags, several species feed on fungi and lichens. Some groups of Tenebrionidae are known as best adopted to dry conditions; so, they are richly presented in arid and semi-arid habitats.

As a result, **deserts and semideserts** inhabit more than 40 species of darkling beetles. Semideserts of different types occupy *Pachyscelis musiva* (Faldermann in Ménétriés, 1832) *Pimelia cursor* Ménétriés, 1832 (Fig. 5.25), *P. persica* Faldermann, 1837, *Trachyderma christophi christophi* (Faust, 1875), *Dichillus araxidis* Reitter,

Fig. 5.25 *Pimelia cursor*

1889, *D. rugatus rugatus* Baudi di Selve, 1874, *Oogaster piceus* (Ménétriés, 1832), *Platamodes dentipes dentipes* Ménétriés, 1849, *Stenosis armeniacus* (Motschulsky, 1849), *Tagenostola pilosa* (Motschulsky, 1839), *Gnathosia modesta* (Faldermann, 1837), *Zophosis* (*Septentriophosis*) *rugosa* Faldermann, 1837, *Cossyphus tauricus* Fischer von Waldheim, 1832, *Laena ferruginea* Küster, 1846, *Gonocephalum pubiferum* Reitter, 1904, *Opatroides punctulatus parvulus* (Faldermann, 1837), *Penthicus* (*Discotus*) *dilectans* (Faldermann, 1836), *Scleropatroides hirtulus* (Baudi di Selve, 1876), *Leichenum mucronatum* Küster, 1849, *Dissonomus picipes* (Faldermann, 1837), red-listed *Ectromopsis bogatschevi* (Iablokoff-Khnzorian, 1957) (Fig. 5.26), *Adelphinus* (*Adelphinops*) *ordubadensis* Reitter, 1890, *Catomus antoniae* Reitter, 1890, Armenian endemic *Entomogonus* (*Delonurops*) *clavimanus* Reitter, 1903, *Hedyphanes mannerheimi* Faldermann, 1837, *H. tagenioides* Faldermann, 1832, *Scaurus araxinus* Richter, 1945, *Phtora hauseriana* (Reitter, 1895), *Ph.plagiocnema* (Iablokoff-Khnzorian, 1956), *Allecula divisa* Reitter, 1889, *Isomira stricta* Khnzorian, 1976, *Cteniopus elegans* (Faldermann, 1837), *Omophlus flavipennis* Küster, 1849, etc. Several species prefer sandy desert, for example, *Adesmia maillei* Solier, 1835, red-listed *Cyphostethe semenovi* Bogatchev, 1947, *Arthrodosis globosus* (Faldermann, 1837), *Blaps araxicola* Seidlitz, 1893, *B. ominosa* Ménétriés, 1832, etc.

In **phryganoid vegetation** are distributed about 20 species of darkling beetles, some of which can be found also in steppe and light forest habitats, for example, *Calyptopsis pulchella pulchella* (Faldermarm, 1837), *Omophlus emmae* (Khnzorian, 1959) *Mycetocharina riabovi* Iablokoff-Khnzorian, 1959, *M. orientalis* (Faust,

Fig. 5.26 *Ectromopsis bogatschevi* (Red Book: CR)

1876), *Hymenalia basalis* (Faust, 1876), *Omophlus laciniatus* Seidlitz, 1896, *O. caucasicus caucasicus* Kirsch, 1869, etc.

Mainly in **light forests** live 20 species including red-listed Armenian endemic *Entomogonus amandanus* (Reitter, 1902), *Eustenomacidius* (*Caucasohelops*) *svetlanae araxi* Nabozhenko, 2006, *Euboeus kalashiani* Nabozhenko, 2022, *Pentaphyllus chrysomeloides* (Rossi, 1792), *Myrmechixenus picinus* (Aubé, 1850), *Isomira antennalis* Reitter, 1884, *Omophlus ochraceipennis* Faldermann, 1837, *O. obscurus* Reitter, 1890, etc.

Rich is fauna of **broad-leaved forest** which consist of about 40 species, including numerous fungivorous and lichenophagous species. These are *Laena lederi* Weise, 1878, *L. wanensis* Schuster, 1940, *Lagria hirta* (Linnaeus, 1758), *Alphitobius diaperinus* (Panzer, 1796), *Diaclina testudinea* (Piller et Mitterpacher, 1783), *Bolitophagus reticulatus* (Linnaeus, 1767), *Eledona agricola* (Herbst, 1783), *Eledonoprius serrifrons* (Reitter, 1890), *Palorus ratzeburgii* (Wissmann, 1848), *Neatus subaequalis* Reitter, 1920, *Cryphaeus cornutus* (Fischer von Waldheim, 1823), *Alphitophagus bifasciatus* (Say, 1824), *Diaperis boleti* (Linnaeus, 1758), *Neomida quadricornis* (Motschulsky, 1873), *Platydema tristis* Laporte et Brullé, 1831, *P. violacea* (Fabricius, 1790), *Corticeus basalis* Reitter, 1884, *C. longulus* (Gyllenhal, 1827), *C.suberis* (Lucas, 1846), *C. unicolor* Piller et Mitterpacher, 1783, *Scaphidema metallica metallica* (Fabricius, 1792), *Odocnemis aurichalcea* (Adams, 1817), red-listed *Isomira armena* Khnzorian, 1976, *I. caucasica* Reitter, 1890, etc.

Few species occur in **pine forests:** *Corticeus bicolor* (Olivier, 1790), as well as *Bolitophagus reticulatus* (Linnaeus, 1767) and *Alphitobius diaperinus* (Panzer, 1796) connected also with broad-leaved forest.

Steppe is another belt with rich fauna with about 40 species of darkling beetles; these fauna share some species with semideserts and meadows, and also include numerous species more typical for this habitat type. These are, for example, *Calyptopsis caucasica* (Kraatz, 1865), *C. emarginata* Reitter, 1889, *Dailognatha caraboides* (Eschscholtz, 1831), *D. pumila* (Baudi di Selve, 1874), *Tentyria striatopunctata* Ménétriés, 1832, *T. tessulata tessulata* Tauscher, 1812, *Blaps lethifera pterotapha* Fischer von Waldheim in Ménétriés, 1832, *B. mortisaga* (Linnaeus, 1758), *B. pudica* Ballion, 1888, *Dendarus crenulatus* (Ménétriés, 1832) *D. extensus* (Faldermann, 1837), *Gonocephalum granulatum pusillum* (Fabricius, 1792), *G rusticum* (Olivier, 1811) *O. geminatum* Brullé, 1832, *O. sabulosum sabulosum* (Linnaeus, 1758) *P. strabonis* Seidlitz, 1893 *P. femoralis femoralis* (Linnaeus, 1767), red-listed *Armenohelops armeniacus* Nabozhenko, 2002 (Fig. 5.27), *Nalassus faldermanni* (Faldermann, 1837), *Crypticus quisquilius quisquilius* (Linnaeus, 1761), *Cyphogenia* (*Lechriomus*) *lucifuga* (Adams, 1817), *Mycetochara linearis* (Illiger, 1794), *Podonta elongata* (Ménétriés, 1832), *Omophlus curtulus* Kirsch, 1878, etc.

Meadows occupy some species of *Cylindrinotus* (*C. gibbicollis* Faldermann, 1837, *C. femoratus* (Faldermann, 1837), red-listed *C. erivanus* (Reitter, 1902) (Fig. 5.28), *Nalassus diteras* (Allard, 1876), *Odocnemis allardi* Nabozhenko et

Fig. 5.27 *Armenohelops armeniacus* (Red Book: EN)

Fig. 5.28 *Cylindrinotus erivanus* (Red Book: EN)

Keskin, 2016, *Zophohelops humeridens* (Reitter, 1902), etc. Among these *C. erivanus* is Armenian endemic, and others are endemics of the Caucasus ecoregion.

Rather few species in Armenia are known mainly from **wetlands,** for example, *Palorus ratzeburgii* (Wissmann, 1848), invasive *P. orientalis* Fleischer, 1900, *Alphitophagus bifasciatus* (Say, 1824), *Pentaphyllus testaceus* (Hellwig, 1792), *Diaclina testudinea* (Piller et Mitterpacher, 1783), red-listed *Cteniopus persimilis* Reitter, 1890, *Omophlus nitidicollis* Seidlitz, 1896, etc. in tugai and *Scleropatroides breviusculus* (Reitter, 1889), *Centorus filiformis* Motschulsky, 1872, *C. trogosita* Motschulsky, 1872, *Cheirodes sardous* Gené, 1839, *Ch brevicollis* Wollaston, 1864 in salt marshes.

5.1.3.5 Family Longhorn Beetles – Cerambycidae

Longhorn beetle fauna of Armenia was revised by N.N. Plavilstshikov (1948) who reported from the country 214 species; in Catalogue are listed 256 species and subspecies (Drumont et al. 2020).

More than 100 species and subspecies are endemics of the Caucasian ecoregion, among them 30 taxa are endemic for Armenia: *Cortodera colchica dilizhanica* Danilevsky, 2014, *C. c. colchica erevanica* Danilevsky, 2014, *C. c. kalashiani* Danilevsky, 2000, *C. kazaryani* Danilevsky, 2014, *Anoplistes agababiani* (Danilevsky, 2000), *Dorcadion artemi* Lazarev, 2019, *D. bistriatum* Pic, 1898, *D. cineriferum* Suvorov, 1910, *D. daratshitshagi* Suvorov, 1915, *D. gorbunovi gorbunovi* Danilevsky, 1985, *D. g. rubenyani* Lazarev, 2014, *D. indutum* Faldermann, 1837, *D. khosrovi* Lazarev, 2019, *D. laeve vladimiri* Danilevsky & Murzin, 2009, *D. megriense* Lazarev, 2009, *D. nigrosuturatum* Reitter, 1897, *D. scabricolle gegarkunicum* Lazarev, 2020 (Fig. 5.29), *D. s. pseudosevangense* Lazarev, 2020,

Fig. 5.29 *Dorcadion scabricolle*

D. s. s. sevangense Reitter, 1889, *D. s. tavushense* Lazarev, 2020, *D. s. tekhense* Lazarev, 2020, *D. s. vaykense* Lazarev, 2020, *D. semilucens* Kraatz, 1873, *D. sevliczi* Danilevsky, 1985, *D. sisianense* Lazarev, 2009, *D. sulcipenne goktshanum* Suvorov, 1915, *Conizonia kalashiani* Danilevsky, 1992, *Phytoecia dantchenkoi* Danilevsky, 2008, *Ph. kazaryani* Danilevsky, 2020, *Ph. marki* Danilevsky, 2008.

Like jewel beetles, the representatives of the family are exclusively phytophagous with larvae developing in the wood of trees and shrubs, stems and roots of grassy plants and some are developing freely in the soil; in opposition to jewel beetles, there are now leaf miners among longhorns. Besides they are less demanding of temperature conditions and can occupy some high mountain habitats. Another factor determining the distribution of longhorns is the presence of host plants, thus, many species can be found in different habitats together with their hosts.

Unlike other families, longhorn beetles are relatively poorly represented in **deserts and semideserts** of Armenia. Here are distributed about 20 species of the family, for example, *Apatophysis vedica* Danilevsky, 2008 (Fig. 5.30), *Chlorophorus faldermanni* (Faldermann, 1837), *Certallum ebulinum* (Linnaeus, 1767), *Oxylia argentata argentata* (Ménétriés, 1832), *Oberea erythrocephala erythrocephala* (Schrank, 1776), *Phytoecia vittipennis pravei* Plavilstshikov, 1926, *Ph. armeniaca* Frivaldszky, 1878, *Ph. scutellata* (Fabricius, 1793), *Ph. kurdistana* Ganglbauer, 1884, *Ph. coerulescens coerulescens* (Scopoli, 1763), etc. Majority of them, except *Apatophysis*, can be found also in steppes, especially in their low part.

Also about 20 species of longhorns are typical for **phryganoid** habitats. These are red-listed endemic *Anoplistes agababiani* (Danilevsky, 2000) (Fig. 5.31) on *Ephedra*, several Lamiinae, connected with different Asteraceae, for example, *Mallosia armeniaca* Pic, 1897, *M. brevipes* Pic, 1897 (Fig. 5.32), *M. scovitzii* (Faldermann, 1837), *Phytoecia erivanica* Reitter, 1899, red-listed *Ph. pici* Reitter, 1892, *Ph. diademata* Faldermann, 1837, *Ph. antoniae* Reitter, 1889 (Fig. 5.33),

Fig. 5.30 *Apatophysis vedica*

Fig. 5.31 *Anoplistes agababiani* (Red Book: CR)

Fig. 5.32 *Mallosia brevipes*

Agapanthia dahli walteri Reitter, 1898 (Fig. 5.34), etc., as well as with other semi-shrubs and grasses: *Pteromallosia albolineata* (Hampe, 1852) (Fig. 5.35), *Ph. millefolii* Adams, 1817, *Ph. plasoni* Ganglbauer, 1884, *Ph. praetextata praetextata* Steven, 1817, *Ph. faldermanni* Faldermann, 1837, red-listed endemic of South Caucasus *A. korostelevi* Danilevsky, 1985, etc.

Fig. 5.33 *Phytoecia antoniae*

Fig. 5.34 *Agapanthia dahli walteri*

The habitats of **light forests** of different types occupy about 30 species. Among these *Semanotus russicus russicus* (Fabricius, 1777) develops on *Juniperus*, several species are connected with deciduous trees and shrubs, namely, *Cerambyx dux* Faldermann, 1837, *Anaglyptus danilevskii* Miroshnikov, 2000, *Xylotrechus antilope bitlisiensis* S.Marklund & D.Marklund, 2013, *X. arvicola lazarevi* Danilevsky, 2016, *Hesperophanes sericeus* (Fabricius, 1787), invasive species *Trichoferus campestris* (Faldermann, 1835) recently penetrated into natural habitats, *Purpuricenus budensis* (Götz, 1783), *P. kaehleri menetriesi* Motschulsky, 1845,

Fig. 5.35 *Pteromallosia albolineata*

Molorchus kiesenwetteri hircus Abeille de Perrin, 1881, *M. umbellatarum umbellatarum* (Schreber, 1759), *Callimoxys gracilis* (Brullé, 1832), *Callimus femoratus* (Germar, 1823), *Stenopterus rufus geniculatus* Kraatz, 1863, *Clytus schneideri schneideri* Kiesenwetter, 1879, *Deroplia genei genei* (Aragona, 1830), *Tetrops gilvipes gilvipes* (Faldermann, 1837), *T. praeustus praeustus* (Linnaeus, 1758), etc. Mainly in these habitats can be found *Phytoecia cylindrica* (Linnaeus, 1758) living on different Apiaceae. Some of the species above-mentioned can penetrate into typical forests.

Most diverse is the fauna of **broad-leaved forests,** which includes more than 50 species. These are mainly polyphagous species developing of different kinds of broad-leaved trees (*Quercus, Carpinus, Fagus*, etc.) at different stages of drying up, for example, *Aegosoma scabricorne* (Scopoli, 1763), *Prionus coriarius* (Linnaeus, 1758) (sometimes lives also on coniferous trees), *Rhagium caucasicum caucasicum* Reitter, 1889, *Rh. fasciculatum* Faldermann, 1837, *Stenocorus quercus aureopubens* (Pic, 1908), *S. insitivus insitivus* (Germar, 1823), *Leptorhabdium caucasicum* (Kraatz, 1879), *Dinoptera collaris* (Linnaeus, 1758), *Stenurella novercalis* Reitter, 1901, *S. bifasciata bifasciata* (Müller, 1776) (Fig. 5.36), *S. jaegeri* Hummel, 1825, *Anastrangalia dubia melanota* (Faldermann, 1837), *Grammoptera ustulata tibialis* Kraatz, 1886, *Leptura quadrifasciata lederi* Ganglbauer, 1882, *Pachytodes erraticus* (Dalman, 1817), *Rutpela maculata manca* (Schaufuss, 1863), *Stictoleptura cordigera cordigera* (Fuessly, 1775), *Strangalia attenuata* (Linnaeus, 1758), *Fallacia elegans* (Faldermann, 1837), *Necydalis ulmi* (Chevrolat, 1838), *Anaglyptus arabicus* Küster, 1847, *A. mysticoides mysticoides* Reitter, 1894, *Paraclytus sexguttatus* (Adams, 1817), *Chlorophorus figuratus* (Scopoli, 1763), *Ch. sartor* (Müller, 1766), *Clytus rhamni temesiensis* (Germar, 1823), *Isotomus comptus comptus* Mannerheim,

Fig. 5.36 *Stenurella bifasciata bifasciata*

Fig. 5.37 *Cerambyx cerdo acuminatus*. (Red Book: VU)

1825, *Plagionotus arcuatus lugubris* Ménétriés, 1832, *Ropalopus macropus* (Germar, 1823) *R.clavipes* (Fabricius, 1775, *Cerambyx scopolii scopolii* Fuessly, 1775), as well as red-listed and included in the Annex II of Bern Convention *C. cerdo acuminatus* Motschulsky, 1853 (Fig. 5.37) and *Rosalia alpina alpina* (Linnaeus, 1758), *Gracilia minuta* (Fabricius, 1781), *Obrium cantharinum cantharinum* (Linnaeus, 1767) *Phymatodes testaceus* (Linnaeus, 1758), *Molorchus monticola* Plavilstshikov, 1931, *Aegomorphus clavipes* (Schrank, 1781), *Mesosa curculionoides* (Linnaeus, 1760), *Leiopus femoratus* Fairmaire, 1859, *L. nebulosus caucasicus*

Ganglbauer, 1887, *Morimus verecundus verecundus* (Faldermann, 1836), *Anaesthetis testacea testacea* (Fabricius, 1781), etc., etc. Mainly in the forests live *Phytoecia affinis boeberi* Ganglbauer, 1884 and *Agapanthia chalybea* Faldermann, 1837 developing on grassy plants.

Fauna of **pine forests** consists of about ten species developing on *Pinus*, majority of them are considered the forest pests. These are *Rhagium inquisitor schtschukini* Semenov, 1898, recently found in North Armenia, as well as more or less widely distributed in the country *Arhopalus rusticus rusticus* (Linnaeus, 1758), *Asemum striatum* (Linnaeus, 1758), *Spondylis buprestoides* (Linnaeus, 1758), *Hylotrupes bajulus* (Linnaeus, 1758), *Callidium violaceum* (Linnaeus, 1758), *Acanthocinus aedilis* (Linnaeus, 1758), *Pogonocherus fasciculatus fasciculatus* (DeGeer, 1775), *Monochamus galloprovincialis transitivus* Lazarev, 2017, etc.

Steppe belt is characterized by rich fauna as well including more than 40 species and subspecies. Mainly here live several species and subspecies of *Cortodera* living in roots of different grasses (*C. colchica erevanica* Danilevsky, 2014, *C. alpina armeniaca* Pic, 1898 (Fig. 5.38), *C. a. umbripennis* Reitter, 1890, *C. pseudomophlus* Reitter, 1889, *C. transcaspica lobanovi* Kaziutshitz, 1988), as well as numerous taxa of *Dorcadion* developing freely in the soil: *D. apicerufum* Breuning, 1943, *D. dimidiatum dimidiatum* Motschulsky, 1838, *D. mniszechi* Kraatz, 1873 (Fig. 5.39), *D. nitidum* Motschulsky, 1838, *D. striolatum* Kraatz, 1873, *D. sulcipenne argonauta* Suvorov, 1913, *D. s. caucasicum* Küster, 1847, *D. s. goktshanum* Suvorov, 1915, red-listed *D. kasikoporanum* Pic, 1902, *D. gorbunovi gorbunovi* Danilevsky, 1985, *D. g. rubenyani* Lazarev, 2014 (the species is incuded in the Red Book of Armenia without subspecies specification), *D. laeve vladimiri* Danilevsky & Murzin, 2009, *D. megriense* Lazarev, 2009, etc. Majority of the species abovementioned have rather restricted range being endemics of the Caucasian Ecoregion, *C. colchica erevanica* and last seven taxa of *Dorcadion* are endemics of Armenia.

Fig. 5.38 *Cortodera alpina armeniaca*

Fig. 5.39 *Dorcadion mniszechi*

Fig. 5.40 *Echinocerus floralis floralis*

Tragant plots inhabit *Xylotrechus sieversi sieversi* Ganglbauer, 1890. On cereals develop *Calamobius filum* (Rossi, 1790), *Theophilea cylindricollis* Pic, 1895, on different kinds of grasses live *Pseudovadonia livida bicarinata* (Arnold, 1869), *P. l. desbrochersi* (Pic, 1891), *Vadonia unipunctata unipunctata* (Fabricius, 1787), *V. bitlisiensis* Chevrolat, 1882, *Certallum ebulinum* Linnaeus, 1767, *Chlorophorus varius varius* (Müller, 1766), *Echinocerus floralis floralis* (Pallas, 1773) (Fig. 5.40), *Neoplagionotus bobelayei huseyini* Lazarev, 2016 (Fig. 5.41), *Pilemia hirsutula hirsutula* (Frölich, 1793), *Phytoecia astarte lederi* Pic, 1899 *Ph. caerulea caerulea* (Scopoli, 1772) *Ph. virgula virgula* (Charpentier, 1825), *Ph. pustulata pustulata* (Schrank, 1776), *Agapanthia lederi lederi* Ganglbauer, 1884, *A. persicola* Reitter, 1894, *A. kirbyi valandovensis* Sláma, 2015, etc.

Fig. 5.41 *Neoplagionotus bobelayei huseyini*

Some of the species above-mentioned (*Dorcadion, Cortodera*) can penetrate to meadow-steppes and low part of alpine and subalpine **meadows,** and some taxa of these genera are more typical for this belt, for example, *C. kazaryani* Danilevsky, 2014, *C colchica colchica* Reitter, 1890, *C. c. kalashiani* Danilevsky, 2000, *C. kaphanica* Danilevsky, 1985, *Dorcadion apicerufum* Breuning, 1943, *D.wagneri wagneri* Küster, 1846, *D. sisianense* Lazarev, *D. indutum* Faldermann, 1837, *D. nigrosuturatum* Reitter, 1897, *D. semilucens* Küster, 1846, *D. sevliczi* Danilevsky, 1985, *D. bistriatum* Pic, 1898, *D. cineriferum* Suvorov, 1910, etc. *C. kazaryani* and *C. c. kalashiani*, as well as last 7 species of *Dorcadion* are endemics of Armenia, and last three are included into Red Book of Armenia.

Like jewel beetles, **wetlands** inhabit several species living on Salicaceae (*Populus, Salix*), *Aromia moschata ambrosiaca* (Steven, 1809) (Fig. 5.42), *Xylotrechus rusticus* (Linnaeus, 1758), *Saperda populnea populnea* (Linnaeus, 1758), *S. octopunctata* (Scopoli, 1772), *S carcharias* (Linnaeus, 1758), *Lamia textor* (Linnaeus, 1758) *Oberea oculata* (Linnaeus, 1758), etc., as well as *Mesoprionus asiaticus* (Faldermann, 1837) on *Tamarix.*

5.1.3.6 Rare and Threatened Species

So far 57 species of beetles belonging to 11 families are included in the Red Book of RA. These are ground beetles (Carabidae): *Carabus (Procerus) scabrosus fallettianus* Cavazzutti, 1997 – VU B 1ab(iii)+B 2ab(iii), *Dyschirius sevanensis* Khnzorian, 1962 – EN B 1a+B2a, *Trechus infuscatus* Chaudoir, 1850– CR B2a,

Fig. 5.42 *Aromia moschata ambrosiaca*

Duvalius stepanavanensis Khnzorian, 1963 – CR B2a, *Duvalius yatsenkokhmelevskii* (Khnzorian, 1960) – CR B2a, *Deltomerus khnzoriani* Kurnakov, 1960 – CR B2a, *Chylotomus alexandri* Kalashian, 1999 – CR B2a, *Poecilus festivus* (Chaudoir, 1868) – VU B 1b(iii)+B 2b(iii), *Pristonychus arenicus* Kalashian, 1979 – CR B2a; Leiodidae (in Red Book as Cholevidae) *Philomessor kalashiani* Khnzorian, 1988 – CR B1a+ B2a; superfamily Scarabaeoidea from families bumble bee scarab beetles (Glaphyridae) (in the Red Book mentioned as Scarabaeidae): *Glaphyrus calvaster* Zaitzev, 1923 – EX, *G. caucasicus* Kraatz, 1887 – EN B 1ab(iii)+B2ab(iii) and scarabs themselves (Scarabaeidae): *Adoretus rubenyani* Kalashian, 2002 – EN B1a+ B2a, *Anisoplia reitteriana* Semenov, 1903 – EN B 1ab(iii)+B2ab(iii), *Pharaonus caucasicus* (Reitter, 1893) – CR (B1ab(ii, iii, v), B2ab (ii, iii, v), *Tanyproctus araxidis* Reitter, 1901 – EN B 1a+B2a, *T. vedicus* Kalashian,1999 – CR B1a+ B2a, *Pseudopachydema medvedevi* Khnzorian, 1971 – CR B1a+ B2a; click beetles (Elateridae): *Aeoloides figuratus*(Germar, 1844) – VU B2a, *Drasterius atricapillus* (Germar, 1824) – EN B2a, *Ctenicera pectinicornis* (Linnaeus, 1758) – VU B 1b (Fig. 5.43), *Cardiophorus araxicola* Khnzorian, 1970 – CR (B1ab(ii, iii, v), B2ab (ii, iii, v) (Fig. 5.44), *C. pseudogramineus* Mardjanian, 1977 – EN B1a (Fig. 5.45), *C. somcheticus* Schwarz, 1896 – EN B1+ B2a, *Craspedostethus permodicus* (Faldermann, 1835) – VU B 1b(iii) + B 2b(iii); jewel beetles (Buprestidae): *Acmaeoderella pellitula* (Reitter, 1890) – EN B 1a + B 2a, *Anthaxia breviformis* Kalashian, 1988 – CR B1a+B2a, *A. superba* Abeille de Perrin, 1900 – CR B1a, *Sphaerobothris aghababiani* Volkovitsh & Kalashian, 1998 – CR Ba +B2a, *Sphenoptera geghardica*Kalashian & Zykov, 1994 – EN B 1ab(iii), *S. khnzoriani* Kalashian, 1996 – EN B2a; Blister beetles (Meloidae): *Mylabris sedilithorax*

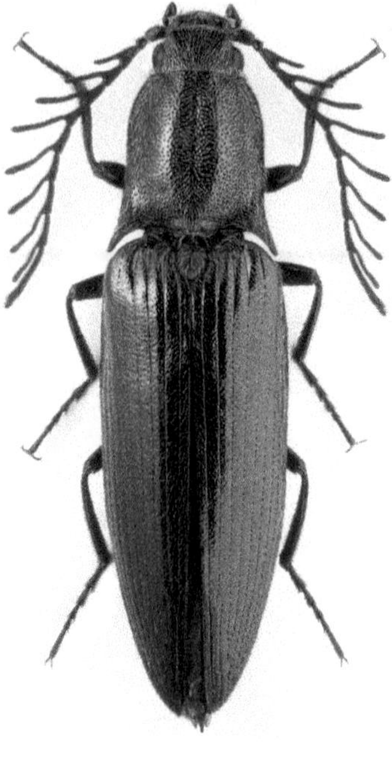

Fig. 5.43 *Ctenicera pectinicornis* VU (Elateridae)

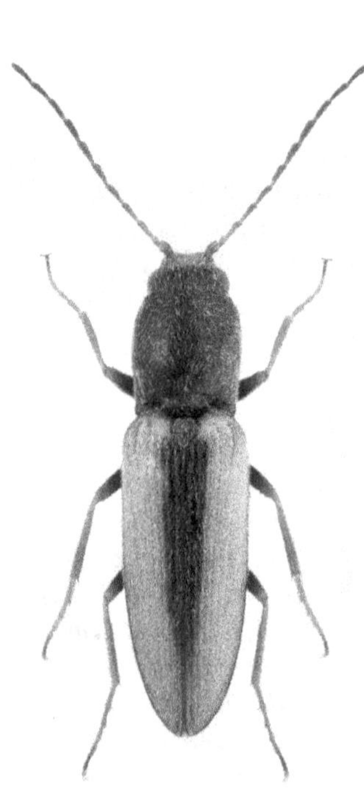

Fig. 5.44 *Cardiophorus araxicola* CR (Elateridae)

Fig. 5.45 *Cardiophorus pseudogramineus* EN (Elateridae)

Sumakov, 1924 – EN B 1a; darkling beetles (Tenebrionidae): *Adelphinus ordubadensis* Reitter 1890 –EN B1a, *Armenohelops armeniacus* Nabozhenko, 2002 – EN B1a, *Cylindronotus erivanus* (Reitter, 1901) – EN B1a, *Ectromopsis bogatchevi* (Khnzoryan, 1957) – CR B2a b(iii), *Entomogonus amandanus* Reitter, 1901 – EN B 1a, *Laena constricta*Khnzorian, 1957 – EN B 1ab(iii)+B2ab(iii), *Cyphostethe semenovi* Reitter, 1895 – EN B 1ab(iii)+B2ab(iii), *Cteniopus persimilis* Reitter, 1890 – EN B1a, *Isomira armena* Khnzorian, 1976 – EN B1a (the last two as Alleculidae); longhorn beetles (Cerambycidae): *Cortodera kaphanica* Danilevsky, 1985 – EN B 1a, *Cerambyx cerdo acuminatus* Motschulsky, 1852 – VU B 1a+B 2a; IUCN (ver 2.3) VU A1c+2c; *Rosalia alpina* (Linnaeus, 1758) – EN B 1a+B2a; IUCN (ver 2.3) VU A1c+2c; *Anoplistes aghababiani* Danilevsky, 1999 – CR Ba +B2a (mentioned as *Asias*), *Dorcadion kasikoporanum* Pic, 1902 – CR B1a, *D. bistriatum* Pic, 1898 – EN B 1a+B2a, *D. gorbunovi* Danilevsky, 1985 – EN B 1a, *D. semilucens* Kraatz, 1873 – EN B 1a+B2a, *D. cineriferum* Suvorov, 1909 – EN B 1a+B2a, *D. sevliczi* Danilevsky, 1985 – CR B1a, *Conizonia kalashiani* Danilevsky, 1992 – CR B1a, *Phytoecia pici* Reitter, 1892 – EN B 1a, *Agapanthia korostelevi* Danilevsky, 1985 – EN B 1a+B2a; leaf beetle (Chrysomelidae) *Cryptocephalus araxicola* Khnzorian, 1968 – EN B 1ab(iii) (Fig. 5.46) and weevils (Curculionidae): *Cyclobaris richteri* Ter-Minasian, 1955 – CR B 1ab(iii)+B2ab(iii) (Fig. 5.47) and *Baris mirifica* Khnzorian, 1958 – CR B 1ab(iii)+B2ab(iii) (Fig. 5.48).

The list must be updated, taking into consideration numerous Armenian endemics, sometimes inhabiting vulnerable habitats, as well as representatives of the

Fig. 5.46 *Cryptocephalus araxicola* EN (Chysomelidae)

Fig. 5.47 *Cyclobaris richteri* CR (Curculionidae)

families not presented or insufficiently represented in the book (e.g. Curculionidae, Chrysomelidae, Histeridae, Staphylinidae, etc.).

Fig. 5.48 *Baris mirifica* CR (Curculionidae)

5.1.4 Butterflies (Lepidoptera)

In Armenia, the butterflies are represented by 232 species, which belong to five families, among those the most species-rich one is *Lycaenidae* with 85 species, followed by *Nymphalidae* with 83 species, then *Hesperidae* – 32 species, *Pieridae* – 27 species, and *Papilionidae* – 5 species (Annex 3).

The fauna of butterflies in Armenia is mainly represented by two major groups: European and Iranian species, while the minor part is composed by endemics of the Caucasus ecoregion.

Among the European species there are such species as *Thymelicus sylvestris* and *Colias hyale*, among Iranian species – such typical representatives as *Satyrium hyrcanica* and *Coenonympha saadi*. Endemics of the Caucasus ecoregion include such species as *Colias chlorocoma*, *Polyommatus damonides*, and *Satyrus effendi*. At the same time, a number of species have wider Palearctic distribution, for example, *Pyrgus sidae*, *Papilio machaon*, *Colias erate*, and *Argynnis paphia*.

During last decades, we did not observe any latitudinal shifts in butterfly distribution, but instead there is a certain trend in change of vertical distribution of several species. Thus, since the period of 1995–2001, when the first rather comprehensive survey of butterflies was implemented, there are several cases of vertical movements have been documented. For example, *Anthocharis grunerii* (Fig. 5.49), a typical species of semideserts and juniper woodlands was observed expanding its upper border of distribution further up, penetrating into deciduous woodlands and forests. The host plant of *Anthocharis grunerii* is *Microthlaspi umbellatum*, which

Fig. 5.49 *Anthocharis grunerii*

is growing in arid habitats. Therefore, such a penetration indicates an aridization of the deciduous woodlands. The same pattern was observed for *Colias aurorina*, which in a similar manner expands its distribution range to forests and meadows, following expansion of its host plant, the tragacanth astragals. Another, but not last, example is a semidesert species *Pontia chloridice*, which was recently observed at about 1800 m above sea level, colonizing mountain steppes.

Most probably, this expansion is related to such factors of the climate change as increase of average temperature and decrease of humidity and precipitations and is a subject of further study of the influence of climate change on the biodiversity.

5.1.4.1 Landscape Distribution

As it was mentioned already, the ecosystems of Armenia are described in detail in Chap. 2 of the current book. Therefore, we will classify the communities of butterflies related to the certain habitats, rather than describe the ecosystems of Armenia. The butterflies can be divided into several communities, some of those can be related to the vertical zones, while the others are intrazonal. The zonal communities are butterflies of semideserts, shrublands (shiblyak), juniper open forests, deciduous forests, tragacanth mountain steppes, grassy mountain steppes, meadows, and alpine carpets. The intrazonal communities include butterflies of wetlands (predominantly marshes), rocks and screes, butterflies of riparian shrubs and woodlands, and butterflies of arable lands.

Butterflies of Semideserts

Semideserts of Armenia occupy elevation range from 400 m till about 1200 m a.s.l. and are mainly characterized by perennial and annual herbs, semishrubs, and shrubs. The semideserts are inhabited by 74 butterfly species. Most of those species occur in other arid habitats as well, and only four species: *Pontia chloridice*, *Melitaea ornata*, *Melitaea abbas*, and *Melitaea vedica* inhabiting the semidesert only.

Butterflies of Shrublands (Shiblyak)

The habitat is dominated by *Paliurus spina-christi* and is mainly characterized by rather dense coverage of shrubs. The shrublands begin at the elevation range from about 600 m a.s.l. in the southeastern Armenia and at about 800 m a.s.l. in the northeastern Armenia. The shrublands are inhabited by 61 butterfly species. There are no inhabitants of this habitat only, all the species live in other arid habitats too (e.g. in semideserts or juniper woodlands). Nevertheless, some species, associated with *Paliurus spina-christi*, like *Callophrys chalybeitincta*, or *Rhamnus pallasii*, like *Satyrium spini* and *Gonepteryx farinosa* (Fig. 5.50) can be considered quite typical for this habitat.

Butterflies of Juniper Woodland

This type of habitat is dominated by Juniper tree species *Juniperus polycarpos* and is often combined with some other elements, such as shrubs, deciduous woodlands, or tragacanth-dominated mountain steppes. In the southern and southeastern

Fig. 5.50 *Gonepteryx farinosa*

Armenia this habitat gradually replaces shrublands, while in Central Armenia its elevation range starts from 1500 m a.s.l., and in Eastern Armenia around Lake Sevan it begins at about 2000 m a.s.l. Juniper woodlands of Armenia are inhabited by 94 butterfly species. Again, there are exclusive inhabitants of Juniper woodlands, as the butterflies of this habitat live either in next altitudinal zone – deciduous woodlands of in previous zone of semidesert and shrubland. Nevertheless, Probably, the *Limenitis reducta* a species that is linked to various *Lonicera* sp. could be considered as one of the typical inhabitants. At the same time, the density of such species as *Satyrus amasinus* and *Melanargia larissa* is higher here than in other habitats.

Butterflies of Deciduous Forests

Deciduous forests in Armenia start at various elevations in Southern, Central, and Northern Armenia, and pretty much vary in the species composition. While the Central and Southern forests are represented by oak-hornbeam forests, dominated by *Quercus* spp. and *Carpinus betulus*, the forests of northern and northeastern regions are dominated by *Quercus* spp. and *Fagus orientalis*. In southeastern Armenia, the forests' belt starts at about 600 m a.s.l., while in northeastern regions it begins at about 500 m a.s.l. The forests are inhabited by 65 butterfly species, among those there are 20 species, which live only in forests. Those are mainly presented by representatives of Nymphalidae family, such as *Satyrus dryas* (Fig. 5.51), *Argynnis paphia*, and *Mellicta caucasogenita*, but also include species from other families, such as *Carcharodus flocciferus*, *Gonepteryx rhamni*, and *Favonius quercus*. Thus, the forests have quite a big number of specialist species.

Fig. 5.51 *Satyrus dryas*

Fig. 5.52 *Callophrys paulae*

Butterflies of Tragacanth Mountain Steppes

Tragacanth mountain steppes are occurring at the elevations from 1200 to about 2400 m a.s.l. and are dominated by various legumes, with most abundant species of *Astragalus* sp. and *Onobrychis cornuta*. Tragacanth mountain steppes are inhabited by 99 butterfly species, among those there are 18 species, which live only in this habitat. Vast majority of those species is presented by representatives of Lycaenidae, like *Polyommatus (Agrodiaetus) huberti*, *P. vanensis*, and *Callophrys paulae* (Fig. 5.52), which is not surprising as many of Lycaenidae using legumes as host plants. At the meantime, some Pieridae, like *Leptidea duponcheli* and Hesperiidae, like *Spialia phlomidis*, are also specialists of this habitat.

Butterflies of Grassy Mountain Steppes

Grassy mountain steppes and meadow-steppes are occurring at the elevations from 1400 to about 2200 m a.s.l. and are dominated by various cereals. Grassy mountain steppes are inhabited by 57 butterfly species. None of the species are specialized to live in this habitat only, but there are several butterflies, which have higher abundance here than in other habitats, and therefore can be conditionally considered as typical species of grassy mountain steppes. Mainly those are from Nymphalidae family, specifically, from the subfamily Satyrinae, such as *Arethusana arethusa* and *Chazara briseis*.

Butterflies of Subalpine Meadows

The meadows in Armenia are occurring at the elevations from 2000 m a.s.l. in northern Armenia and from 2200 m a.s.l. in southern parts of the country. This type of habitat is dominated by mesophylllic herbaceous plants from *Poaceae* family and includes also wide variety of other herbs. The high-level moisture of this habitat allows it staying green throughout the summer including the driest period from late August till mid-September. The meadows of Armenia are inhabited by 76 butterfly species. Among those there are 17 species specialized in living in this habitat only. They include representatives of all families, such as *Mellicta aurelia*, *Coenonympha leander*, *Lycaena virgaurea*, *Colias hyale*, *Parnassius apollo*, *Carterocephalus palaemon*, and others.

Butterflies of Alpine Carpets

This habitat is dominated by short perennial herbaceous vegetation and is always alternated by various screes and rocks. The lower border of alpine meadows and carpets starts at about 2700 m a.s.l. The habitat is characterized by very short summer and high level of precipitations. The alpine zone in Armenia is inhabited by 24 butterfly species, including 14 specialists. Among the specialists, there are representatives of all families except Papilionidae, for example, *Boloria caucasica*, *Erebia graucasica*, *Polyommatus (Agrodiaetus) altivagans*, *Lycaena thetis* (Fig. 5.53), *Colias thisoa*, *Pyrgus jupei*, and others.

Fig. 5.53 *Lycaena thetis*

Butterflies of Rocks and Screes

This is another type of intrazonal habitat, which is represented from about 400 to 4000 m a.s.l. There are two major types of these formations: volcanic and sedimentary. The rocks and screes are inhabited by 14 butterfly species. Among those species there is only one specialist that lives only in this type of habitat – *Pieris krueperi*, while some other species are quite linked to the rocks and screes, located in juniper woodlands and other arid lands, for example, *Pieris ergane* and *Polygonia egea*.

Butterflies of Riparian Zone

To some extent, the riparian shrubs and woodlands can be considered as intrazonal; however, it is true only for lower parts of riparian shrubs – from 375 m to about 1800 m a.s.l., where the riparian zone contrasts with the surrounding arid habitats. Above that elevation the area is usually occupied be deciduous forests and the plant composition of riparian zone does not significantly differ from the main zonal habitat. The riparian shrubs and woodlands are dominated by trees *Juglans regia*, *Celtis australis*, and *Populus* spp., wide variety of bushes with domination of *Rubus* spp., and among herbs – by *Phragmites australis, Typha* spp., etc. The riparian zone is inhabited by 44 butterfly species including 3 specialists of this habitat: *Plebeius christophi* (Fig. 5.54), *Eogenes alcides*, and *Gegenes nostrodamus*. Many other species move to riparian shrubs and woodlands from deciduous forests, like *Nymphalis xanthomelas*, or from juniper woodlands, like *Libythea celtis*.

Fig. 5.54 *Plebeius christophi*

Butterflies of Wetlands

Under wetlands in the context of butterflies, we consider grassy and brackish marshes. The grassy marshes typically occur on the plateaus and in close proximity of Lake Sevan at the elevations from 1900 to 2400 m a.s.l., and are characterized by domination of various species of *Scírpus*, *Carex*, *Rumex*, *Polygonum*, etc. Brackish marshes are located in Ararat Plain and in adjacent clay semideserts at the elevation range from 800 to about 1200 m a.s.l. and are dominated by species of *Phragmites* and *Typha*. The wetlands of Armenia are inhabited by nine butterfly species, among those two species considered to occur only in the marshes – *Thymelicus acteon* and *Lycaena dispar*, and other two occur also in the meadows – *Maculinea nausithous* and *Lycaena candens*.

Butterflies of Arable Lands

In Armenia, there are three major types of arable lands: (1) grain fields; (2) gardens; and (3) orchards. The grain fields are mostly located at the higher elevations from 1500 m up to 2200 m a.s.l., while the gardens and orchards occupy the lower warmer areas from 500 m to 1500 m a.s.l. The grain fields and the gardens present fully transformed monocultural habitats, where only edges of the arable lands remain for the original type of habitat – mainly mountain steppes and semideserts. The orchards usually host much more of original habitat because the herbs which grow in orchards are used as a hay for the livestock. In rigorous terrain, such as in southeastern Armenia, the orchards have the most mosaic structure and are combined with original riparian shrublands, which allows occurrence of higher number of wild plants, and therefore, the insect species. The butterfly fauna, which inhabits the arable landscape, mostly consists of the most adaptable species of the original habitat, which was transformed. Thus, in the grain fields, the *Pontia edusa* and *Colias crocea* can be found, in the gardens – the *Pieris pseudorapae* and *Pieris brassicae*, and in the orchards – the *Carcharodus alceae*, *Iphiclides podalirius* (Fig. 5.55), and *Lysandra bellargus*.

It is worth to mention that during last 20 years, a change of upper or lower borders of the species distribution was determined. It includes alpine species, for which the lower border of their range was shifted up, and species of arid habitats, which have expanded the upper border of their vertical distribution towards degraded parts of forest and through mountain steppes. Butterflies are very sensitive and quite mobile and react to the transformation of the habitats at the very early stages. Most of the observed shifts are related to the economic activities, which in combination with the factors of climate change cause habitat degradation.

Fig. 5.55 *Iphiclides podalirius*

5.1.4.2 Threatened Species and Threats

The last assessment of the conservation statuses of the butterflies of Armenia was conducted in 2009 in the frames of Red Book of Animals of Armenia (Aghasyan and Kalashyan 2010). According to that publication, in Armenia there are three species classified as Critically Endangered, eight species were assessed as Endangered, 13 as Vulnerable, and none of the species were classified as Data Deficient. The category Near Threatened was not considered for the mentioned edition of the Red Data Book.

The recent studies undertaken in frames of European Butterfly Monitoring and National Butterfly Monitoring Scheme (the publications are specified below) demonstrated a need for revision of the 2010 edition of the Red List of butterflies of Armenia, as many of the species should be included in the list, as well as the Near Threatened category should also be considered.

The major threats for the butterflies in Armenia are mostly related to the transformation and destruction of habitats, chemical pollution, overharvest of the biological resources, and the climate change, while the direct persecution does not appear to play any significant role.

Transformation and destruction of habitats is true for both terrestrial and freshwater ecosystems. The forests of Armenia are suffering from the unsustainable forestry management practices, which lead to habitat fragmentation and removal of snag and hollow trees. It affects the specialists of deep forests, which need some shade and a certain level of humidity, like *Satyrus dryas, Pararge aegria* (Fig. 5.56), and *Argynnis paphia* (Aghababyan et al. 2010). At the same time forest fragmentation opens space for such generalist species as *Colias crocea* and *Colias alfacariensis* (Aghababyan et al. 2010).

Fig. 5.56 *Pararge aegria*

The semideserts are being occupied by horticulture, which completely transforms the habitat under orchards and vineyards. Such a transformation destroys the bushes and scrubs, affecting such species as *Satyrium ledereri* and *Satyrium hyrcanica*, but moreover important destroys herbal vegetation, causing decline of such specialized species, as *Coenonympha saadi* and *Satyrus amasinus*.

The mountain steppes and meadows used for pasture suffer from the overgrazing and uncontrolled mowing. Overgrazing destroys the habitat in several phases. At the earlier stages, it results in change of the plant composition, mainly because the livestock feeds selectively, leaving out the poisonous and thorny plants, which occupy the space of grasses and legumes. Thus, it decreases the plant diversity of the mountain steppes and meadows, which affects such specialized species as *Polyommatus (Agrodiaetus) ninae* (Fig. 5.57), *Melanargia russiae* and *Coenonympha leander*. At the later stages, it results in soil erosion, affecting most of the species of the habitat, even such common ones, as *Lasiommata megera* (van Swaay et al. 2015, 2016).

The freshwater ecosystems also suffer from transformation. The brackish marshes of Ararat Plain and the grassy marshes of the mountain plateaus are being consistently drained to be used for agriculture or for water acquisition, thus destroying habitats for such specialized species, like *Maculinea nausithous* and *Lycaena candens*. The fragmentation of the rivers due to construction of small hydropower plants leads to decrease of humidity of the area, and this process is being exacerbated by the climate change casing, often critical consequences for such specialists like *Polyommatus myrrha* via influencing their host plants (Aghababyan and Khanamirian 2018).

In the meantime, there are some anthropogenic activities which are affecting various ecosystems. Those are related to open-pit mining and urban development,

Fig. 5.57 *Polyommatus (Agrodiaetus) ninae*

Fig. 5.58 *Thaleropis ionia*

which are completely destroying the ecosystems, causing extermination of up to 80% of the butterfly fauna.

The pollution can be conditionally divided into mechanical and chemical, but in the context of butterflies it is worth to observe the chemical pollution, and specifically pollution of the environment by persistent pesticides. This is especially true for the species, which live in arable lands, like *Cupido argiades* and *Gegenes nostrodamus*, which predominantly occur in orchards, also for the species, which live next to arable lands, like *Thaleropis ionia* (Fig. 5.58), as well as for the species of

Fig. 5.59 *Parnassius apollo*

deciduous forests, which are being processed with insecticides in frames of pest control.

The overharvest of biological resources can be related to removal of butterflies' food items – their host plants. Usually, such harvest plays a minor role in butterflies' population, because the collection of the edible plants is made by local inhabitants, by hand. However, in some cases it can cause the decline of the specialized species. For example, collection of *Ferula* sp., the plant that is widely used for marinade is made mainly in its early stage, and the people usually collect the softest central branches, which are supposed to give flowers later. It decreases the number of *Ferula* flowers in the habitat, and thus is depriving the species *Callophrys danchenkoi* of its host plant, because the larvae of this carpophagous species feed only on flowers and later on seeds of the host plant.

The last threat is related to the consequences of the climate change, which are influencing the ecosystems and related butterfly fauna through several factors, such as warming of the mid temperatures (annual, spring-summer, winter) and decrease of precipitations and humidity. Those factors accelerate aridization of the habitats, influencing the change of plant composition and the rate of their transformation. While many different species suffer from these factors, the most vulnerable ones are probably the community of alpine and forest species. Thus, among alpine species there are such specialized ones, as *Parnassius apollo* (Fig. 5.59) and *Satyrus effendi*. The mentioned factors result in shifting of the range of their host plants, but these butterflies need also some other conditions, like steepness of the slopes, level of areas' coverage by rocks, and so on. Among forest species, there are such examples as *Nymphalis antiopa*, which demonstrates steep decline of its population in the absence of other visible threats. The factors of climate change can cause change of temperature and humidity optimum for its larval or pupal stage, or can cause an outbreak of some parasitic species, which kill the eggs or larvae of this butterfly.

Negative influence of climate change is visible on other specialized butterfly species of alpine zone, deciduous forests, and other habitats.

For conservation of butterflies of Armenia, there are 32 potential PBAs – Prime Butterfly Areas (van Swaay and Warren 2006) have been identified, and among those, 15 PBAs have been assessed (Khanamirian et al. 2014; Khanamirian and Aghababyan 2019; Aghababyan et al. 2022a). The PBAs do not have state protection status, but their identification supplies in assessment of national protected areas or emerald sites (Fayvush et al. 2016).

5.2 Vertebrate Animals

5.2.1 Fisch

Bardukh Gabrielyan

We witness changes in fish species communities due to the expansion or reduction of their habitats, introductions (accidental or scientific acclimatization), destruction of spawning grounds, obstructions to migrating routes, climate change. As a result, fish species composition may undergo certain changes over relatively short periods of time.

In determining species composition, many researchers tend to rely on the results of a single research method, for example, molecular or caryological or morphological. However, a combination of these methods would provide the most reliable basis for determining fish species composition.

Systematic ichthyological investigations on Lake Sevan and other reservoirs and rivers of Armenia began in 1923, when Sevan Lake Station (now called the Institute of Hydroecology and Ichthyology of the National Academy of Sciences of Armenia) was established. Since then the ichthyofauna of reservoirs and rivers which may have commercial importance has been investigated and fish ecology and morphology studied.

This annotated list of the fish of Armenia is based on as thorough as possible analysis of the main scientific data about the ichthyofauna of Armenia, in which available data sources were carefully chosen to take into account not only data obtained by modern taxonomy methods (e.g. molecular), but also historical data on the presence of particular species, their catch, natural reproduction events, availability of suitable habitats and other factors.

ORDER I. SALMONIFORMES

Family 1. SALMONIDAE

1. ***Salmo trutta fario* (Linnaeus 1758)**

Found in almost all rivers of Armenia and the Akhurian and Arpi Lich reservoirs. At present, this fish species has almost disappeared from some rivers due to

Fig. 5.60 *Salmo ischchan*

poaching and water pollution. It can live up to 12 years and the maximum body length can reach 25 cm. Males mature at body length 12.5 cm (1st–3rd year); females at 14.7 cm (3rd–4th year). On growing, the fish food spectrum shifts from benthos to small-size fish (Smoley 1979; Gabrielyan 2001).

2. ***Salmo ischchan* (Kessler 1877)**

The endemic of Lake Sevan was represented by four ecological races (subspecies), which differed in the period of spawning and ground, growth rate and morphometric features (Barach 1940; Dadikyan 1986; Savvaitova et al. 1989; Gabrielyan 2010a, b). Before lowering of the water level of Lake Sevan, individuals with body length 65 cm could be met (Fig. 5.60). Some individuals were introduced into Lakes Akna, Kari, Sev, the Aparan and Mantash reservoirs, as well as Lake Issyk Kul (Kyrgyzstan).

(a) *Salmo ischchan gegarkuni* (Kessler 1877)

However, the spawning population had been poached in the river mouth by traps. In Lake Sevan, the feeding population had also been poached. As a result, the population has currently dramatically declined. In Lake Sevan, the population is currently being sustained via artificial reproduction (in fish hatcheries). According to IUCN criteria, it is categorized as Critically Endangered (CR A2cd). Listed in the Red Book of Armenia (2010).

Females mature between the 5–6 years, although some females may mature by the third year. Males mainly mature by the 2nd year (Dadikyan, 1986). Young fish mainly feed on benthos and zooplankton, while adults mainly prey (Pivazyan 1984). The fish have commercial importance.

(b) *Salmo ischchan aestivalis* (Fortunatov 1927)

Spawn in rivers (at slow occurs of a river). This is the unique subspecies among *Salmonidae* family, because they spawn in spring – beginning of summer, in contrast to others which spawn in autumn-winter period. The population of this subspecies has dramatically declined due to two factors: drying up of the spawning grounds because of the lowering of the lake's water level and poaching during spawning period in rivers. According to IUCN criteria, they are categorized as Critically Endangered (CR A2cd). Listed in the Red Book of Armenia (2010). Maturity age and food spectrum are similar to *S. ischchan gegarkuni* (Gabrielyan, 2010a, b). The fish have commercial importance.

(c) *Salmo ischchan ischchan* (Kessler 1877)

Spawned in Lake Sevan only, in the littoral. As a result of the lake's water level lowering, spawning grounds were dried up. Since the middle of the 1980s the subspecies has not been met in the lake. Currently, it is considered as disappeared. According to IUCN criteria, it is categorized as Extinct (EX). Listed in the Red Book of Armenia (2010). The fish was the most fast-growing from all subspecies and could reach body mass 8 kg (one of the examples is stored at the zoological museum of the Scientific Center of Zoology and Hydroecology). The fish have commercial importance.

(d) *Salmo ischchan danilewskii* (Yakowlew 1888)

Spawned in Lake Sevan. With the lake's water level lowering, the natural reproduction was affected leading to disappearance of this subspecies. According to IUCN criteria. They are categorized as Extinct (EX). Listed in the Red Book of Armenia (2010). The fish was the most slow-growing of all subspecies, and due to this factor it was of a small commercial importance (Gabrielyan 2010a, b).

3. ***Parasalmo mykiss* (Walbaum 1792)**

Bred in fish farms throughout Armenia, it is a commercially important species. Annual production reaches 500 t. Due to escapes from fish farms the fish can be met in different rivers and even in Lake Sevan. However, the data on their natural reproduction in Lake Sevan are absent. Body length may reach 50 cm. By taste, it is inferior to both Sevan trout (*S. ischchan*) and brook trout (*S. trutta*).

Family 2. COREGONIDAE

4. ***Coregonus lavaretus* (Linnaeus 1758)**

This species is a hybrid of *Coregonus lavaretus ludoga* [referred to as *C. lavaretus lavaretus* by some authors, for example, Reshetnikov et al. (1980)] and *C. lavaretus marineoidis* acclimatized into Lake Sevan in the 1920s. Whitefish is found only in Lake Sevan. Individuals with body length 80 cm were met in the beginning of 1980s. Currently, individuals of even 40 cm are rarely met. Lowering of the water level of the lake created favourable conditions for an increase of its food base (zooplankton and zoobenthos), which resulted in the increase of the fish population.

Since the middle of 1960s, the fish has become the main commercial species in Lake Sevan (Gabrielyan 2010a, b). The fish is the food competitor with trout by feeding on zoobenthos (Smoley 1964) but this competition did not have influence on the sharp reduction of the trout population in Lake Sevan.

ORDER II. CYPRINIFORMES

Family 3. CYPRINIDAE

5. *Acanthalburnus microlepis* (De Fillipi 1863)

Found in the Arpa, Vorotan, Araks, Akhuryan, Metsamor and Hrazdan rivers, as well as in the Kechut reservoir and fish farms of the Ararat Valley. Body length reaches 25 cm (Gabrielyan 2001; Pipoyan 2012). It is an object of recreational fishing. The species is not protected.

6. *Alburnus filippi* (Kessler 1877)

Found in most of Armenia's rivers, except for the basin of Lake Sevan. Maximum body length is 12 cm (Gabrielyan 2001). It is an object of recreational fishing. The species is not protected.

7. *Alburnus hohenackeri* (Kessler 1877)

Found in the Araks and Metsamor rivers and fish farm ponds of the Ararat Valley. Body length reaches 10 cm (Gabrielyan 2001; Pipoyan et al. 2018). It is an object of recreational fishing. The species is not protected.

8. *Alburnoides eichwaldii* (Bloch 1782)

Found in almost all rivers and reservoirs, including Lake Sevan basin. It matures by 2–3 years. Spawning period is from the end of April to August, depending on the altitude and water temperature. Body length reaches 10 cm (Gabrielyan 2001; Pipoyan et al. 2018). It is not an object of commercial and even recreational fishing.

9. *Aristichthys nobilis* (Richardson 1945)

Introduced into Armenia as an aquaculture object. At present, some individuals have been observed in the Akhurian and Araks rivers (Pipoyan et al. 2018).

10. *Aspius aspius* (Linnaeus 1758)

Rare species, with the numbers and habitats being reduced. Some individuals were recorded from the Akhurian, Arpa, Metsamor and Vorotan rivers and the Akhurian and Arpi Lich reservoirs (Gabrielyan 2001). The ecology and biology of the fish is poorly studied in Armenia due to small population size on the territory of Armenia. Among the factors causing the fish population reduction are river flow regulation, pollution, disruption of natural reproduction and poaching. It is an object of recreational fishing. The species is not protected. According to IUCN criteria

Fig. 5.61 *Barbus goktschaicus*

categorized as Vulnerable (VU Blab(iii)). Listed in the Red Book of Armenia (2010), as well is included into Resolution 6 to Bern Convention.

11. ***Barbus lacerta cyri* (De Filippi 1865)**

Widely distributed in Lake Akna, almost all reservoirs of Armenia and the Akhurian, Debed and Metsamor rivers. Spawning occurs from June to August. Eggs develop at a wide temperature range, which explains its wide distribution. Feed on various invertebrates. Body length reaches 18 cm (Gabrielyan 2001). It is an object of recreational fishing. The species is not protected.

12. ***Barbus goktschaicus* (Kessler 1877)**

Vulnerable species, endemic to Lake Sevan and its tributaries. Has three biotypes: lake, lake-river, and river. Body length reaches 30 cm (Fig. 5.61). Spawning period is from the first half of June till the first half of August. Males mature between 3 and 4 years, while females mature between 5 and 6 years. Feed on benthos, zooplankton and eggs of other fish (including trout's); the river biotype feed also on flying insects (Gabrielyan 2010a, b). The main factors of the population reduction are lowering of the water level of Lake Sevan, disturbed natural production and poaching. The fish had small commercial importance. Currently, fishing is prohibited and the species is protected. According to IUCN criteria categorized as Vulnerable (VU A 2 cd). Listed in the Red Book of Armenia (2010).

13. ***Luciobarbus capito* (Güldenstädt 1772)**

Rare species with the numbers and habitats being reduced. The ecology and biology of the fish is poorly studied in Armenia due to small population size. Found in the Akhurian, Metsamor and Hrazdan rivers and the Akhurian and Arpi Lich reservoirs. Body length reaches 40 cm (Gabrielyan 2001). The fish body mass may exceed 2 kg (Pipoyan et al. 2018). The main factors of the population reduction are regulation of river flow, pollution, disturbed natural production. It is an object of recreational fishing. The species is included into Resolution 6 to Bern Convention (as *Barbus capito*).

14. ***Luciobarbus mursa* (Güldenstädt 1772)**

Reduced population mainly due to habitat destruction; rare species. Some individuals were found in the Akhurian, Arpa, Metsamor and Vorotan rivers and the Arpi Lich reservoir (Gabrielyan 2001; Pipoyan et al. 2018). It is a small-bodied species, but in the Araks river, the body length may reach 22 cm. The ecology and biology of the fish is poorly studied in Armenia due to small population size. Other factors of the population reduction are river flow regulation, pollution, deterioration of natural reproduction conditions, impact of invasive fish species and poaching. It is an object of recreational fishing. The species is not protected.

15. ***Blicca bjoerkna transcaucasica* (Dadikyan, 1970)**

Reduced population; locally distributed, endemic subspecies. Inhabit the Metsamor river. Body length reaches 12 cm. Spawns in April–May, in several potions due to a gradual maturation of eggs (Dadikyan 1986; Gabrielyan 2001). The main factors of the population reduction are river flow regulation, pollution, competition with invasive fish species. It is an object of recreational fishing. The species is not protected.

16. ***Carassius auratus gibelio* (Bloch 1782)**

Found in almost all ponds, lakes and rivers of the Ararat Valley, as well as in most of the reservoirs of Armenia. Introduced into Lake Sevan in the early 1980s and nowadays is the second commercially important fish (after whitefish *C. lavaretus*). Body length can reach 30 cm, mass up to 1 kg; can live up to 8 years (Fig. 5.62). Mature fish feed on detritus and benthos. In the Ararat Valley, it matures by the 1st year (body length 7.7 cm and mass 16.3 g). In the lake, it matures by the 3rd year, at body length 14.5 cm and mass 87.2 g. Spawn during July–September three to five times per year in the reservoirs of the Ararat Valley, while in Lake Sevan it occurs two to three times (Gabrielyan 2010a, b). In the past, catches could reach 300–400 t/year. Currently, the population is depleted. It is an object of recreational and commercial fishing. The species is not protected.

17. ***Chondrostoma cyri* (Kessler 1877)**

Some individuals were found in the Akhurian, Debed, Marmarik and Metsamor rivers and related lakes and the Arpi Lich reservoir. Feed on plants, as well as flying

Fig. 5.62 *Carassius auratus gibelio*

insects and larvae of aquatic insects. Body length reaches 20 cm (Gabrielyan 2001). It is an object of recreational fishing. The species is not protected.

18. ***Ctenopharyngodon idellus* (Valenciennes 1844)**

Introduced into natural and artificial ponds of the Ararat Valley for weed control and for commercial purpose. Maximum body length is 40 cm. Aquaculture production could reach 1–2 t/year (Dadikyan 1986). The fish is an important aquaculture object.

19. ***Cyprinus carpio* (Linnaeus 1758)**

Inhabit the Akhurian and Metsamor rivers and some reservoirs of Armenia (Akhurian, Arpi Lich, Gokhas and Spandarian). The fish is a warmwater species. Feed on benthos, zooplankton and aquatic plants. Maximum body length is 50 cm (Berg 1949; Dadikyan 1986; Gabrielyan 2001). Introduced into Lake Sevan in the 1980s but the lake's population is small due to unfavourable habitat and reproduction conditions (low water temperature) (Gabrielyan, 2010). The fish is bred in fish farms; in the 1980s, farm production reached 1000–1500 t/year. It is also an object of recreational fishing. The species is not protected.

20. ***Romanogobio macropterus* (Kamensky, 1901)**

Disappearing species. Previously, it was common to all rivers of Armenia, except for the basin of Lake Sevan. In recent years, some individuals were found in the

Akhurian and Metsamor rivers and the Arpi Lich reservoir. Biology is practically not studied. The main factor of the population reduction is the effect of invasive fishes (Pipoyan et al. 2018; Kuljanishvili et al. 2020). The species is not protected.

21. ***Hypophthalmichthys molitrix*** **(Valenciennes 1844)**

Bred in fish farms. Body length reaches 45 cm (Dadikyan, 1986). In the 1980s, farm production reached 3000–4000 t/year. Recently, it was 200–300 t/year.

22. ***Leucaspius delineates*** **(Heckel 1843)**

Widespread in the lower reach of the Hrazdan River and the reservoirs of Armavir region (Pipoyan 1996; Gabrielyan 2001). The fish does not have commercial importance; biology and ecology is not studied.

23. ***Squalius cephalus orientalis*** **(Nordmann 1840)**

Inhabit almost all reservoirs of Armenia (Aparan, Arpi Lich, Gokhas, Spandarian, etc.), except for the basin of Lake Sevan and several small mountain lakes. Body length reaches 18 cm (Gabrielyan 2001; Pipoyan et al. 2018). It is an object of recreational fishing. The species is not protected.

24. ***Mylopharyngodon piceus*** **(Richardson 1846)**

The fish is bred in fish farms. Some individuals were found in the Akhurian and Araks rivers (Pipoyan et al. 2018).

25. ***Rhodeus amarus*** **(Bloch 1782)**

Widespread in all reservoirs of the Ararat Valley. Some individuals were found in the Metsamor and Araks rivers and related lakes. Reproduce in portions; spawning period lasts from late March to September (Gabrielyan 2001; Pipoyan 2012). It is an object of recreational fishing. The species is included into Resolution 6 to Bern Convention (*as Rhodeus sericeus amarus*).

26. ***Rutilus rutilus schelkovnikovi*** **(Derjavin, 1926)**

Decreasing population. Locally distributed, endemic subspecies. Found in the Sevdgur river basin, a small tributary of Araks River and Lake Aygerlich. Reproduce in portions from March–April. The fatness of this species is the greatest among Armenia's fish. Body length reaches 15 cm (Gabrielyan 2001). The main factors of the population reduction are river flow regulation, pollution, disturbance of natural reproduction, poaching and effect of invasive fishes. Listed in the Red Book of Armenia (2010). According to IUCN criteria categorized as Endangered (EN B1ab(iii)+2ab(III)).

27. ***Capoeta capoeta sevangi*** **(De Filippi 1865)**

Population-reducing endemic subspecies in Lake Sevan. Spawn from late May to early August in Lake Sevan and its tributaries. Due to lowering of the lake's water level, spawning grounds in the lake dried up. The fish is poached in rivers. Females mature by the 6th year, males by 2–3 years. Body length reaches 42 cm (Fig. 5.63).

Fig. 5.63 *Capoeta capoeta sevangi*

This fish prefers to feed on detritus. Juveniles (3 years and below) feed on zooplankton and zoobenthos (Gabrielyan 2010a, b). Before population reduction, the fish was an important object of commercial fish. According to IUCN criteria categorized as Vulnerable (VU A 1 cd). Listed in the Red Book of Armenia (2010).

28. ***Capoeta capoeta* (Güldenstädt 1772)**

Found in almost all reservoirs of Armenia and the Arpa, Debed and Vorotan rivers. The biology of this species in the water bodies of Armenia is not sufficiently studied. Body length reaches 30 cm (Gabrielyan 2001). It is an object of recreational fishing. The species is not protected.

29. ***Abramis brama* (Linnaeus, 1758)**

Invasive species and poorly studied in Armenia. Some individuals were found in the Araks and Metsamor rivers. According to Rubenyan and Rubenyan (2003), local populations exist in some reservoirs of the Ararat Valley and the Debed river. Assumingly, they were established during annual migrations of semi-anadromous populations to the Kura and Araks rivers. Body length reaches 40 cm. It is a common species for recreational fishing. The species is not protected.

Family 4. COBITIDAE

Only one species is present in Armenia.

30. ***Sabanejewia aurata* (De Filippi 1863)**

Endangered, poorly studied species. Some individuals were found in all rivers of Armenia. Body length reaches 10 cm (Gabrielyan 2001). The factors affecting the population on the territory of Armenia are not known. The fish is not important for both commercial and recreational fishing. According to IUCN criteria categorized as Data Deficient (DD). Listed in the Red Book of Armenia (2010), as well is included into Resolution 6 to Bern Convention.

Family 5. GOBIONIDAE

31. ***Pseudorasbora parva* (Temminck and Schlegel 1846)**.

Introduced into the ponds of the Ararat Valley in the 1960s and shortly spread to all reservoirs and rivers of Armenia. Currently, it is also found in Lake Sevan, where it is considered as harmful to local ichthyofauna due to being a food competitor with young fish of local endemics. Small non-commercial fish, not of interest even for recreational fishing. The fish is not protected.

32. ***Gobio gobio* (Linnaeus, 1758)**

Inhabit the rivers Araks, Metsamor, Hrazdan, Debed, Tashir, canals of Armavir region and the ponds of Armash and Masis fish farms (Pipoyan 2012). Can reach 15 cm in length. Non-commercial, widespread species, object of recreational fishing. The species is sensitive to water pollution but does not need special protection measures. Taxonomic status on the territory of Armenia needs further clarification.

Family 6. BALITORIDAE

33. ***Oxynoemacheilus angorae* (Steindachner 1897)**

Inhabit all rivers. Body length reaches 8 cm. It represents a danger to valuable fish species by feeding on their eggs and destroying their spawning grounds (Gabrielyan 2001; Pipoyan et al. 2018). The species is not protected.

34. ***Oxynoemacheilus brandtii* (Steindachner 1897)**

Poorly studied on the territory of Armenia, non-commercial species. Inhabit the Aghstev, Azat, Arpa, Debed, Dzoraget, Hrazdan rivers. Small fish up to 10 cm (Pipoyan, 2012). The fish is not protected.

ORDER III. CYPRINODONTIFORMES

Family 7. POECILIIDAE

35. ***Gambusia holbrooki* (Girard 1859)**

Introduced into Armenia for mosquito control. Feed on mosquito larvae. Body length reaches 6 cm (Gabrielyan 2001). Small non-commercial fish, not of interest even for recreational fishing. The species is not protected.

ORDER IV. SILURIFORMES

Family 8. SILURIDAE

36. ***Silurus glanis* (Linnaeus 1758)**

Found in the Akhuryan, Arpa, Metsamor, Vorotan rivers and the Akhuryan reservoir. Predator fish, which can reach up to 150 cm body length in Armenia (Gabrielyan 2001). It is an object of recreational fishing. The species is not protected.

ORDER V. PERCIFORMES

Family 9. PERCIDAE

37. ***Perca fluviatilis*** **(Linnaeus, 1758)**

Invasive species, found on the territory of Armenia recent years. Adults are predators feeding on other fish species, which makes them undesirable for water bodies of Armenia. Currently, some individuals have been observed in the Hrazdan river and in the water bodies of the Ararat Valley. In Armenia, the biology and ecology of this species is not studied. Body length reaches 13–15 cm (Kuljanishvili et al. 2020). It is an object of recreational fishing. The species is not protected.

ORDER VI. GOBIIFORMES

Family 10. GOBIIDAE

38. ***Knipowitschia caucasica*** **(Berg 1916)**.

Introduced and widespread in the Metsamor river and the reservoirs of the Ararat Valley (Gabrielyan 2001). Later was also introduced into other reservoirs of Armenia. It is an object of recreational fishing. The species is not protected.

39. ***Neogobius fluviatilis*** **(Pallas 1814)**

Introduced and widespread in almost all reservoirs and fish ponds of the Ararat Valley. Later was also introduced into other reservoirs of Armenia (Gabrielyan 2001). It is an object of recreational fishing. The species is not protected.

5.2.2 Amphibia and Reptilia of Armenia

Marine Arakelyan

5.2.2.1 Amphibia

A checklist of the Amphibia of Armenia contains only seven species belonging to four families and six genera (Arakelyan et al. 2011a, b). Mountain cold and dry climate of the country, scarcity of wetlands are the reasons for the small number of suitable habitats for amphibians, which include four frog species, two toad species and one newt species. Among amphibians, the most common in Armenia species is Mash frog, which occurs close to various freshwater habitats in pristine and urbanized area. The next species with high probability of meeting in Armenia is green toad. Prefers open landscapes with moderately dry climate on different altitudes and habitat types. The Caucasian brown frog occurs in all part of Armenia on high elevation from 1200 up to 3000 m above sea level. The ranges of the two tree frog species are divided into the northern part of Armenia for *Hyla orientalis* and the southern part for *H. savigny*. Syntopy between the two species has been reported, but no hybrids are recognized (Stepanyan 2021). The southern and northern of tree frog species differ by preferred habitats, where the first one prefers riparian forest, open deciduous forest close to ponds, while the second one lives in drier habitats including dry steppe and semidesert. Two species of Armenian amphibians are included into the Armenian Red Book: newt *Ommatotriton ophryticus* (estimated as CR) and spadefoot toad *Pelobates syriacus* (VU) (Aghasyan and Kalashyan 2010). Newt was first recorded in Armenia in 1990th (Danielyan et al. 1998) in few ponds in northern areas, and their habitats are continuously degraded under climate change impact (Aghasyan et al. 2009). Spadefoot toad has few locations in Ararat Valley and Southern Armenia. The main threats to this species are high pressure of human impact on habitats which are mainly located around the agricultural landscapes. No endemic for Armenia or Caucasus species are in Armenia.

List of amphibians of Armenia

1. *Ommatotriton ophryticus* (CR)
2. *Pelobates syriacus* (VU)
3. *Bufotes viridis*
4. *Hyla orientalis*
5. *Hyla savignyi*
6. *Pelophylax ridibundus*
7. *Rana macrocnemis*

5.2.2.2 Reptilia

The fauna of Armenia is especially rich by reptiles. Armenia is home to 51 different species of turtles, lizards and snakes (Arakelyan et al. 2011a, b). A large variety of climate and habitats, suitable for reptiles occurs in Armenia. Armenian highlands are rich in terms of species due in large part of hybrid speciation in groups of rock lizards of genus *Darevskia*. This group is particularly interesting because of the presence of parthenogenetic lizard species whose populations consist only of females.

Turtles

The spur-thighed tortoise *Testudo graeca* is widespread throughout the Caucasus region, including Armenia and surrounding territories (Chernov 1939; Darevsky 1957). *Testudo graeca* is listed in the Red Data Book of Armenia, which provides them with protection as threatened species (VU). A study on the geographic distribution of Armenia and NKR turtles (Arakelyan and Parham 2018) shows that *T. graeca* distribution is defined by elevation, usually below 1500 m. Therefore, this species is restricted to the two lowland areas of the Kura in the northern part of Armenia and Arax River drainages in the south. Two very different looking forms of

T. graeca are known from Kura and Arax river drainages. The populations in northeast Armenia (Kura drainage) are very close to the type locality of *T. g. ibera* Pallas, 1814, and referred to this taxon. The southern (Arax) populations of *T. graeca* have been referred to taxon *T. g. armeniaca* Chkhikvadze and Bakradze 1991, which is sometimes elevated to full species status. The tortoises from Arax river valley are the most distinct race of *Testudo graeca* complex found anywhere. *T. g. armeniaca* appears to be highly adapted for a burrowing behavior, has a low-domed shell and a rigid plastron, that distinguish the taxon *armeniaca* from all other *T. graeca* (Arakelyan et al., 2018). The typical habitat of this form is arid regions in semidesert and dry steppe zones. The conservation status of *T. g. armeniaca* is CR according to Red Book of Armenia (2010) and IUCN. In Armenia, the two tortoise areas are disjunctive. *T. g. ibera* also usually found in arid zone with xerophilous vegetation, however in northern Armenia it inhabits mesophilic forests and grasslands. In this area, *T. g. ibera* inhabits rocky hillsides and flat peaks of oak light forest and fringe of the oak-hornbeam forest at an elevation of up to 950 m. We found *T. graeca* in agricultural fields, gardens, and even a cemetery. However, these populations are not allopatric since they are connected through that occupy the area between to the Kura and Arax drainages in Azerbaijan and Nagorno-Karabakh. Speciation in the genus *Testudo* are still highly debated (Parham et al. 2006; Fritz et al. 2007). In this respect, the Nagorno-Karabakh populations of *T. graeca* are especially interesting since they occupy the Kura-Arax lowland where two different taxa apparently come into contact and hybrids are reported.

Two species of freshwater turtles are found on the territory of Armenia: the pond turtle *Emys orbicularis* (Emydidae) and the Caspian turtle *Mauremys caspica* (Geoemydidae). The pond turtle is listed in the Red Book of Armenia as a rare and narrow-range species. In Armenia, *E. orbicularis* is extremely rare, unlike the Caspian tortoise, which is widespread in the north and south of Armenia (Arakelyan et al. 2011a, b). A distinctive external sign of the marsh turtle is yellow spots on the neck, legs and carapace, while the Caspian turtle has longitudinal yellow stripes on the head, neck and legs, a notch on the back of the plastron and no mobility of the plastron.

The pond turtle in northern Armenia inhabits small canals and swampy areas in the floodplain of the Aghstev River. According to our data (Arakelyan et al. 2011a, b), the main limiting factor in the distribution of the pond turtle in the territory of Armenia is the altitude of their habitat up to 1000 m above sea level (Arakelyan and Parham 2018). In general, the bog turtle species has a complex structure. According to Fritz, the pond turtle in northern Armenia is assigned to the subspecies *Emys orbicularis iberica* Eichwald, 1831 (Fritz et al. 2007).

Mauremys caspica is common in rivers that flow through semidesert habitats (Arakelyan et al. 2011a, b), such as the Arax River and the lower reaches of the Akhuryan River. These observations were echoed by Darevsky (1957), who also noted that, in southern Armenia, *M. caspica* occupies a variety of riparian and stagnant aquatic habitats and are very often found in irrigation canals of Ararat Valley (Arakelyan and Parham 2008). In some areas, *M. caspica* occupies mountain springs where they settle in small ponds or pools. In periods of drought, these turtles can be completely isolated from the main drainage. According to our data, *M. caspica* occurs at an altitude up to 1800 m.

List of turtles of Armenia

1. *Emys orbicularis*
2. *Mauremys caspica*
3. *Testudo graeca* (VU)

Lizards

Twenty-five species of lizards belong to five families: Agamidae (2 species), Anguidae (2), Scincidae (4), Gekkonidae (1), Lacertidae (16) (Arakelyan et al. 2011a, b).

For the first time a discovery of parthenogenetic mode of reproduction in vertebrates was made in Armenia in a group of rock lizards (Darevsky 1957), which opens a number of general biological questions. The most characteristic ecological feature of rock lizards of genus *Darevskia* is their habitat on various kinds of rocks and outcrops of rocks along the roads, mountain steppe, as well as on all kinds of stone buildings and walls (Darevsky 1967). Living within the same mountainous country at the same time several species of rock lizards inhabiting similar habitats is largely associated with the specifics of the vertical distribution of individual forms. The general distribution of rock lizards has a well-defined altitude character. The ranges of different species located at different levels may overlap to some extent, which is a specific feature of the distribution of rock lizards throughout the Caucasus. The sensitivity of different forms of rock lizards to a certain humidity regime also affects the zoning of the distribution of the same species on the northern and southern slopes of the ridges (Darevsky 1967). In general, each species of rock lizards in Armenia is characterized by its specific biotopes, where this form reaches the greatest abundance. There are four parthenogenetic species of lizards in Armenia. Of these species, *Darevskia armeniaca* is the most widespread, whose range covers the Armenian Highlands within northern Armenia, northeastern Turkey, southern Georgia, and northwestern Azerbaijan. Inhabits rock outcrops, heaps of large stones, stony placers and slopes of gorges in the forest and mountain-steppe zones at an altitude of 1400–2000 m a.s.l. m. The range of the species *D. unisexualis* is mainly represented by small populations isolated from each other in Central Armenia. The boundaries of the vertical distribution of this species lie within 1700–2100 m a.s.l. Inhabits rock outcrops, heaps of large stones, rocky slopes in the mountain-steppe zone. *D. dahli* occurs mainly in the inner part of the Armenian Highlands within Northern Armenia, Southern Georgia, at an altitude of 900–1800 m a.s.l. Inhabits moderately dry slopes of gorges, outcrops of rocks in the forest zone, in some places also penetrating into rocky areas of the mountain steppe. *D. rostombekowi* is distributed in the Armenian Highlands in Northern Armenia and Northwestern Azerbaijan. A small isolated population has survived on the eastern shore of Lake Sevan. Lives on dry and moderately dry rocks and outcrops of sedimentary rocks in the forest and, less often, mountain-steppe zones.

D. raddei, sexual species and it is maternal form for *D. unisexualis* and *D. rostombekowi*, is widely distributed within southern Armenia, southwestern, western, and southeastern Azerbaijan, and adjacent regions of northwestern Iran. It occurs along river banks, on various outcrops of hard parent rocks, among heaps of large fragments of stones and stony placers both in the mountain-forest and mountain-steppe zones at an altitude of 300–1000 m above sea level. It has two subspecies: nominal and *D.r.narirensis*. *D. valentini* is paternal species for *D.unisexualis* and *D.armeniaca*, and found in Central and Northwestern Armenia, Southern Georgia, Upper Adjara, and Eastern and Southeastern Turkey. The general range of the species is broken into several isolated areas of different sizes. Adheres to rock outcrops, individual stone blocks and heaps of stones in the mountain-steppe and mountain-meadow zones at an altitude of 1800–3000 m above sea level. It is also found among moderately dense herbaceous vegetation on open slopes away from stones and rocks. It uses crevices in rocks, voids under rocks, and rodent burrows as shelters. *D. portschinskii* is paternal species for *D. dahli*, distributed mainly within the right-bank valley of the middle reaches of the Kura River, from where it penetrates into the territory of Northern Armenia along the gorges of the Debet and Agstev rivers, forming several isolated populations here. Inhabits dry and moderately dry rocks and stony cliffs along river banks and on slopes. The boundaries of vertical distribution are within 300–1400 m above sea level (Arakelyan et al. 2011a, b).

In area of overlaps between parthenogenetic and sexual species mainly the triploid and even tetraploid hybrids are originated as a result of interspecific hybridization. The process of microevolution in framework of reticulate or hybridogeneous speciation is ongoing in group of this outstanding reptiles (Arakelyan et al. 2022). Rock lizards currently continue to attract the attention of researchers due to the presence of natural parthenogenetic reproduction and polyploidy in some of them.

Three species of *Lacerta: L. agilis, L. media* and *L. strigata*, are found almost throughout Armenia, where their ranges often overlap. The Armenian subspecies *L. agilis* belongs to the subspecies *L. a. brevicaudata* Peters, 1958, *L. media* and *L. strigata* are nominative subspecies (Arakelyan et al. 2011a, b).

Among three species of *Eremias*, *Eremias a. transcaucasica* was described for Armenia by Darevsky in1953 (Darevsky, 1957) Now it is survived only in one population in mountain steppe near to Sevan Lake. It listed in the Red Book of Armenia as CR. The range globally threatened species (CR in IUCN) *Eremias pleskei* is restricted to the Ararat Plain. In Armenia, currently it is known only from desert part of Gorovan Sand Sanctuary. *Eremias strauchi* is not threatened and found in dry steppe and smidesert zones of Araks River valleys and foothills.

Parvilacerta parva inhabits stony sites in xerophytic mountain steppe at elevations ranging from 700 to 2300 m mainly in northern parts of Armenia. It is rare species for Armenia and listed in the Red Book of Armenia (2010) as CR.

Snake-eyed lizard, *Ophisops elegans,* is widely spread in semidesert zone of Armenia, inhabiting different landscapes. The nominative subspecies lives in Armenia.

Among Scincus, *Ablepharus bivittatus* usually occurs on dry, rocky slopes covered by traganth (*Astragalus, Onobrychis, Acantholimon*) or other xerophyte vegetation between 600 and 2400 m both in northern and southern parts of Armenia. *Ab. chernovi* was first time described in Armenia, it is extremely rare and close to extinction; indeed, since its discovery in 1953, only six individuals have been found in the country (Szczerbak 2003; Arakelyan et al. 2019). *Eumeces schneideri princeps* is widespread mainly along the Araks River valley and surrounding mountain slopes and valleys. It also occurs in a few areas in North Armenia in the Kura River basin (Tavush Province). This species has a limited distribution and is generally considered endangered. Similar, in Armenia, *Trachylepis s. transcaucasica* is mainly found on mountain slopes and valleys facing the Araks River valley (Arakelyan et al. 2011a, b).

Two species are in family Agamidae. *Laudiacia caiscia* is common and widespread. It is rock-dwelling species found in various habitats at elevations ranging from 400 to 2500 m, it is most often seen on rocky slopes covered by xerophytes as well as on rocky slopes bordering mountain roads, among stony heaps. The next species is extremely rare and listed in Armenian Red book and IUCN as CR species. *Phrynocephalus persicus horvathi* is endemic to the middle part of Araks River valley. Currently, the species has disappeared from most of the localities from which formerly it had been reported, and it is now threatened in the few places where it is still present, in large part because of habitat loss resulting from cultivation of the semidesert as well as urbanization.

Legless two species of lizards are in Armenia. *Anguis colchica* – slow worm inhabits edges and glades of deciduous and mixed forests. It also is found in secondary bush associations or in tall grass, as well as in forest-steppe zones in north of country on altitudinal range: 766–2706 m (Darevsky 1957), while *P.apodus* species inhabits a wide variety of habitats between 350 and 2388 m mostly in southern parts (Darevsky 1957). It occurs in gardens, vineyards, open forests, thickets, bushlands, and grasslands.

One species of gecko *Cyrtopodion caspium* introduced species in Armenia, probably from Azerbaijan or North Iran, possibly passively transported by train or automobile transport; naturalized populations and currenly distributed mostly in cities: Meghri Armavir, and Yerevan (Arakelyan et al. 2011a, b).

List of lizards of Armenia

1. *Cyrtopodion caspium*
2. *Laudakia caucasia*
3. *Phrynocephalus persicus* (CR)
4. *Anguis colchica*
5. *Pseudopus apodus*
6. *Darevskia armeniaca*
7. *Darevskia dahli* (EN)
8. *Darevskia portschinskii*
9. *Darevskia praticola* (VU)
10. *Darevskia raddei*

11. *Darevskia rostombekowi* (EN)
12. *Darevskia unisexualis* (VU)
13. *Darevskia valentini*
14. *Eremias arguta* (CR)
15. *Eremias pleskei* (CR)
16. *Eremias strauchi*
17. *Lacerta agilis*
18. *Lacerta media*
19. *Lacerta strigata*
20. *Ophisops elegans*
21. *Parvilacerta parva (CR)*
22. *Ablepharus bivittatus*
23. *Ablepharus chernovi* (CR)
24. *Eumeces schneideri (VU)*
25. *Trachylepis septemtaeniata (VU)*

Snakes

Twent-two species of snakes belong to four families: Typhlopidae (1 species), Biodae (1), Colubridae (16) and Viperidae (4). Among 22 species of snakes, only one family Viperidae with four species are venomous (Arakelyan et al. 2011a, b).

Different species of vipers are spread on different altitudes. The lowland of Ararat valley, and along Arax river as well as the semidesert and steppe zone in north are occupied by *Macrovipera lebetina.* In the southern part of dry mountain steppe inhabited by Armenian viper (*Montivipera raddei*). It is included in IUCN and Red Book of Armenia as VU. The mountain steppe of southern and central parts of Armenia is habitat for *Vipera eriwanensis*. Only on few point in the most northern part of Armenia on elevation more than 2900 are met *Vipera darevskii*, which is rare and included in IUCN and Red Book of Armenia as CR.

Family Calubridae is most diverse. Among them, the common species are *Natrix tessellata* and *N. natrix,* which are widespread along Armenia. Two subspecies of *N. natrix* found in Armenia: *N .n. scutata* and *N. n. persa*. The last subspecies is the most common (Arakelyan et al. 2011a, b).

Coronella austriaca is widespread and common snake in Armenia. It is usually found in mountain forest edges, stony slopes with grasses and bushes. *Dolichophis schmidti* prefers semidesert zone, rocky slopes with xerophytic vegetation, and river banks, bush thickets, and reeds. It is also often found among ruins, and in gardens and vineyards. In some places in Southern Armenia, *Eirenis modestus* is syntopic with *E. punctatolineatus* and *E. collaris*. Among these three species *Eirenis collaris* is common. All of them prefer dry, stony sites or stony slopes covered with xerophytic grass and bushes. In Armenia, *E. persicus* is an extremely rare species, reaching the northwestern limit of its range in the valley of the Araks River.

Elaphe urarticus mainly found in the Araks Valley foothills in Central, South Armenia and North Armenia (Tavush Province) close to the Georgian border.

Inhabits rocky sites overgrown with bushes and grasses, on slopes and in mountain steppe and meadows. In Armenia, *Hemorrhois nummifer* occurs only in valleys and foothills of the Araks River valley, while *Hemorrrhois raviergiery* is widespread in various habitats. *Malpolon monspessulanus* is one of the most thermophilous snakes found in Armenia; typical habitats include stony semideserts characterized by fragmented rocks as well as sandy soils with scarce vegetation (vicinity of Gorovan, Ararat Province), dry open forests (Khosrov Reserve), and occasionally in gardens and vineyards mainly in the Ararat Valley and along the Araks River in South Armenia (Syunik Province), it has also been reported from one locality in Aragatsotn Province. *Platyceps najadum* known from the Araks and Kura River valleys. *Rhynchocalamus melanocephalus satunini* occurs in Armenia (Chernov 1939; Darevsky 1970). It is rare species and known from a few localities on mountain slopes bordering the Araks Valley in Ararat and Syunik Provinces. *Telescopus fallax* is also rare species and mainly found in dry zones of Central and South Armenia along the Araks River Valley and neighbouring slopes and valleys. *Zamenis hohenackeri* is mainly found on foothills in the Araks River Valley in Central and South Armenia and it also occurs in North Armenia in Tavush Province (Arakelyan et al. 2011a, b).

One species of Typhlopidae family, *Xerotyphlops vermicularis* inhabits rocky slopes with xerophilous vegetation in the semidesert zone. The worm snakes spend most of their time underground. Also one species have in family of Biodae, *Eryx jaculus* may found along the Araks River valley and surrounding mountain slopes and valleys in Armenia.

List of snakes of Armenia

1. *Xerotyphlops vermicularis*
2. *Eryx jaculus*
3. *Coronella austriaca*
4. *Dolichophis schmidti*
5. *Eirenis punctatolineatus*
6. *Eirenis collaris*
7. *Eirenis modestus*
8. *Eirenis persicus (CR, as Pseudocychlophis)*
9. *Elaphe urartica*
10. *Hemorrhois nummifer*
11. *Hemorrhois ravergieri*
12. *Malpolon monspessulanus*
13. *Natrix natrix*
14. *Natrix tessellata*
15. *Platyceps najadum*
16. *Rhynchocalamus melanocephalus (VU)*
17. *Telescopus fallax (VU)*
18. *Zamenis hohenackeri (VU)*
19. *Macrovipera lebetina*
20. *Montivipera raddei* (VU)

21. *Vipera darevskii* (CR)
22. *Vipera eriwanensis* (VU)

5.2.2.3 Geographical Analysis of Armenian Amphibia and Reptilia

Armenia belongs to Palearctic herpetogeographical division and arid Mediterranean-Asiatic subregion and with predominance of arid territories and rich herpetofauna (Szczerbak 2003). In Mediterranean province and Caucasian subprovince, which cover the whole territories Armenia three districts are occur: Caucasian – Mountain Wooded district (indicator species are *Rana macrocnemis, Darevskia portschinskii, D. valentini, D. armeniaca, D. dahli, D. rostombekowi*); Caucasian Alpine and Subalpine district (indicator species are *Darevskia unisexualis, Parvilacerta parva* and *Vipera darevskia*) and the Eastern-Caucasian Desert Steppe district which contains many Turanian species (indicators are *Phrynocephalus persicus, Ablepharus chernovi, Eremias strauchi, E. arguta transcaucsica, E. pleskei, Darevskia raddei, Hemorrhois ravergieri, Zamenis urarticus, Vipera eriwanensis, Montivipera raddei*).

The entire territory of Armenia can be considered as a vast interfluve of the Araks-Kura, where Arax is the southern border of Armenia. The southern part of Armenia are greatly differ by herp species from northern and eastern. It has a great impact of Iranian group of species with preferring of xerophyte habitats. The southern and northern parts of the herpetofauna are separated by western slopes of Aragats mountains up to Zangezur range. The high occurrence of European mesophyll species are found in north part of Armenia, where climate is more cold and humid.

Due to the special location of Armenia at the junction of two floristic provinces and four zoogeographic ones, here there is a unique opportunity to combine various forms as a result of secondary intergradation and, as a result, favourable conditions for the of hybrid zones. Thus, the formation of hybrid zones is possible when geographic ranges (for allopatric populations) or ecological niches (for sympatric populations) overlap. However, studies of the hybrid zones of Armenia and adjacent territories can also contribute to research devoted to the study of the ways of formation and development of the modern vertebrate fauna of Armenia and the restoration of their biogeography.

The valleys of the Araks and Kura rivers are not only one of the important centres of endemism, but also one of the largest centres of contact between various zoogeographic, phytogeographic and geological regions in Asia (Darevsky 1957, 1967; Tuniyev 1995; Chkhikvadze and Bakradze 1991). The valley of the Araks River, passing in the arid and subarid regions of the three highlands of Western Asia (Asia Minor, Armenian and Iranian), is characterized by rich and unique flora and fauna. One of the most active centres of speciation in all of Western Asia is located here. Due to the commonality of formation and the special historical path of development of the regions, the endemism noted above for the flora fully applies to the fauna as well. Even Satunin (1916) notes that the “zoogeographical areas” identified by them almost completely correspond to phytogeographical areas. Similarly, reptiles also have a lower dispersal capacity than birds and mammals, so this circumstance better provides a certain isolation of species populations and promotes the effects of gene flow.

The distribution of the fauna and its origin can be judged from the relationships of the constituent elements with the faunas of adjacent territories (Darevsky 1957). Many species of reptiles in Armenia have extensive ranges, covering several natural zones with different landscapes, in which they occupy a variety of biotopes. We found data on the zoogeographic distribution of reptiles in Armenia in the works of Darevsky (1956) and Tuniyev (1995).

Summarizing the literature data, we identified five eco-geographical groups of Armenian amphibians and reptiles.

CAUCASIAN Group

Amphibia: *Rana macrocnemis; Hyla orientalis*;

Lizards: *Lacerta agilis brevicaudata, Parvilacerta parva*, 4 parthenogenetic and 5 bisexual species of *Darevskia* (*D. unisexualis, D. armeniaca, D. dahli, D. rostombekowi, D. portschinskii, D.valentini D. raddei, D. nairensis, D. praticola*);

Snakes: *Vipera darevskii, V. eriwanensis, Montivipera raddei.*

MEDITERRANEAN Group

Amphibia*: Bufotes viridis, Pelobates syriacus, Pelophylax ridibundus;*

Turtles*: Mauremys caspica, Testudo graeca;*

Lizards*: Pseudopus apodus, Ophisops elegans, Ablepharus chernovi, Lacerta strigata, L. media;*

Snakes*: Xerotyphlops vermicularis, Eryx jaculus, Natrix natrix, N. tessellata, Zamenis hohenackeri Platyceps najadum, Dolichophis schmidti, Eirenis modestus, Telescopus fallax, Malpolon monspessulanus, Macrovipera lebetina.*

IRANIAN Group

Lizards: *Laudakia caucasia, Trapelus lessonae, Trachylepis septemtaeniata, Eremias pleskey, E. strauchi, Phrynocephalus persica, Eumeces shneidery, Ablepharus bivittatus*;

Snakes: *Hemorrhois nummifer, H. ravergieri, Eirenis collaris, E. punctatolineatus, Rhynchocalamus melanocephalus satunini.*

It is large group of species in Armenia and most of them inhabit only the southern parts of the country in Arax River Valley, but some of them penetrate to northern part along the Kura River Valley like (*Macrovipera lebetina* and *Hemorrhois raviergiery*). All of them prefer dry habitats like steppe and semidesert.

ASIA MINOR Group

Lizards*: Parvilacerta parva, Lacerta strigata*

Snakes: *Eirenis modestus, Zamenis hohenackeri, Elaphe urartica.*

This group of animals penetrate to Armenia from Balkans and eastern parts of areas of Mediterranean Sea. In contrast to Iranian group, they escape from the semidesert and very dry habitats and may live in northern parts with exception of

Parvilacerta parva, for which Armenia is the northern limit of their distribution.

EUROPEAN Group
Turtles*: Emys orbicularis*
Lizards*: Anguis colchica*
Snakes*: Coronella austriaca, Natrix natrix scutata*
This is small, but very distinctive group of reptiles, which are living mainly in northern part of Armenia and belong to mesophyllic species with typical habitats in broadleaf forest and steppes.

TURANIAN Group
Lizards*: Cyrtopodion caspius, Phrynocephalus persicus, Eremias arguta*
Snakes: *Eirenis persicus.*

These four species have origin from Middle East and prefer sandy, dry habitats located in shore of Sevan Lake and southern parts of Armenia.

The most of the reptiles are of Mediterranean origin and are mainly associated with the valleys of the Araks and Kura rivers, where they inhabit arid ecosystems. Mediterranean species in the Caucasus are mainly autochthonous, ancient and relic. In Armenia, most species of Iranian origin are confined to the semidesert zone of the Araks River valley, where they live on the extreme northern border of their ranges. At the same time, some of them are also found in the north, where they penetrate bypassing along the valley of the Kura river. There is no doubt that in the past the valleys of the Araks and Kura rivers were wide migratory routes for fauna penetrating Transcaucasia from the south (Darevsky 1957; Arakelyan et al. 2011a, b). In particular, many species of reptiles of Iranian and Eastern Mediterranean origin, spreading along the Kura valley and further along the valleys of its tributaries, originating from the ridges of the Lesser Caucasus, penetrated into northern Armenia, regardless of their existence in the Araks valley in the south. Thus, during the period of repeated transgressions of the sea, some species of reptiles (*Testudo graeca, Laudakia caucasia, Eumeces shneideri, Eirenis collaris, Eryx jaculus, Platyceps najadum, Hemorrhois ravergieri*) could survive along foothills of the Lesser Caucasus in northern Armenia. This circumstance allows us to assume the presence of different genetic linages in the north and in the south in various species of reptiles in Armenia and adjacent territories. Consequently, in the places of their contacts (e.g. in the territory of the Kura-Arax lowland), zones of hybridization are possible, where hybridogenic speciation is not excluded, and can proceed according to the following scheme. The initial population breaks up into isolated small populations, each of which has its own evolutionary destiny. Each population adapts to its habitat conditions independent of other populations of the same species. Individuals from different populations get the opportunity to interbreed with each other when the primary isolating barriers are removed. If in primary isolated populations there was no divergence in traits that determine reproductive isolation, then when primary isolation is eliminated, a simple merger of populations occurs. If the divergence of populations in many traits has gone so far that, according to most criteria, they can be considered different species, then secondary intergradation can be considered as the process of formation of a new species by hybridogenesis. The further fate of hybrid individuals will depend either on the ability to pass to reticular evolution or on the viability of hybrids.

European species of reptiles are mainly associated with the mountain-steppe and mountain-forest zones and do not descend into the valley of the Araks River. Turanian species could penetrate through the Caspian Isthmus into the valleys of the Kura and Araks rivers. The species of this group are generally psammophytes, living in sandy deserts.

The group of autochthonous species of the Lesser Caucasus includes endemic species and subspecies. According to Satunin (1916), all endemics of Transcaucasia undoubtedly have a southern origin. The center of formation of a number of autochthonous species of Caucasian reptiles was the Armenian Highlands. According to Darevsky (1956), the center of this territory lies in a triangle between Lake Sevan in the north and lakes Van and Urmia in the west and southeast. It was here that such species as *Montivipera raddei, Ablepharus chernovi, Darevskia brandtii, D. chlorogaster, Vipera kaznakovi, Telescopus fallax iberus* appeared.

Some elements of the herpetofauna of the Lesser Caucasus, such as the group of rock lizards *Darevskia, Ablepharus chernovi, Montivipera raddei*, are neoendemics, the occurrence of which is associated with the formation of mountain landscapes.

Thus, the zoogeographic analysis showed that the fauna of herps of the Armenian Highlands and the Lesser Caucasus is represented by species that have a different origin, and most of them are located on the border of their distribution area. Moreover, it was noted that at the border of their range, reptiles do not become stenotopic, but, on the contrary, penetrate into neighboring zones (Darevsky 1957). Such, for example, typically semidesert species as the *Ophisops elegans* and *Eremias shtrauchi*, rising to the mountains, at the border of their range in the foothills of the Araks River valley, penetrate into the mountain steppe, into juniper open forests and are found far in the forest (Arakelyan et al. 2011a, b). The trend towards active dispersal of species leads to an attempt to develop new biotopes and, consequently, to a wide eurytopy of species at the border of their ranges (Darevsky 1957). This distribution of animals creates conditions for the meeting of different close forms, which also contributes to the formation of hybrid zones.

The paleontological data for a zoogeographical analysis of the modern distribution of the fauna in the valleys of the Araks and Kura rivers is scarce. If we consider the geological past of the territory of interest, we can distinguish the following stages. The land between the Kura-Araks rivers began to form at the end of the Lower Oligocene. The modern fauna of amphibians of Armenia began to form from the late Miocene, which was due to the beginning of the terrestrial development of the modern territory of the country (Vasilyan 2009). It has also been established that the probable penetration of the ancestral forms of vertebrates of the modern fauna into the territory of Armenia occurred from the southern and southeastern parts of the adjacent territories (Darevsky 1957). Representatives of many modern genera of lizards and snakes entered the Caucasus from the south already from the middle of the Tertiary period. In the middle of the Miocene, some xerophile species from the

Mediterranean, such as the Mediterranean tortoise, penetrate into Transcaucasia. By the end of the Middle Miocene, the territories of Southeastern Anatolia and Central Armenia are rising and the sea waters are receding. At the beginning of the Late Miocene (Sarmatian), shallow reservoirs with brackish (approaching freshwater) waters appeared in Central Armenia, where, apparently, from time to time, the sea waters of the Kura Bay penetrated (Vasilyan 2009). Later, in the late Miocene, these reservoirs disappear. The land connection between the Caucasus and the south and southeast, established since the Miocene, has never been interrupted again, which led to the unhindered penetration of the Near East faunistic elements into the Caucasian peninsula. Together with changes in the flora, the xerophytic genera as *Laudakia, Eumeces, Eremias, Trachilepis, Xerotyphlops, Eryx*, and *Eirenis* penetrate into Transcaucasia. The northern continental connection is established when the Caspian first separated from the Black Sea basin, which contributed to the settlement of the Caucasus in the Pliocene by migrants from the north. It is possible that at this time some forest species, such as *Anguis colchica*, penetrated into the Caucasus. In the opposite direction, the meadow lizard *Darevskia praticola* migrated to the Balkans from the Caucasus. The main core of the herpetofauna of Transcaucasia had already been formed by the end of the Pliocene. The modern picture of the distribution of a number of reptile species in Armenia was formed under the influence of volcanic processes experienced by its territories in the recent geological past, which, in particular, is due to a gap in the distribution area of some species (Darevsky 1957). In the Pleistocene, the Caucasus was subjected to glaciation twice, although there were refugia for the fauna of the northern and middle regions of the Russian Plain, pushed back by the glacier and migrating south. Extensive areas of refugia cover the foothills and the middle belt of the mountains of Eastern Transcaucasia, located in a semicircle around the Kura-Arks lowland with areas extending to the foothills of Talysh, as well as to the south to the southern slopes of the Lesser Caucasus. Refugia are also located on the Armenian Highlands along the banks of the river Araks (Tuniyev 1995). During the epoch of glaciations, such species as *Emys orbicularis, Lacerta agilis, Coronella austriaca* and some others penetrate the territories occupied by refugia from the north. From the south, in addition to the ancient settlement route of the Turanian elements bypassing the Caspian Sea, their penetration to the west from the deserts of Central Asia could be carried out along the so-called Caspian Isthmus. Such species as *Cyrtopodion caspium, Phrynocephalus persicus, Eremias arguta* could settle along the Caspian Isthmus from the south.

On the territory of Armenia, among amphibians, possible speciation due to hybridization can occur in various forms of the marsh frog *Pelophylax ridibundus*, tree frogs *Hyla savignyi* and *H. orientalis*. In the terrestrial tortoises *Testudo graeca* and the freshwater tortoises *Emys orbicularis*, we have for the first time found zones of hybridization. It has been repeatedly noted that a specific feature of the distribution of rock lizards of the genus *Darevskia* within the Caucasus is the partial overlap of their ranges observed in them (Darevsky 1967; Danielyan 1989). At the same time, in one case, the zones of common habitat are most often associated with the specifics of the vertical distribution of lizards and are usually found in the gorges of mountain rivers, along which forms common in the foothills, as if wedged into the area of distribution of forms, the range of which is located higher. In the next case, the zones of sympatry are quite wide, and both closely related species are jointly distributed over a considerable extent of their ranges (Darevsky 1967). Questions of hybridization in this group have attracted attention for more than a decade, since parthenogenetic species have been formed as a result of their hybridization. Quite wide overlap zones exist between green lizards of the genus *Lacerta* (*L. agilis, L. strigata, L. media*) (Aslanyan 2004). Based on the data of morphological and molecular genetic analysis, it was shown that two forms of *Phrynocepahlus horváthi* and *Ph. persicus* are common in the Araks river basin (including Lake Urmia) on the territory of Armenia, NKR, Turkey, Nakhichevan, Northwestern and Central Iran, where their hybridization is also possible (Melnikov et al. 2013). Among snakes, sympatric zones exist between two forms of snakes (*Natrix natrix persa* and *N. n. scutata*), as well as between species of *Eirenis* (*E. collaris, E. punctatolineatus, E. modestus*). In the Caucasus and Central Asia, there are zones of sympatry of the snakes *Hemorrhois ravergiery* and *H. nummifer*. The rare species of Darevsky's viper *Vipera darevskii* also seems to be the result of hybridization .

Thus, due to the location of the valleys of the Kura and Araks rivers at the crossroads of historical migration routes of flora and fauna, complex topography, tectonic activity, the presence of ancient refugia in the territory, hybridization and the creation of new species by sympatric speciation provided a rich biodiversity of Armenia.

Habitats of Armenian Amphibia and Reptilia

Semidesert with fragments of desert. Nearly 30 species of reptiles live in the semidesert belt. The dominant species *Testudo graeca*, *Eremias pleskei*, *E. strauchi*, *Phrynocephalus persicus horvathi*, *Trachylepis septemtaeniata*, *Eumeces schneideri*, *Eryx jaculus*, and *Malpolon monspessulanus*, whereas *Mauremys caspica*, *Xerotyphlops vermicularis*, *Dolichophis schmidti*, *Eirenis collaris*, *E. punctatolineatus*, *Hemorrhois nummifer*, *Hemorrhois ravergieri*, *Platyceps najadum*, *Rhynchocalamus melanocephalus*, *Macrovipera lebetina* are less common. Among amphibians, *Bufotes varidis* is more often found, rarely *Hyla savignyi*, and only in a few places *Pelobates syriacus*. Eurytopic species, are distributed across nearly all of the habitats, whereas reptiles such as *Laudakia caucasia*, *Pseudopus apodus*, *Darevskia raddei*, *Lacerta strigata*, and *Ophisops elegans* have more limited distributions and are found only in a vertical range of 400–2200 m and then only in the semidesert, steppe, and forest zones. Desert patches are found as little islets among semidesert vegetation in the Ararat Plain. Here are found (1) In psammophytous communities (on sandy soils) following lizard species: *Phrynocephalus persicus*, *Eremias strauchi*, *E. pleskei*, *Lacerta strigata*, and the snakes *Eryx jaculus*, *Malpolon monspessulanus*, and *Platyceps najadum*. (2) , In halophitous communities which tolerate saline soils, lizards and snakes are scarce: *Lacerta strigata* often occurs in the mesophile parts, whereas *Natrix natrix* and *N. tessellata* are found near water reservoirs. *Eremias strauchi* and *Ophisops elegans* as well as *Malpolon monspessulanus* can be observed at the edge of the desert, whereas *Bufotes viridis* is the only amphibian species found in this kind of habitat. Most of the nonirrigated Armenian Araks River valley is characterized by stony dry landscape. The most common snakes are *Eryx jaculus*, *Malpolon monspessulanus*, and *Macrovipera lebetina*, and more rarely *Dolichophis schmidti*, *Hemorrhois ravergieri*, and *Platyceps najadum*. On slopes, *Xerotyphlops vermicularis*, *Eirenis collaris*, and *E. punctatolineatus* can be found. The most frequently encountered lizards are *Laudakia caucasia*, *Pseudopus apodus*, *Eumeces schneideri*, *Darevskia raddei*, *Ophisops elegans*, and *Trachylepis septemtaeniata*. (3) In gypsophilous communities which tolerate soils rich in gypsum most common reptiles are *Pseudopus apodus*, *Ophisops elegans*, *Eumeces schneideri*, *Eryx jaculus*, *Eirenis collaris*, *Hemorrhois ravergieri*, and *Macrovipera lebetina*.

Mountain steppe. Elevations of belt vary from 800 to 2000 m in the north to 2400–2500 m in the south. *Dry mountain steppe* represents a transition from semi-deserts to mountain steppes. It includes steppes in the northeastern areas of Armenia up to 900 m, as well as those in the Araks Depression in the south (1200–1400 m). The dry mountain steppe is inhabited mostly by *Eremias strauchi*, *Lacerta media*, *Eumeces schneideri*, *Trachylepis septemtaeniata*, *Ablepharus bivittatus*, whereas *Ophisops elegans* occasionally penetrates it from the semidesert zone. Among snakes *Xerotyphlops vermicularis*, *Eryx jaculus*, *Dolichophis schmidti*, *Eirenis collaris*, *E. punctatolineatus*, *Elaphe urartica*, *Hemorrhois nummifer*, *Platyceps najadum*, *Telescopus fallax*, and *Macrovipera lebetina* are found here as well. *Hyla savignyi* is common in the valley of the Araks River. *Testudo graeca*, *Laudakia caucasia*, *Pseudopus apodus*, *Lacerta media*, *L. strigata*, *Ophisops elegans*, *Xerotyphlops vermicularis*, *Eryx jaculus*, *Dolichophis schmidti*, *Eirenis collaris*, *E. modestus*, *Elaphe urartica*, *Hemorrhois nummifer*, *H. ravergieri*, *Natrix natrix*, *N. tessellata*, *Telescopus fallax*, and *Macrovipera lebetina* are found here. In Southern Armenia, snakes are common in both the semidesert and dry mountain steppe. Due to the distribution of their habitat (respectively in Northeastern Armenia in the Kura River drainage basin and along the Araks River in the south), widely separated by high elevations and forested mountains, northern populations of the above-mentioned reptiles are isolated from the southern ones.

Real mountain steppe. This is the most widespread vegetation zone in Armenia from 1400 m to 1800 m in the north, and to 2000 m in the south. The most common lizards seen on the gentle, vegetation-covered slopes are *Lacerta agilis*, *L. media*, *L. strigata*, and *Pseudopus apodus*, whereas *Laudakia caucasia*, *Darevskia armeniaca*, *D. dahli*, *D. nairensis*, *D. portschinskii*, *D. raddei*, *D. rostombekovi*, *D. unisexualis*, and occasionally *D. valentini* predominate on rocky outcrops and talus accumulations. Other lizards occur here too, but they are less common and rarely seen: *Parvilacerta parva*, *Eremias arguta*, *Ablepharus bivittatus*, and *Ablepharus chernovi*. With respect to snakes, the most common is *Vipera eriwanensis*. In some places, especially those characterized by herb-bunchgrass, it occurs sympatrically with *Montivipera raddei*. *Eryx jaculus* and *Telescopus fallax* are also encountered, but less frequently, as are *Coronella austriaca*, *Hemorrhois nummifer*, *H. ravergieri*, *Platyceps najadum*, and even more rarely *Elaphe urartica* and *Zamenis hohenackeri*. *Natrix natrix* and *N. tessellata* are to be seen close to water, in or near ponds and streams. Among the testudinids, *Testudo graeca* penetrates the steppe from subjacent zones. Lastly, amphibians, for the most part, are rare; the commonest species in arable land is *Bufotes viridis*, whereas *Hyla orientalis shelkownikowi* and *Rana macrocnemis* occur in humid places along ravines of small rivers and streams and *Pelophylax ridibundus* only in ponds or lakes.

Forest. Among the reptiles, *Anguis colchica* appears to be the most typical reptile of this habitat, although *Darevskia armeniaca, D. dahli, D. portschinskii*, and *D. praticola* are found here as well. Snakes, on the other hand, are only occasionally seen, but of those found there, *Coronella austriaca* is the most common; however, *Elaphe urartica, Hemorrhois ravergieri*, and *Platyceps najadum* are also found here as well as *Natrix natrix* and *N. tessellata*, almost always in the vicinity of water. *Testudo graeca* occurs along forest edges. Among amphibians, *Ommatotriton ophryticus* occurs only in northern Lori Province, whereas *Hyla orientalis* is more widely distributed.

Arid open forests are found in both northern and southern parts of the country, at elevations of 900 m to 1800 m. Among the reptiles encountered in the juniper open forests, *Laudakia caucasia* and *Darevskia raddei* are common; other species found here include *Ablepharus bivittatus, Eumeces schneideri, Lacerta media, L. strigata, Pseudopus apodus*, and *Trachylepis septemtaeniata* among the lizards, and *Eirenis punctatolineatus, Hemorrhois ravergieri, Zamenis hohenackeri*, and *Montivipera raddei* among the snakes, and, more rarely, four additional species of snakes, *Dolichophis schmidti, Elaphe urartica, Telescopus fallax*, and *Macrovipera lebetina*.

Subalpine and alpine belt occupy the highest altitudes up to 3600 m (up to 4095 m – Mount Aragats, but reptiles do not climb to this elevation). Only a few species of reptiles occur in this belt; among these are *Darevskia armeniaca, D. raddei, Lacerta agilis, L. media*, and *Darevskia valentini*, the latter of which appears to be restricted to this belt, as well as *Vipera darevskii* and *Vipera eriwanensis* (up to 2800 m). Other reptiles found here include two species of *Natrix, Elaphe urartica, Hemorrhois ravergieri*, and *Zamenis hohenackeri* (the latter up to 2600 m). *Rana macrocnemis* is the most widespread of the amphibians, but *Pelophylax ridibundus* and *Bufotes viridis* occur here as well.

Wetlands. This habitat is intrazonal. Among reptiles, the most common are the freshwater turtles *Mauremys caspica*, and, locally, *Emys orbicularis*. The lacertid lizard, *Lacerta agilis*, is usually found along the fringes of the wetlands. This habitat is typical for *Natrix tessellate* and *N. natrix*. Frogs are especially well represented in the wetlands, notably *Pelophylax ridibundus, Rana macrocnemis*, as well as *Bufotes viridis* and *Hyla orientalis*.

Agricultural landscapes. Many species of reptiles are found in the cultivated areas. For example, *Testudo graeca* is often found in gardens and vineyards. A number of snakes are commonly found in these areas as well and include *Hemorrhois nummifer, H. ravergieri, Natrix natrix, N. tessellata, Platyceps najadum, Macrovipera lebetina*, and, somewhat less frequently, *Dolichophis schmidti, Elaphe urartica, Telescopus fallax*, and *Zamenis hohenackeri*. The lizards *Lacerta media, L. strigata, Pseudopus apodus* are regularly observed in cultivated areas, whereas

Eumeces schneideri and *Ophisops elegans* are less often seen (Arakelyan et al. 2011a, b).

5.2.3 *Birds*

Karen Aghababyan

In Armenia, the birds are represented by 376 species, among which, the 241 are breeding in the country, while 135 occur during migration and wintering periods (Annex 2). The birds belong to 67 families, among those the most species-rich ones are Ducks (*Anatidae*) with 30 species, Hawks and Eagles (*Accipitridae*) with 30 species, Old World Flycatchers (*Muscicapidae*) – 26 species, and Sandpipers (*Scolopacidae*) – 25 species.

The fauna of breeding birds in Armenia is mainly represented by two major groups: European and Asian species, while the minor part is composed by endemics of the Caucasus and the species, which have almost cosmopolitan distribution. Among the European species there are such species as blackbird (*Turdus merula*) and nuthatch (*Sitta europaea*), among Asian species – such typical representatives as Finsch's wheatear (*Oenanthe finschii*) and chukar (*Alectoris chukar*) (Fig. 5.64). Endemics of the Caucasus count only three species: Caucasian grouse (*Lyrurus mlokosiewiczi*), mountain chiffchaff (*Phylloscopus sindianus*), and Armenian gull (*Larus armenicus*). Cosmopolitan ones include such species as peregrine falcon (*Falco peregrinus*) and, in some extent, golden eagle (*Aquila chrysaetos*) (Fig. 5.65).

The rich fauna of migratory birds is formed due to the unique position of Armenia, which is situated on the crossroads of three global migration flyways: Black Sea – Mediterranean, Central Asian, and West Asian – East African (Kirby

Fig. 5.64 *Alectoris chukar*

Fig. 5.65 *Aquila chrysaetos*

et al. 2008). The rigorous terrain of the country creates various gorges, which host a number Passerines, like willow warbler (*Phylloscopus trochilus*) to get a rest and the shelter in the thickets. The flatter parts of mountain plateaus attract a number of raptors such as pallid harrier (*Circus macrourus*) to get the rest during migration and even to hunt on the rodents, when possible. Lake Sevan, the grassy marshes of Shirak and Lori Plateaus, and the brackish marshes of Ararat Plain provide wide opportunities for stopover of the waterbirds, like green-winged teal (*Anas crecca*). While most of the migrants cross Armenia on the way to Africa and Southern Asia, the others stay in Armenia over the winter, like the tundra swan (*Cygnus columbianus*) and hen harrier (*Circus cyaneus*).

During last decades, there is a trend of change of the species' distribution in Eurasia, which influences Armenia as well. Thus, since 1995 when the first more or less comprehensive inventory was implemented, the new species have been found in Armenia at breeding ground. For example, the see-see partridge (*Ammoperdix griseogularis*) appeared in Armenia in 2000s, and since that time became a regular breeder (Ananyan 2004; Aghababyan et al. 2021). Most probably the species expanded its distribution to the north from Iran.

Similarly, we observe a number of other species, which have been breeding in southern and warmer regions, but now are expanding their distribution to the North, and start occurring in Armenia. The number of summer registrations of masked shrike (*Lanius nubicus*) increases (Ter-Voskanyan and Aghababyan 2014), indirectly indicating a possible breeding of the species in the country.

The spur-winged plover (*Vanellus spinosus*), first recorded in the country as occasional visitor, became a year-round resident, most probably with regular breeding (Aghababyan 2019b) (Fig. 5.66).

Fig. 5.66 *Vanellus spinosus*

Desert finch (*Rhodospiza obsoleta*), which first was recorded in Armenia in 2013 (Ananian et al. 2013), now became a regular breeder in the semidesert areas of Central Armenia.

The regular monitoring, implemented in the country, makes it possible to document not only penetration of the new species into Armenia, but also less obvious shifts of Armenian breeders. Thus, the sombre tit expanded its breeding range to the North during last decade, occupying not only southern slopes of Meghri ridge, but now also the slopes of Bargusht ridge (located to the north). The latitudinal shifts are important to document for several reasons. At first, invasion of new species, could have negative effect on the indigenous ones, casing their decline. At second, the enrichment of Armenian fauna opens new opportunities for incoming tourism, helping to market birdwatching in Armenia for Europe. Also, it might have an input into the study of influence of climate change on the Armenian biodiversity.

5.2.3.1 Landscape Distribution

The ecosystems of Armenia are described in detail in Chap. 2 of the current book. Here we will classify the bird communities, rather than provide a description of the ecosystems of the country. Thus, the birds can be divided into several communities, some of those can be related to the vertical zones, while the others are intrazonal. The zonal communities are birds of semideserts, shrublands (shiblyak), juniper open forests, deciduous forests, mountain steppes, meadows and alpine carpets. The intrazonal communities include birds of wetlands (in a broad sense), rocks and screes, birds of riparian zone (including riparian shrubs and woodlands), and birds of arable lands.

Fig. 5.67 *Oenanthe isabellina*

Birds of Semideserts

Semideserts of Armenia occupy elevation range from 400 m till about 1200 m a.s.l. and are mainly characterized by perennial and annual herbs, semishrubs, and shrubs. The semideserts are inhabited by 60 breeding species. The most typical inhabitants of semideserts are Wheatears and Larks. Thus, Isabelline Wheatear (*Oenanthe isabellina*) and Crested Lark (*Galerida cristata*) are one of the most numerous species in the semideserts (Fig. 5.67).

Birds of Shrublands (Shiblyak)

The habitat is dominated by *Paliurus spina-christi* and is mainly characterized by rather dense coverage of shrubs. The shrublands begin at the elevation range from about 600 m a.s.l. in the Southeastern Armenia and at about 800 m a.s.l. in the Northeastern Armenia. The shrublands are inhabited by 57 breeding species. The most typical inhabitants of the shrublands are the Sylviid Warblers and Buntings. For example, lesser whitethroat (*Curruca curruca*) and black-headed bunting (*Emberiza melanocephala*) are the most common species in this habitat.

Birds of Juniper Woodland

This type of habitat is dominated by Juniper tree species *Juniperus polycarpos* and is often combined with some other elements, such as shrubs, deciduous woodlands, or tragacanth dominated mountain steppes. In the Southern and Southeastern Armenia, this habitat gradually replaces shrublands, while in Central Armenia its elevation range starts from 1500 m a.s.l., and in Eastern Armenia around Lake Sevan it begins at about 2000 m a.s.l. Juniper woodlands of Armenia are inhabited by 50 breeding species. The most typical inhabitants of Juniper woodlands are red-fronted serin (*Serinus pusillus*) and sombre tit (*Poecile lugubris*).

Birds of Deciduous Forests

Deciduous forests in Armenia start at various elevations in Southern, Central, and Northern Armenia, and pretty much vary in the species composition. While the Central and Southern forests are represented by oak-hornbeam forests, dominated by *Quercus* spp. and *Carpinus betulus*; the forests of Northern and Northeastern regions are dominated by *Quercus* spp. and *Fagus orientalis*. In Southeastern Armenia, the forests' belt starts at about 600 m a.s.l., while in Northeastern regions it begins at about 500 m a.s.l. The forests are inhabited by 63 breeding species, among those the most typical ones are flycatchers, leaf warblers, tits, and woodpeckers. Thus, the most common species in deciduous forests are green warbler (*Phylloscopus nitidus*), chiffchaff (*Phylloscopus collybita*), and coal tit (*Periparus ater*) other typical species include semicollared flycatcher (*Ficedula semitorquata*), great spotted woodpecker (*Dendrocopos major*), and green woodpecker (*Picus viridis*).

Birds of Mountain Steppes

Steppes are occurring at the elevations from 1400 to about 2200 m a.s.l. and are dominated by various cereals. The breeding species diversity encounters 46 species, including such typical groups, like kestrels, quails and partridges, and some larks. Thus, the common quail (*Coturnix coturnix*) inhabits all types of steppes, the Eurasian skylark (*Alauda arvensis*) is one of the commonest birds of this habitat, and the common kestrel (*Falco tinnunculus*) occurs in the steppes throughout the country, although is less abundant than nonpredatory species like larks.

Birds of Meadows

The meadows in Armenia are occurring at the elevations from 2000 m a.s.l. in Northern Armenia and from 2200 m a.s.l. in southern parts of the country. This type of habitat is dominated by mesophyllic herbaceous plants from *Poaceae* family and includes also wide variety of other herbs. The high-level moisture of this habitat allows it staying green throughout the summer including the driest period from late August till mid-September. The meadows of Armenia are inhabited by 38 breeding species, including such typical species, like corn crake (*Crex crex*), grasshopper warbler (*Locustella naevia*), and common rosefinch (*Carpodacus erythrinus*).

Birds of Alpine Meadows

This habitat is dominated by short perennial herbaceous vegetation and is always alternated by various screes and rocks. The lower border of alpine meadows and carpets starts at about 2700 m a.s.l. The habitat is characterized by very short summer and high level of precipitations. The alpine zone in Armenia is inhabited by 17 breeding species, including such typical ones, as horned lark (*Eremophila alpestris*), alpine accentor (*Prunella collaris*), and white-winged snowfinch (*Montifringilla nivalis*).

Birds of Wetlands

The lakes and marshes are intrazonal type of habitat that does not depend on elevation; however, the lower marshes of Ararat Plain have brackish nature, while the grassy marshes located at higher zone of Northern Armenia can be considered as freshwater ones. Wetlands of Armenia include the second large high-mountain lake in the world – Lake Sevan. The fauna of breeding birds of wetlands encounters 78 species, most of which are highly specialized for living in this habitat, and include various waders, ducks, herons and egrets, cormorants, pelicans, rails and coots, and grebes. The most typical species that occupies variety of wetlands are little grebe (*Tachybabtus ruficollis*), green sandpiper (*Tringa ochropus*), mallard (*Anas platyrhynchos*), little bittern (*Ixobrychus minutus*), and common coot (*Fulica atra*).

Birds of Rocks and Screes

This is another type of intrazonal habitat, which is represented from about 400 to 4000 m a.s.l. There are two major types of these formations: volcanic and sedimentary. The rocks and screes are inhabited by 25 species, among which the most typical species are long-legged buzzard (*Buteo rufinus*), red-billed chough (*Pyrrhocorax pyrrhocorax*), and black redstart (*Phoenicurus ochruros*). Of course, there are number of other species, which breed on the rocks and cliffs, like the blue rock-thrush (*Monticola solitarius*), which breeds on the cliffs of semideserts, or the wallcreeper (*Tichodroma muraria*), which occurs in subalpine zone, but the birds, which are encountered in the mentioned 25 species are those which occupy rocks and screes regardless to elevation.

Birds of Riparian Zone

To some extent the riparian shrubs and woodlands can be considered as intrazonal; however, it is true only for lower parts of riparian shrubs – from 375 m to about 1800 m a.s.l., where the riparian zone contrasts with the surrounding arid habitats. Above that elevation the area is usually occupied by deciduous forests and the plant composition of riparian zone does not differ from the main zonal habitat. The riparian shrubs and woodlands are dominated by trees *Juglans regia*, *Celtis australis*, and *Populus* spp., wide variety of bushes with domination of *Rubus* spp., and among herbs – by *Phragmites australis, Typha* spp., etc. The habitat is also characterized by small areas of sandy and gravelly beaches, which play important role in the life of some waders and wagtails. The breeding birds, which breed in this habitat encounter 49 species, among which the most typical ones are white-throated dipper (*Cinclus cinclus*), kingfisher (*Alcedo atthis*), and common sandpiper (*Actitis hypoleucos*) (Fig. 5.68).

Birds of Arable Lands

In Armenia, there are three major types of arable lands: (1) grain fields; (2) gardens; and (3) orchards. The grain fields are mostly located at the higher elevations from 1500 m up to 2200 m a.s.l., while the gardens and orchards occupy the lower warmer areas from 500 m to 1500 m a.s.l. The grain fields and the gardens present fully transformed monocultural habitats, where only edges of the arable lands remain for the original type of habitat – mainly mountain steppes and semideserts. The orchards usually host much more of original habitat, because the herbs which grow in orchards are used as a hay for the livestock. In rigorous terrain, such as in Southeastern Armenia, the orchards have the most mosaic structure and are combined with original riparian shrublands, which allows occurrence of higher number of wild plants, and therefore the insect species. The bird fauna, which inhabits the arable landscape, mostly consists of the most adaptable species of the original habitat, which was transformed. Thus, in the grain fields, the Eurasian skylark (*Alauda arvensis*) and common quail (*Coturnix coturnix*) can be found, in the gardens – the

Fig. 5.68 *Alcedo atthis*

greater whitethroat (*Curruca communis*), and in the orchards – the Eurasian blackbird (*Turdus merula*) and great tit (*Parus major*).

It is worth to mention that during last 20 years, a change of upper or lower borders of the species distribution was determined on 26 species. It includes alpine species, for which the lower border of their range was shifted up, and species of arid habitats, which have expanded the upper border of their vertical distribution towards degraded parts of forest and through mountain steppes. Birds are highly mobile animals and quickly react to the changes of habitat conditions. Mostly those shifts are related to the habitat transformation caused by human, exacerbated by the factors of climate change (Figs. 5.69, 5.70, 5.71 and 5.72).

5.2.4 Mammals

Astghik Ghazaryan

Armenia, located between Europe and Asia, can be regarded as one of the major hotspots of biodiversity interlinking the faunas of the Mediterranean, the Middle East, the Caucasus and the Caspian areas. The Caucasus region is one of the world's most biologically diverse regions. Hotspots inside its territory are globally significant centres of biological richness and also for cultural diversity. Armenia is known as biodiversity hotspot in the Caucasus (Myers et al. 2000).

The mammalian fauna of Armenia is diverse and belongs to the Palearctic zoogeographic range and represents the Caucasian global biodiversity hotspot. It is in the crossroads of two subregions of Palearctic zone – European and Mediterranean. Mammals are very mobile animals and had the opportunity to spread widely, due to

Fig. 5.69 *Gypaetus barbatus*

Fig. 5.70 *Himantopus himantopus*

Fig. 5.71 *Vanellus leucurus*

which their origin is still under the research of scientists. According to the data provided by the IUCN, most of the animals distributed in Armenia are of European and Mediterranean origin. We have species that are widely distributed, such as the jerboa, rat, and wolf. European Topic Center on Biological Diversity (ETC/BD) presented the Geographical map for Europe in 2016; according to it Armenia is in crossroads of Alpine, Mediterranean, Anatolian and Iranian subregions (https://www.eea.europa.eu/data-and-maps/figures/biogeographical-regions-in-europe-2).

The different landscapes create a diverse habitat for a lot of species (Dahl, 1954). There are 94 species of mammals distributed in Armenia; they are represented by

Fig. 5.72 *Vanellus vanellus*

six orders: Rodentia (34 species), Logomorpha (1), Chiroptera (30), Eulipotyphla (11), Carnivora (15), and Artiodactyla (5).

5.2.4.1 Rodentia

The rodents are one of the biggest and richest orders of mammals not only for the world also for Armenia. Thirty-four species of this order are distributed not only in wildlife also in human areas. Six species (18%) of rodents are listed in the Red Data Book of Animals of Armenia (Aghasyan & Kalashian 2010) in different categories. Armenian birch mouse is endemic species which have very small area of distribution. Dahl's jird and small five-toed jerboa have restricted area of distribution. Two species (ondatra and nutria) are invasive, and are very well distributed in Armenia. 57.6% of rodents dwelling in Armenia have large area of distribution. And the spread of the remaining 24.4% of species is mainly related to the parable of a certain landscape. The activity of five species of rodents is mainly related to human activity. They are synanthropic species and live in or near human buildings. The majority of field mice and hamsters are considered pests for grain crops.

One of the endangered species is the Asia Minor ground squirrel (*Spermophilus xanthoprymnus*) (Fig. 5.73). In Armenia, its preferred altitude is 1100–2400 m a.s.l. in Pambak Valley and Shirak Plateau, and it also lives in fields, on the edges of gardens, and in the ruins of abandoned villages. Another endangered species is *Hystrix indica*, which expands its distribution area from the southern part of Armenia to the north. *Dryomis nitedula* and *Sciurus anomalus* are common species for Armenian forests. Among the rodents that cause more damage to agricultural areas are voles, hamsters and mole voles. Among the synanthropic species of rodents are mice (*Mus musculus, Mus macedonicus*) and rats (*Rattus norvegicus, Rattus rattus*), in some

Fig. 5.73 *Spermophilus xanthoprymnus*

rural communities also *Apodemus witherbyi. Apodemus ponticus and Apodemus uralensis* are widespread forest species. *Arvicola amphibious, Ondatra zibethicus, Miocaster coypus* are common and very well distributed animals that prefer wetlands. There are four species of jirds in Armenia, one of which *Meriones dahli* is critically endangered in Armenia.

List of species

1. *Hystrix indica VU*
2. *Dryomys nitedula*
3. *Myoxus glis*
4. *Spermophilus xanthoprymnus EN*
5. *Sciurus anomalus*
6. *Ondatra zibethicus*
7. *Nannospalax nehringi*
8. *Ellobius lutescens*
9. *Scarturus williamsi*
10. *Scarturus elater EN*
11. *Sicista armenica EN*
12. *Apodemus ponticus*
13. *Apodemus uralensis*
14. *Apodemus witherbyi*
15. *Arvicola amphibius*
16. *Mus macedonicus*
17. *Mus musculus*
18. *Rattus norvegicus*
19. *Rattus rattus*
20. *Cricetulus migratorius*

21. *Chionomys nivalis*
22. *Mesocricetus brandti*
23. *Microtus arvalis*
24. *Microtus levis*
25. *Microtus schidlovskii EN*
26. *Microtus socialis*
27. *Microtus (Terricola) daghestanicus*
28. *Microtus (Terricola) majori*
29. *Microtus (Terricola) nasarovi*
30. *Meriones dahli EN*
31. *Meriones persicus*
32. *Meriones tristrami*
33. *Meriones vinogradovi*
34. *Miocaster coypus*

5.2.4.2 Logomorpha

This order is represented with one species – European hare (*Lepus europeus*). This species is very well distributed in Armenia.

5.2.4.3 Chiroptera

The extremely various landscapes providing many types of habitats for bats. While the biodiversity of several vertebrate species is well studied, the bat biodiversity remains largely unstudied. Bats are the second-rich order of mammals. This order is represented with four families of Microchiropteran bats: Rhinolophidae, Vespertilionidae, Molosidae, Miniopteridae. There are several species, which are belonging to one species complex and need deep molecular investigation for identification of species. Bat detector method helps in research of bats and their monitoring, but at the same time there are a lot of difficulties to make an identification of bat calls of very close species. In this order there are species which are living in forests, in caves and human buildings. 30% of bats species are listed in the Armenian Red book. There are several species, which are not registered in Red Book but have very small area of distribution. Armenia is very unique place where breeding and wintering colonies of bats were registered.

An endemic bat species has been described from the Sevan basin: the Armenian whiskered bat *Myotis hajastanicus* (Dietz et al., 2016). It was described as a subspecies of the widely distributed *Myotis mystacinus* by Argyropulo (1939). A detailed review of the Western Palearctic whiskered bats pointed out the unique morphological characters of Lake Sevan specimens and raised them to species rank: *M. hajastanicus* (Benda and Tsytsulina 2000). Referring to that publication, *M. hajastanicus* was also given species rank by Simmons (2005) in the reference list of mammal species of the world. Genetic analysis of a museum specimen placed the species within the *Myotis brandtii* clade, surprisingly distant to the morphologically more similar *M. mystacinus* clade (Tsytsulina and Masuda 2004). The whiskered bat population of Lake Sevan basin still persists and at least two nursery colonies exist in the terra typica of the taxon "hajastanicus". However, neither morphological nor genetic criteria support the distinctness of the taxon from closely related other species of the *M. mystacinus* clade. Morphological and genetical characters place *hajastanicus* clearly together with *aurascens*, a taxon lately regarded to be a synonym of *Myotis davidii* (Benda et al. 2012, 2016). Therefore, the validity of the species *M. hajastanicus* have been rejected.

Myotis nattereri sensu lato are distributed in Northwest Africa, Europe, and parts of the Middle East In parts of the Caucasus, the nominal form is sympatric with a larger form, *M. schaubi*. This larger form is clearly distinct both in morphology and in genetics (Ruedi and Mayer 2001; Salicini et al. 2011). *M. nattereri* species group has a complex and reticulate evolutionary history involving multiple cases of hybridization between divergent genetic lineages. We identified unknown cryptic diversity in the eastern ranges. All of the identified genetic lineages evolved in distinct glacial refugia and later on expanded their ranges during the interglacial periods. In some of these expansions, neighbouring lineages got into contact and hybridized with each other.

There are five species of horseshoe bats in Armenia, three (*Rh. Mehely, Rh. blasii, Rh. euryale*) of which are common. Two species of pipistrelle bats (*Pipistrellus pipistrellus, Pipistrellus kuhli*) are widespread and two are rare species in Armenia (*Pipistrellus nathusi, Pipistrellus pygmaeus*). *Plecotus macrobularis* is widespread species in the southern parts of Armenia, while *Plecutus auritus* is rare species dwelling in forestry areas. Another forestry species *Nyctalus noctula, Nyctalus leisleri* and *Barbastella barbastellus* are common for forests of Armenia, but *Barbastella caspicus* is cave-dwelling species known only from one locality.

List of species

1. *Rhinolophus ferrumequinum* (Fig. 5.74)
2. *Rhinolophus euryale* (VU)
3. *Rhinolophus blasii* (EN)
4. *Rhinolophus hipposideros*
5. *Rhinolophus mehelyi* (VU)
6. *Miniopterus pallidus* (VU)
7. *Pipistrellus pipistrellus*
8. *Pipistrellus nathusii*
9. *Pipistrellus kuhlii*
10. *Pipistrellus pygmaeus*
11. *Hypsugo savii*
12. *Plecotus auritus* (VU)
13. *Plecotus macrobullaris*
14. *Nyctalus lasiopterus*
15. *Nyctalus leisleri*
16. *Nyctalus noctula*

Fig. 5.74 *Rhinolophus ferrumequinum*

17. *Myotis aurascens*
18. *Myotis bechsteinii* (VU)
19. *Myotis blythii*
20. *Myotis daubentonii*
21. *Myotis emarginatus*
22. *Myotis mystacinus*
23. *Myotis nattereri*
24. *Myotis schaubi* (DD)
25. *Vespertilio murinus*
26. *Barbastella barbastellus*
27. *Barbastella caspica* (VU)
28. *Eptesicus bottae*
29. *Eptesicus serotinus*
30. *Tadarida teniotis* (DD)

5.2.4.4 Eulipotyphla

Eulipotyphla in Armenia are represented by 11 species. They distributed from 400 m to 2500 m a.s. Two species of the Eulipotyphla are endangered species, but the population of *Neomys teres* has started to recover. The remaining species of the order are widespread in almost all landscapes. Levant mole prefers forest areas and surroundings. Some of the species prefer to live with humans in cities. There is very few information on Suncus Estrucus which is listed in Armenian Red Data Book. Two species: southern white breasted hedgehog and long-eared hedgehog are dwelling in Armenia. First of them is very well-distributed species in Armenia, while second is rare species and dwelling in southern part of Armenia.

List of species

1. *Talpa levantis*
2. *Neomys teres* (EN)
3. *Suncus etruscus* (VU)
4. *Sorex minutus*
5. *Sorex raddei*
6. *Sorex satunini*
7. *Sorex volnuchini*
8. *Crocidura leucodon*
9. *Crocidura suaveolens*
10. *Erinaceus concolor* (Fig. 5.75)
11. *Hemiechinus auritus* (EN)

Fig. 5.75 *Erinaceus concolor*

5.2.4.5 Carnivora

The largest terrestrial species in the order Carnivora are some of the world's most admired mammals; however, they are also among the most vulnerable group of species (Wolf and Ripple 2018). They are often in immediate need of conservation planning and action (Treves and Ullas 2003), as humans rapidly dominating landscapes across the world, causing problems for large carnivores, which require extensive and well-connected areas of good habitat (Ghoddousi et al. 2020). During the last two centuries, most of the large carnivores have had significant population losses and range constriction globally (Ripple et al. 2014). The order Carnivore in Armenia contain 15 species of large and medium size animals. Fifty percent of the species of this order are rare and need protection. The lynx is widely distributed in Armenia, but the number of its population is not large. The Caucasian lynx (*Lynx lynx dinniki*) although registered as a species of Least Concern on the IUCN Red List, the status and trend of lynx populations in Armenia is worrying because of the rare encounters in the field during our last year's observations. The Caucasian subspecies of lynx is poorly known and its status is uncertain on the whole area.

The leopard is one of the rarest species of the Armenian Highlands and the Caucasus Ecoregion. In the USSR, before the 1970s, leopards and other large predators had been considered as vermin and intensively wiped out under state support. In 1972, the leopard was officially declared a protected species, and in 1987, it was included in the Red Data Book as "endangered" (Khorozyan et al. 2007, 2011). In 2008, the Caucasian or Persian leopard (*Panthera pardus ciscaucasica* = *P.p. saxicolor* = *P.p.tulliana*) living in the region was designated by the 2008 IUCN Red List as "endangered" (EN C2a(i); Khorozyan 2008) and its updated status assessment is pending now. The leopard is included also in the Red Data Book of Armenia (Aghasyan and Kalashyan 2010) as "critically endangered"(CR C2a(i);D). The leopard range in Armenia is located in the country's south within the provinces of Ararat, Vayots Dzor and Syunik. Before the 1970s, this big cat also used to be recorded in northeastern Armenia, but then disappeared there (Khorozyan et al. 2007, 2011). It occurs from "Khosrov Forest" State Reserve southwards to the Armenia-Iran state border on the Geghama, Zangezur, Vayk, Bargushat and Meghri ridges. Gray wolf (*Canis lupus*), Eurasian lynx (*Lynx lynx*) and brown bear (*Ursus arctos*) are the main competitors of leopards.

Brown bears are classified as Least Concern internationally, since their global population is high and distributed over three continents. However, there are isolated populations that are considered as Threatened (McLellan et al. 2016), and one of them is the Caucasian subpopulation. Brown bear is listed in the Red Data Book of Armenia as a "Vulnerable" species, although they are recorded in almost all regions of Armenia (Malkhasyan and Kazaryan 2012). Moreover, species is included in the National Red Data Books or Red Lists of other South Caucasus countries, namely, Azerbaijan and Georgia and assessed as being at higher risk (Zazanashvili et al. 2020).

Mustelids are fur-bearing carnivores, which are well distributed in Armenia despite Marbled polecat and Eurasian otter. Pallas' cat is recorded in Armenia in two places: Urts ridge and in Meghri region. According to Red Data Book of RA, the latest record was made in 1935. There is a unpublished data about the pallas' cats, recorded in 1991.

There were some records of striped hyena before 1940. On 2010, one dead animal was found near Nrnadzor village (Meghri region).

List of species

1. *Felis silvestris* (VU)
2. *Felis chaus*
3. *Lynx lynx*
4. *Otocolobus manul* (RE)
5. *Panthera pardus* (CR)
6. *Ursus arctos* (VU)
7. *Canis aureus*
8. *Canis lupus*
9. *Vulpes vulpes* (Fig. 5.76)
10. *Hyaena hyaena* (RE)
11. *Meles canescens*
12. *Mustela nivalis*
13. *Martes foina*
14. *Lutra lutra EN*
15. *Vormela peregusna* (VU)

Fig. 5.76 *Vulpes vulpes*

5.2.4.6 Artiodactyla

Artiodactyla order contains five species which are dwelling in Armenia. Mountain ungulates are often particularly threatened. These species often naturally occur in fragmented populations, because they depend on specific altitudinal belts, such as alpine meadows, or landscape features such as cliffs on which they rely as refuges to be safe from predators (Acevedo et al. 2007; Bleich et al. 1990; Gavashelishvili 2004; Gross et al. 2002). Mouflon and bezoar goat belong to the Western Asian faunistic complex and the mouflon has penetrated into the Caucasus isthmus. Only a small portion of mouflon range extends to the north of the Arax River and the stronghold of the species is located in Iran (Ziaie and Gutleb 1997).

Bezoar goats (Fig. 5.77), ancestors of domestic goats, once ranged from Balkans to Pakistan, but today only occupy small pockets of habitat within this region (Naderi et al. 2008; Weinberg et al. 2008). In the Caucasus ecoregion, bezoar goats only occur in small, isolated populations scattered across the region (Weinberg et al. 2008). This species is tightly linked to steep cliffs and rocky outcrops, on which they critically depend for escaping from predators (Weinberg 2001), making their populations naturally patchy.

The Caspian or Caucasian Red Deer is a subspecies of the red deer. Until 1954, the red deer had been commonly spread in the forests of Northern, Eastern and Southern Armenia. Poaching and careless cutting of forest had gradually led to the disappearance of this animal from the area of Armenia, and today the red deer is recorded in the Red Data Book of Armenia (2010) as "critically endangered." Within the framework of the Caucasian Red Deer Reintroduction Project made by WWF Armenia, 14 individuals of the animal were brought to Armenia from the Islamic Republic of Iran. The Red Deer were transported to the Dilijan National Park, where the Red Deer Breeding Center has been built in advance.

Fig. 5.77 *Capra aegagrus*

Fig. 5.78 *Sus scrofa*

Roe deer and wild boar are well-distributed animals in Armenia but they are under the treat because of illegal hunting.

List of species

1. *Ovis orientalis* (EN)
2. *Capra aegagrus* (VU)
3. *Capreolus capreolus*
4. *Sus scrofa* (Fig. 5.78)
5. *Cervus elaphus* (CR)

5.2.4.7 Distribution of Mammals Among Habitats

Semidesert and Desert

This habitat is characteristic of the dry and rocky lowlands of the Ararat valley, Zangezur, Meghri, and Vaik regions, at altitudes of 800–1500 m. These habitats are found in the Ararat valley and adjacent hills, in extremely arid climatic conditions. Soils are characterized by gypsum, clay, and sand. This zone is a habitat for a number of mammals. If the number of other vertebrates (amphibians, reptiles, birds) in the semidesert zone corelated to different seasons of the year, then the number of mammal species is relatively stable. In the winter, mainly the bats, some rodents and insectivores go into hibernation. Forty-five percent of the mammal species distributed in the country can be found in this zone. Indicator for this zone is *Hemiechinus auritus*, which is very well distributed in Armavir region. Among the bats, *Rhinolophus mehely, Rhinolophus euryale, Rhinolophus ferrumequinum, Miniopterus pallidus, Pipistrellus kuhli, Pipistrellus pipistrellus*, *Myotis blythi* as well as the *Eptesicus serotinus* and *Eptesicus bottae* are found here. It is possible to register Serotine bat during late spring and early summer. There are a large number of rodents living in the semidesert zone. *Meriones vinogradovi, Scarturus williamsi* and several species of voles are common animals for this zone. Two very rare rodents: *Scarturus elater* and *Meriones persicus* were registered in this zone. There are not new registrations about presence of *Meriones dahli* in Armenia more than 5 years. In Goravan sandy areas, *Scarturus elater* have been found. *Microtus levis* and *M. socialis* are common for this zone. Because of big number of rodents, foxes, jackals, badgers and other carnivores are also registered here. Two rare carnivores *Felis chaus* and *Otocolobus manul* are characterized for semidesert zone. In southern part of Armenia, another very rare animal *Hyaena hyaena* is registered. This zone can be visited also by *Mustela nivalis*, sometimes, *Ellobius lutesncese* from mountain steppes which is usual for them moving to the upper parts of the semideserts. *Canis lupus* and *Vulpes vulpes* mainly are observed during winter.

Mountain Steppe

Mountain steppes are the dominant vegetation type for most of the country. Meadow steppes occur in the highlands, while patches of forest also occur on ridge tops among steppes in the northeast and Sjunik regions. Biodiversity is also quite rich in this zone. Arid mountain steppe represents a transition from semideserts to mountain steppes. Among the insectivores *Crocidura suaveloense* and *C. leucodon* are found here. These animals sometimes share their nests with voles. *Plecotus auritus, Myotis nattereri, M. blythi, M. davidi, Hypsugo savii, Pipistrrelus khuli, P. pipistrellus*, bats' species were recorded in this zone. *Miniopterus pallidus* also possible to be recorded here. An indicator for western mountain steppes of Armenia is widespread species *Vormella peregusna*, and its main food *Spermophillus xanthoprimus*, which is registered in the Red Book of Armenia. *Spermophillus xanthoprimus* is distributed from mountain steppe to the semidesert areas. In case of dry weather, they go into summer hibernation. The species of rodents which can be met in this zone are *Ellobius lutescense, Cricetullus migratorius, Microtus arvalis, M. majori* and *Apodemus witherbyi*. In Armenia from North to South *Hystrix indica* is well distributed. *Lepus europeus* also has wide distribution in this zone, especially in southern parts of Armenia. A big number of carnivores – wolves, foxes, jackals, badgers are widespread in mountain steppe. Among the animals registered in the Red Book, this zone is characterized by bezoar goats and Armenian mouflons, especially for rocky and stony slopes (Fig. 5.79). The population of the *Ovis orientalis* has been preserved in the southern regions of Armenia.

Forests

Forests generally occur at altitudes between 500 m and 2100 m in the north (up to 2500 m in the south). Armenian forests are predominantly broadleaved (97%). This habitat has its own characteristic mammals' species. Among insectivores *Talpa*

Fig. 5.79 Bezoar goat on a rocky slope

levantis prefers broad-leaved, mainly beech forest. Beside moles other insectivores are also registered in this habitat: *Sorex volnuchini* and *S. raddei*. *Sorex volnuchini* prefers the upper edges of forests. In the lower parts of the forest, *Crocidura guldenstaedti* is dwelling. Bats prefer only old, mature forests. Among such species are *Plectotus auritus, Tadarida teniotis, Pipistrellus nathusi Nictalus noctula, N. leisleri, Myotis bechsteini, Barbastella barbastelus*. Some rodents are typical for forest ecosystems. *Squrius anomalus, Dryomis nitedula, Apodemus ponticus* and *A. uralensis* are common species for Armenian forests. In the lower edges of forests, *Microtus majori* is dwelling. Among the carnivores, *Felis sylvestris, Lynx lynx* are distributed here. Recently, in Ijevan forest *Panthera pardus* was detected, which was known from the southern parts of Armenia. Carnivores well dwelled in forest because of richness of prey. *Capreolus capreolus* is common for forest ecosystem. *Cervus elapher* also was common in past for northern part of Armenia. Now there is no native population present in forests. The red deer reintroduced from Iran is being kept in the forests of Dilijan National park for the purpose of release. *Sus scrofa* prefers oak forests. In 2012, due to swine fever in RA, the wild boar population suffered, but now population of wild boar is increasing.

Subalpine and Alpine Meadows

Subalpine meadows are distributed at altitudes from 2200 to 2700 m. Alpine meadows occupy the highest altitudes above subalpine meadows (up to 3800 m). Subalpine and alpine meadows are poor of vertebrates. Here small rodents predominate. This habitat is dominated by the *Chionomys nivalis*, which lives in rocky areas. Beside snow vole, *Microtus arvalis* also registered at this attitude. *Sorex volnuchini* also registered in subalpine belt, from 2000–2400 m a.s.l. *Sicista armenica* which is endemic for Armenia, was found in northern part of Sevan basin, in subalpine meadows. *Lepus europeus* very well dwelled in this habitat. There are no bat habitats in this belt, but bats are crossing it during migration and sometimes for feeding. *Vespertilio murinus, Hypsugo savii, Myotis blythi, M. davidii, Nyctalus noctula* and *Eptesicus serotinus* were recorded here in the end of July and in August. *Canis lupus*, *Vulpes vulpes* and *Ursus arctos* are also registered in subalpine belt, especially in the mountain ranges of Zangezur, Tsaghkunyats, Pambak and Aragats massive. Bezoar goats and Armenian mouflon can be found in the mountain meadows of Mount Aramazd.

Wetlands

Approximately 10% of the country is covered by wetlands and saline and alkaline lands. The latter cover about 25,000 ha, including areas in the Ararat Valley. Wetlands are intrazonal and considered as threatened habitats in the country. Among mammals the most common here are *Arvicola amphibius, Lutra lutra, Nyomis teres.* Climate training in Armenia of *Myocastor coypus* started in Shirak region, now it is also found in Arpi lake and on the banks of Akhuryan river. *Ondatra zibethicus* also is common for wetlands. First time *Ondatra zibethicus* was acclimatized in 1940 in Ayghrlich lake and then well distributed in Araks river valley.

Agricultural Landscapes

Many species of mammals could be found in cultivated areas. Rodents, mainly voles, hamsters and mole voles are common for this ecosystem. *Vulpes vulpes, Canis aureus* and *Ursus arctos* in Vayots dzor region preferred the agricultural landscapes because of richness of food. *Erinaceus concolor* dwells in fruit gardens and vineyards mainly feeding pests.

References

Acevedo P, Cassinello J, Hortal J, Gortázar C (2007) Invasive exotic aoudad (*Ammotragus lervia*) as a major threat to native Iberian ibex (*Capra pyrenaica*): a habitat suitability model approach. Divers Distrib 13:587–597

Aghababian K (2008) The distribution and number of golden eagles in Armenia. In: Research and conservation of the raptors in Northern Eurasia: materials of the 5th conference on raptors of Northern Eurasia. Ivanovo, 4–7 February 2008, pp 169–170

Aghababian K, Tumanyan S (2009) On the breeding peregrines *Falco peregrinus brookei* in some regions of Armenia. In: Sielicki J, Mizera T (eds) Peregrine falcon populations – status and perspectives in the 21st century. Turul, Warsaw, pp 99–108

Aghababyan K (2018) Habitat requirements of the Semicollared Flycatcher *Ficedula semitorquata* in Armenia. A BOU-funded project report. BOU, Peterborough
Aghababyan K (2019a) Recent counts of White-headed Ducks *Oxyura leucocephala* in Armash wetlands, Armenia. Sandgrouse 41:5–6
Aghababyan K (2019b) Summer observations of lesser white-fronted goose *Anser erythropus* and spur-winged lapwing *Vanellus spinosus* in Armenia. Sandgrouse 41:2–4
Aghababyan K (2021) Status and conservation of White-tailed Lapwing *Vanellus leucurus* in Armenia. Wader Study 128(1):87–92
Aghababyan K, Khanamirian G (2018) State of *Polyommatus myrrha cinyraea* in Armenia. Electron J Nat Sci NAS RA 1(30):27–29
Aghababyan K, Khanamirian G (2020) The state of bearded vulture, *Gypaetus barbatus* (Linnaeus, 1758) in Armenia. Bird Census News 32(1–2):11–16
Aghababyan K, Stepanyan H (2020) Booted Eagle *Hieraaetus pennatus* J. F. Gmelin, 1788 in Armenia: update on conservation status. J Life Sci 14:14–21
Aghababyan K, Ananyan V, Kalashyan M, Khanamiryan G (2010) Analysis of forest pests and pestholes exacerbated by climate change and climate variability in Syunik Marz of Armenia and to identification of the most applicable prevention measures. Report under the "Adaptation to Climate Change Impacts in Mountain Forest Ecosystems of Armenia" UNDP/GEF/00051202 Project. http://www.nature-ic.am/en/PR_FA_Reports
Aghababyan KE, Ter-Voskanyan H, Tumanyan S, Khachatryan A (2015) First National Atlas of the birds of Armenia. Bird Census News 28(2):52–58
Aghababyan K, Ter-Voskanyan H, Khachatryan A, Gevorgyan V, Khanamirian G (2017) The state of semi-collared flycatcher (*Ficedula semitorquata* Homeyer, 1885) in Armenia. In: Proceedings of the III international scientific conference "Biological diversity and conservation problems of the fauna – 3", Yerevan
Aghababyan K, Khanamirian G, Gevorgyan V (2019) An update of current situation of Griffon Vultures *Gyps fulvus* (Hablitz, 1783) in Armenia. Tichodroma – J Slovak Ornithol Soci 31:3–10
Aghababyan K, Gevorgyan G, Boyajyan M (2021) New observations of see-see partridge (*Ammoperdix griseogularis*) in Armenia. Sandgrouse 43(1):133–135
Aghababyan K, Khachatryan A, Ghazaryan A, Gevorgyan V (2021a) About the state of Corn Crake *Crex crex* Bechstein 1803 in Armenia. Bird Census News 34(1):9–17
Aghababyan K, Khanamirian G, Ghazaryan A, Gevorgyan V (2021b) About conservation status of Northern lapwing *Vanellus vanellus* in Armenia. J Ecol Nat Resour 5(3):1–9
Aghababyan K, Khachatryan A, Baloyan S, Ghazaryan A, Gevorgyan V (2021c) Assessing the current status of the common Pochard *Aythya ferina* in Armenia. Wildfowl 71:147–166
Aghababyan K, Khanamirian G, Khachatryan A, Martirosyan B, Grigoryan V, Zuerker T, Baloyan S (2022a) Evaluation of importance of Teksar Mountain of Armenia for bird and butterfly protection. Int J Zool Anim Biol 5(5):1–14. https://doi.org/10.23880/izab-16000401
Aghababyan K, Khanamirian G, Manukyan M, Ghazaryan H, Pahutyan N, Khechoyan A, Hambardzumyan K, Sharimanyan M, Avetisyan K, Tsaturyan A, Arabyan A, Sahakyan K, Sayadyan A, Khachatryan A, Martirosyan B, Gevorgyan V, Ghazaryan A, Sundar G (2022b) White Storks *Ciconia ciconia* L. became victims of environmental pollution of Ararat Plain of Armenia. J Ecol Nat Resour 6(1):1–10
Aghababyan K, Khanamirian G, Khachatryan A, Grigoryan V, Tamazyan T, Baloyan S (2022c) Revision of important bird and biodiversity areas of Armenia. Int J Zool Anim Biol 5(1):1–27
Aghababyan K, Aebischer N, Baloyan S (2022d) The modern state of Chukar *Alectoris chukar* J. E. Gray, 1830 in Armenia. Ornis Hungarica 30(1):80–96
Aghasyan A, Kalashyan M (eds) (2010) The red book of animals of the Republic of Armenia. Invertebrates and vertebrates, 2nd edn. Yerevan
Aghasyan AL, Aghasyan LA, Kaloyan GA (2009) Study of the present status of amphibian and reptile populations for updating the red data book of Armenia and the IUCN red list. In: Zazanashvili N, Mallon D (eds) Status and protection of globally threatened species in the Caucasus. Tbilisi, pp 125–130
Akramowskaya EG (1983) Politomicheskiy tsifrovoy i perfocartnyy opredelitel' poluzheskokrylykh nasekomykh semeystva Coreidae (Heteroptera) Armenia (Polytomic digital and punch card guide of heteropteran isects of the family Coreidae (Heteroptea) of Armenia, Yerevan

Akramowski NN (1948) Fauna strekoz Sovetskoy Armenii (The dragonfly fauna of the soviet Armenia). Zoologicheskiy sbornik Akademii Nauk Armyanskoy SSR 5:117–118
Akramowski NN (1976) Mollyuski (Mollusca). Fauna Armyanskoy SSR (Mollusks (Mollusca). Fauna of Armenian SSR, Yerevan
Ananian V (2004) The first see-see partridge *Ammoperdix griseogularis* in Armenia. Sandgrouse 26(2):137–139
Ananian VY, Drovetski SV, Koblik EA, Fadeev IV, Agayan SA (2013) The desert finch *Rhodospiza obsolete*—a new species for the fauna of Armenia. Russ J Ornithol 22(895):1799–1803
Ananyan V, Tumanyan S, Janoyan G, Aghababyan K, Bildstein K (2011) On breeding Levant Sparrowhawk in selected regions of Armenia. Sandgrouse 33(2):114–119
Arakelian GR (1994) Murav'i (Formicidae). Fauna Armyanskoy SSR. (Ants (Formicidae). Fauna of Armenian SSR.), Yerevan
Arakelyan M, Parham JF (2008) The geographic distribution of turtles in Armenia and Nagorno-Karabakh Republic (Artsakh). Chelonian Conserv Biol 7(1):70–77
Arakelyan MS, Danielyan FD, Corti C, Sindarco R, Leviton A (2011a) Herpetofauna of Armenia and Nagorno Karabakh, Ithaca
Arakelyan MS, Danielyan FD, Corti C, Sindaco R, Leviton AE (2011b) The Herpetofauna of Armenia and Nagorno-Karabakh. SSAR, Salt Lake City
Arakelyan M, Türkozan O, Hezaveh N, Parham JF (2018) Ecomorphology of tortoises (*Testudo graeca* complex) from the Araks River Valley. Russ J Herpetol 25(4):245–252
Arakelyan M, Ananian V, Petrosyan R (2019) Rediscovery of Chernov's skink (*Ablepharus chernovi* Darevsky, 1953) in Armenia. Herpetol Notes 12:475–477
Arakelyan M, Spangenberg V, Petrosyan V, Ryskov A, Kolomiets O, Galoyan E (2022) Evolution of parthenogenetic reproduction in Caucasian rock lizards: a review. Curr Zool 36:1–19
Argyropulo AI (1939) Über einige Säugetiere Armeniens. Zool Pap Biol Inst AS Arm SSR 1:27–66
Aslanyan AV (2004) Ecological-faunistic analysis of lizards of Armenia. Thesis, Yerevan
Avagian GD (1968) Saranchovye (Acridoidea) Fauna Armyanskoy SSR. (Locusts (Acridoidea). Fauna of Armenian SSR.), Yerevan
Avagian GD (1981) Kuznechikovye (Tettigonoidea). Fauna Armyanskoy SSR. (Grasshoppers (Tettigonoidea). Fauna of Armenian SSR), Yerevan
Barach GP (1940) Ryby Armenii (iz materialov po ichtiofaune Zakavkazja) (Fish of Armenia. From materials of ichtiofauna of the Transcaucasia). Reports of Sevan hydrobiological station, 6, 1, pp 5–70
Benda P, Tsytsulina KA (2000) Taxonomic revision of *Myotis mystacinus* group (Mammalia: Chiroptera) in the western Palearctic. Acta Soc Zool. Bohem 64:331–398
Benda P, Faizolâhi K, Andreas M, Obuch J, Reiter A, Ševčík M, Uhrin M, Vallo P, Ashrafi S (2012) Bats (Mammalia: Chiroptera) of the Eastern Mediterranean and Middle East. Part 10. Bat fauna of Iran. Acta Soc Zool Bohem 76:163–582
Benda P, Gazaryan S, Vallo P (2016) On the distribution and taxonomy of bats of the *Myotis mystacinus* morphogroup from the Caucasus region (Chiroptera: Vespertilionidae). Turk J Zool 40:1–22
Berg JS (1949) Presnovodnye ryby Irana i sopredel'nykh stran (Freshwater fish of Iran and neighboring countries). Rep Zool Inst AS USSR 8(4):783–858
Bleich VC, Wehausen JD, Holl SA (1990) Desert-dwelling mountain sheep – conservation implications of a naturally fragmented distribution. Conserv Biol 4:383–390
Borkhsenius NS (1949) Opredelitel' chervetsov i shchitovok (Coccoidea) Armenii. (Identification guide of Scale insects of Armenia), Yerevan
Charitonov DE (1936) A supplement to the catalogue of Russian spiders. Uchenye zapiski Permskogo Univ 2:167–222
Charitonov DE (1956) Overview of the spider family Dysderidae in the fauna of the USSR. Uchenye Zapiski Molotovskogo Gosudarstvennogo Universiteta 10:17–39
Chernov SA (1939) Gerpetofauna Armjanskoj SSR i Nkhichevanskoj ASSR (herpetological fauna of Armenian SSR and Nakhichevan ASSR). Zoological sbornik ArmFAN SSSR 1:79–194
Chkhikvadze VM, Bakradze MA (1991) On the systematic position of the recent land turtle from the Araxes valley. Proceedings of the Tbilisi State Universitat, 305

Dadikyan MG (1986) Ryby Armenii (Fish of Armenia), Yerevan
Dahl SK (1954) Fauna Armjanskoj SSR (Fauna of Armenian SSR), v.1, Yerevan
Danielyan FD (1989) Theory of hybrid origin of parthenogenesis in the group of Caucasian rock lizards. Doctorate thesis Yerevan
Danielyan FD, Darevsky IS, Makaryan A (1998) Record of the banded newt (*Triturus vittatus*) on the territory of Armenia. In: Advances in amphibian research in the former soviet union, vol 3, pp 185–186
Darevsky IS (1957) Reptile fauna of Armenia and Nakhichevan and its zoogeographical analysis. Doctorate thesis, Yerevan
Darevsky IS (1967) Skal'nye yashcheritsy Kavkaza: sistematika, ekologiya, filogeniya polimorfnykh yashcherits podroda Arshaeolaserta (Rock lizards of the Caucasus: taxonomy, ecology, phylogeny of polymorphic lizards of the subgenus Archeolacerta), Leningrad
Dietz C, Helverson O, Nill D (2009) Bats of Britain, Europe and Northwest. Africa; A & C Black Publishers Ltd
Dietz C, Ghzaryan A, Papov G, Dundarova H, Frieder M (2016) *Myotis hajastanicus* is a local vicariant of a widespread species rather than a critically endangered endemic of the Sevan lake basin (Armenia). Mamm Biol, Zeitschrift für Säugetierkunde 81(5):518–522
Dilbarian K, Kocharyan M (2014) Kleschi-ploskotelki (Acariformes: Tenuipalpidae) Armenii (Flat mites (Acariformes: Tenuipalpidae) of Armenia), Yerevan
Drumont A, Komiya Z, Danilevsky ML, Adlbauer K, Niisato T, Ohbayashi N, Yamasako J, Jang HK, Lee S, Lim J, Oh S-H, Lazarev MA (2020) Family Cerambycidae. In: Danilevsky M (ed) Catalogue of Palaearctic Coleoptera Revised and updated second edition. Volume 6/1. Chrysomeloidea I (Vesperidae, Disteniidae, Cerambycidae). Brill, Leiden/Boston, pp 104–480
Dunin PM (1984) Fauna and ecology of the spiders (Aranei) of the Apsheron Peninsula (Azerbajanian SSR). In: Utochkin A (ed) Fauna i ekologia paukoobrasnykh. Perm University, pp 45–49
Dunin PM (1988) Cribellate spiders (Aranei, Cribellatae) of Azerbaijan. Entomologicheskoe obozrenie 67:190–203
Dunin PM (1989) New spider species of the Harpactea genus from Armenia (Aranei, Dysderidae). Zool Zhurnal 68(7):142–145
Dunin PM (1991) New spider species of the genus Dysdera from the Caucasus (Arenei, Haplogynae, Dysderidae). Zool Zhurnal 70(8):90–98
Dunin PM (1992) The spider family Dysderidae of the Caucasian fauna (Arachnida Aranei Haplogynae). Arthropoda Selecta 1(3):35–76
Dunin PM, Zacharjan VA (1991) New spider species of genus *Zodarion* from the Caucasus (Aranei, Zodariidae). Zool Zhurnal 70:142–144
Emeljanov AF (1974) Predlozheniya po klassifikatsii i nomenclature arealov (proposals on the classification and nomenclature of areals). Entomologicheskoe obozrenie 53(3):497–522
Eskov KY (1987) The spider genus *Robertus* O. Pickard-Cambridge in the USSR with an analysis of its distribution (Arachnida: Araneae: Theridiidae). Senckenberg Biol 67:279–296
Fayvush G, Arakelyan M, Aghababyan K, Aleksanyan A, Aslanyan A, Ghazaryan A, Oganesyan M, Kalashyan M, Nahapetyan S (2016) The Emerald Network in the Republic of Armenia, Yerevan
First National Report to The Convention on Biological Diversity (1998) Yerevan. https://www.cbd.int/doc/world/am-nr-01-en
Fritz U, Guicking D, Kami H, Arakelyan M, Auer M, Ayaz D, Fernandez CA, Bakiev AG, Celani A, Džuki G, Fahd S, Havaš P, Joger U, Khabibullin VF, Mazanaeva LF, Široky P, Tripep S, Velez AV, Anton GV, Wink M (2007) Mitochondrial phylogeography of European pond turtles (*Emys orbicularis, Emys trinacris*) – an update. Amphibia-Reptilia 28:418–426
Gabrielyan BK (2001) An annotated checklist of freshwater fishes of Armenia. Naga, ICLARM, vol 24, Nos. 3&4, Malaysia, pp 23–29
Gabrielyan BK (2010a) Fishes of Lake Sevan. Publishing House "Gitutyun" of NAS RA
Gabrielyan BK (2010b) Osteichties bony fishes, 2nd edn. The Red Book of the Republic of Armenia, pp 183–194
Gabrielyan BK, Khosrovyan AM (2004) Stock and fishery dynamics of Coregonus Lavaretus in changing ecological conditions of Lake Sevan, Armenia. J Ecohydrol Hydrobiol 4(4):229–235

Gavashelishvili A (2004) Habitat selection by east Caucasian tur (*Capra cylindricornis*). Biol Conserv 120:391–398

Ghoddousi A et al (2020) Mapping connectivity and conflict risk to identify safe corridors for the Persian leopard. Landsc Ecol 35(8):1809–1825

Gross JE, Kneeland MC, Reed DF, Reich RM (2002) Gis-based habitat models for mountain goats. J Mammal 83:218–228

Harutyunian ES (1977) Opredelitel' fitoseyidnykh kleshchey sel'skokhozyaystvennykh kul'tur Armenii (Identification guide of Phytoseid mites of agricultural cultures of Armenia), Yerevan

Harutyunian ES, Dilbarian KP (2006) Parazitiformnye kleshchi (Acarina: Parasitiformes Reutyer, 1909) Respubliki Armeniya i ikh znachenie v razlichnykh cenozakh (Parasitiform mites (Acarina: Parasitiformes Reutyer, 1909) of the Republic of Armenia and their significance in different coenosises), Yerevan

Hayrapetyan TA, Aslanyan AV, Papov GY, Ghazaryan AS (2014) New data on small mammals (Insectivora, Chiroptera, Rodents) in Southern part of Armenia. In: Proceedings of YSU, chemical and beological sciences, vol 2. Yerevan State University, pp 43–47

Heath MF, Evans MI, Hoccom DG, Payne AJ, Peet NB (eds) (2000) Important Bird Areas in Europe: priority sites for conservation. Birdlife International, Cambridge

Hertevtsian EK (1986) Entsirtidy (Encyrtidae). Fauna Armyanskoy SSR (Encyrtids (Encyrtidae). Fauna of Armenian SSR), Yerevan

Huber C, Farkač J, Bousquet Y, Marggi W, Nagel P, Häckel M, Březina B, Putchkov AV, Matalin AV, Goulet H, Valainis U, Balkenohl M, Wrase DW, Zaballos JP, Pérezgonzáles S, Andújar FC, Giachino PM, Toledano L, Neri P, Kopecký T, Belousov IA, Zamotailov A, Robertson JA, Moore W, Hrdlička J, Kirschenhofer E, Kataev BM, Jaeger B, Kabak I, Schmidt J, Hovorka O, Casale A, Hieke F, Serrano J, Andújar A (2017) Carabidae. In: Löbl I, Löbl D (eds) Catalogue of Palaearctic Coleoptera. Revised and updated second edition. Volume 1. Archostemata – Myxophaga – Adephaga. Brill, Leiden/Boston, pp 31–838

Iablokoff-Khnzorian SM (1961) Opyt vosstanovleniya genezisa fauny zhestkokrylykh Armenii (The attempt of restoring the genesis of the coleopteran fauna of Armenia), Yerevan

Iablokoff-Khnzorian SM (1967) Platinchatoussye (Scarabaeoidea). Fauna Armyanskoy SSR. (Lamellicorn beetles (Scarabaeoidea). Fauna of Armenian SSR), Yerevan

Iablokoff-Khnzorian SM (1976) Zhuzhelitsy (Carabidae). Chast' 1. Fauna Armyanskoy SSR. (Ground beetles (Carabidae), Part 1. Fauna of Armenian SSR), Yerevan

Iablokoff-Khnzorian SM (1983) Mayki (Meloidae) i pyltseedy (Alleculidae). Fauna Armyanskoy SSR (Blister-beetles (Meloidae) and comb-clawed beetles (Alleculidae). Fauna of Armenian SSR), Yerevan

IUCN (2022) Habitats classification scheme (Version 3.1). https://www.iucnredlist.org/resources/habitat-classification-scheme

Iwan D, Löbl I, Bouchard P, Bousquet Y, Kamiński MJ, Merkl O, Ando K, Schawaller W (2020) Family Tenebrionidae Latreille, 1802. In: Löbl I, Löbl D (eds) Catalogue of Palaearctic Coleoptera. Revised and updated second edition. Volume 5. Tenebrionoidea. Brill, Leiden/Boston, pp 104–475

Javakhishvili Z, Aghababyan K, Sultanov E, Tohidifare M, Mnatsekanovf R, Isfendiyaroğlu S (2020) Status of birds in the Caucasus. In: Zazanashvili N, Garforth M, Bitsadze M (eds) Ecoregional conservation plan for the Caucasus, 2020 edition. Tbilisi, pp 72–82

Karapetian AP (1985) Zernovki (Bruchidae). Fauna Armyanskoy SSR (Seed beetles (Bruchidae). Fauna of Armenian SSR), Yerevan

Khanamirian G, Aghababyan K (2019) Further development of Prime Butterfly Area network in Armenia. In: Aksu-Zhabagly nature reserve proceedings, 12, Shymkent, pp 69–80

Khanamirian GG, Aghababyan KE, Warren MS, van Swaay CAM (2014) Identification of Prime Butterfly Areas in Meghri District of Armenia. In: Proceedings of international conference "Biological diversity and conservation problems of the fauna of the Caucasus – 2", September 23–26, 2014, Yerevan, pp 202–205

Khanamiryan G, Aghababian K (2012) Habitat distribution of butterflies (Lepidoptera: Rhopalocera) of Meghri region of Armenia. Caucasian Enthomol Bull 8(1):145–148

Khanbekian YR, Kalashian MY (1992) Fauna oribatidnykh kleshchey (Oribatei) Araratskoy ravniny i ispol'zovanie ikh v kachestve bioindikatorov promyshlennykh zagryazneniy sredy (The fauna of the oribatid-mites (Oribatei) of Ararat plain and their use as bioindicators of the industrial pollutions of the environment). Biol J Armenia 45(2):126–130
Khorozyan I (2008) *Panthera pardus* ssp. *saxicolor*. IUCN Red List Threatened Species 2008:e. T15961A5334217. https://doi.org/10.2305/IUCN.UK.2008.RLTS.T15961A5334217.en
Khorozyan I, Cazon A, Malkhasyan AG, Abramov AV (2007) Using thin-layer chromatography of fecal bile acids to study the Leopard (*Panthera pardus*)
Khorozyan I, Malkhasyan AG, Murtskhvaladze M (2011) The striped hyaena rediscovered in Armenia. Folia Zoologica-Praha 60(3):253–261
Kirby JS, Stattersfield AJ, Butchart SHM, Evans MI, Grimmett RFA, Jones VR, O'Sullivan J, Tucker GM, Newton I (2008) Key conservation issues for migratory land- and waterbird species on the world's major flyways. Bird Conserv Int 18:74–90
Koch L (1878) Kaukasische Arachnoideen. In: Schneider O (ed) Naturwissenschaftliche Beiträge zur Kenntniss der Kaukasusländer auf Grund seiner Sammelausbeute. Sitzungs-Berichte der Naturforschenden Gesellschaft ISIS in Dresden, pp 36–71, 159–160
Kubáň V, Volkovitsh MG, Kalashian MJ, Jendek E (2016) Superfamily Buprestoidea Leach, 1815. In: Löbl I, Löbl D (eds) Catalogue of Palaearctic Coleoptera Revised and updated second edition. Volume 3. Scarabaeoidea, Scirtoidea, Dascilloidea, Buprestoidea, Byrrhoidea. Brill, Leiden/Boston, pp 432–574
Kulczyński W (1895) Araneae a Dre G. Horvath in Bessarabia, Chersoneso Taurico, Transcaucasia et Armenia Russica collectae. Természtrajzi Füzetek 18:3–38
Kuljanishvili T, Epitashvili G, Freyhof J, Japoshvili B, Kalous L, Levin B, Mustafayev N, Ibrahimov S, Pipoyan S (2020) Checklist of the freshwater fishes of Armenia, Azerbaijan and Georgia. J Appl Ichthyol. https://doi.org/10.1111/jai.14038
Logunov DV (1998) *Pseudeuophrys* is a valid genus of the jumping spiders (Araneae, Salticidae). Revue Arachnologique 12:109–128
Logunov DV (1999) Two new jumping spider species from the Caucasus (Aranei: Salticidae). Arthropoda Selecta 7:301–303
Logunov DV (2015) Taxonomic-faunistic notes on the jumping spiders of the Mediterranean (Aranei: Salticidae). Arthropoda Selecta 24:33–85. https://doi.org/10.15298/arthsel.24.1.03
Logunov DV, Guseinov EF (2008) A faunistic review of the spider family Philodromidae (Aranei) of Azerbaijan. Arthropoda Selecta 17:117–131
Malkhasyan A, Kazaryan A (2012) Brown bears in Armenia. Int Bear News 21:38–40
Mardjanian MA (1986) Shchelkuni (Elateridae). Fauna Armyanskoy SSR (Click-beetles (Elateridae). Fauna of Armenian SSR), Yerevan
Marusik YM (1989) New data on the fauna and synonyms of the spiders of the USSR (Arachnida, Aranei). In: Lange AB (ed) Fauna i Ekologiy Paukov i Skorpionov. Arakhnologicheskii Sbornik Akademii Nauk SSSR, Moscow, pp 39–52
Marusik YM, Zonstein SL (2021) Description of *Raveniola ambardzumyani* n. sp. from Armenia (Araneae: Nemesiidae). Isr J Entomol 51:93–101
Mcheidze T (1964) Spiders (Araneina). Zhivotnyi Mir Gruzii. Academia Nauk Gruzinskoj SSR 2:48–116
Mcheidze T (1997) Spiders of Georgia: Systematics, ecology, zoogeographic review, Tbilisi
McLellan BN, Proctor MF, Huber D, Michael S (2016) Brown bear (*Ursus arctos*) Isolated populations (Supplementary material to *Ursus arctos* red listing account)
Melnikov D, Melnikova E, Nazarov R, Rajabizadeh M (2013) Taxonomic revision of *Phrynocephalus persicus* de Filippi, 1863 complex with description of a new species from Zagros, Southern Iran. Curr Stud Herpetol 13(1/2):34–46
Mikhailov KG (1986) New species of spiders from the families Clubionidae and Liocranidae from the middle Asia and the Caucasus. Zool Zhurnal 65:798–802
Mikhailov KG (1990) The spider genus *Clubiona* Latreille 1804 in the Caucasus, USSR (Arachnida: Araneae: Clubionidae). Senckenberg Biol 70:299–322
Mikhailov KG (2013) The spiders (Arachnida: Aranei) of Russia and adjacent countries: a non-annotated checklist. Arthropoda Selecta Suppl 3:1–262

Mikhailov KG (2016) Advances in the study of the spider fauna (Aranei) of Russia and adjacent regions: a 2015 update. Vestnik zoologii 50:309–320. https://doi.org/10.1515/vzoo-2016-0038

Mikhailov KG, Otto S, Japoshvili G (2017) A new species from the *Clubiona caerulescens* group from the Caucasus (Araneae: Clubionidae). Zool Middle East 63:362–368. https://doi.org/10.1080/09397140.2017.1361188

Mirumian L, Skuhravá M (2022) The gall midges (Diptera: Cecidomyiidae) of Armenia. 2. New records of occurrence of gall midges in Armenia. Acta Soc Zool Bohem Praha 85:23–34

Mirumyan LS (2011) Phytophagous gall midges (Diptera, Cecidomyiidae) of Armenia. Acta Soc Zool Bohem Praha 75:87–106

Mirumyan L, Skuhrava M (2022) New records of two gall midge species (Diptera: Cecidomyiidae) from Armenia. Acta Soc Zool Bohem Praha 85:67–69

Mirzoyan SA (1977) Dendrofil'nye nasekomye lesov i parkov Armenii (Dendrophilous insects of forests and parks of Armenia), Yerevan

Movsesyan SO (ed) (1987) Krasnaja rniga Armjanskoj SSR. Zhivotnye (Red Data Book of Armenian SSR. Animals), Yerevan

Myers N et al (2000) Biodiversity hotspots for conservation priorities. Nature 403:853–858

Nabozhenko MV, Kalashian MY, Mazmanyan MA (2021) A faunistic review of darkling beetles (Coleoptera: Tenebrionidae; excluding Alleculinae) of Armenia and partly the Nakhichevan Autonomous Republic of Azerbaijan with new records and taxonomic notes. Kavkazskij Entomologiceskij Bulleten 17(2):425–450. https://doi.org/10.23885/181433262021172-425450

Naderi S et al (2008) The goat domestication process inferred from large-scale mitochondrial DNA analysis of wild and domestic individuals. Proc NAS 105:17659–17664

Nentwig W, Blick T, Bosmans R, Gloor D, Hänggi A, Kropf C (2023) Spiders of Europe. Version 2023. https://doi.org/10.24436/1

Nikolajev GV, Král D, Bezděk A, Kon M, Pittino R, Bartolozzi L, Sprecher-Uebersax E, Ballerio A, Nikodým M, Dellacasa M, Dellacasa G, Rakovič M, Ziani S, Ahrens D, Zorn C, Krell FT (2016) Superfamily Scarabaeoidea Latreille, 1802. In: Löbl I, Löbl D (eds) Catalogue of Palaearctic Coleoptera Revised and updated second edition. Volume 3. Scarabaeoidea, Scirtoidea, Dascilloidea, Buprestoidea, Byrrhoidea. Brill, Leiden/Boston, pp 33–412

Otto S (2022) Caucasian spiders. A faunistic database on the spiders of the Caucasus Ecoregion. Database version 02.2022. Internet: caucasus-spiders.info

Parham JF, Türkozan O, Stuart BL, Arakelyan M, Shafei S, Macey JR, Werner YL, Papenfuss TJ (2006) Genetic evidence for premature taxonomic inflation in middle eastern tortoises. Proc Calif Acad Sci 57(3):955–964

Peters G (1958) Die Zauneidechse des Kleinen Kaukasus als besondere Unterart – *Lacerta agilis brevicaudata* ssp.n. Zoologische Jahrbücher. Abteilung für Systematik, Geographie und Biologie der Tiere, Jena 86b(1/2):127–138

Pipoyan SK (1996) A new species in the fauna of Armenia – *Leucaspius delineatus* (Cyprinidae). Voprosy Ichthiologii 36(1):134–137

Pipoyan SK (2012) Ikhtiofauna Armenia (Ichtiofauna of Armenia), Yerevan

Pipoyan S, Gabrielyan B, Tigranyan E, Vardanyan T, Barsegyan N, Khachatryan H (2018) Armenian names taxonomic units of fish of Armenia. Biol J Armenia 3(70):6–12

Pivazyan SA (1984) Pitanie i pishchevye vzaimootnoshenija forelej i siga ozera Sevan (Nutrition and nutritional relationships of trout and whitefish of Lake Sevan). Reports of Sevan hydrobiological station, 19, pp. 161–253

Plavilstshikov NN (1948) Opredelitel' zhukov-drovosekov Armenii (Identification guide of Longhorn beetles of Armenia), Yerevan

Rakov SY, Logunov DV (1997) A critical review of the genus *Heliophanus* C. L. Koch, 1833, of Middle Asia and the Caucasus (Aranei Salticidae). Arthropoda Selecta 5(3/4):67–104

Reshetnikov JS (1980) Ecologia i sistematika sigovykh ryb (Ecology and taxonomy of whitefishes), Moscow

Richter AA (1945) Skorpiony Armenii (Scorpions of Armenia), Yerevan

Ripple WJ et al (2014) Status and ecological effects of the world's largest carnivores. Science 343:6167
Roewer CF (1955) Die Araneen der Österreichischen Iran-Expedition 1949/50. Sitzungsberichte der Österreichischen Akademie der Wissenschaften (I) 164:751–782
Rubenyan A, Rubenyan T (2003) Nekotorye voprosy proniknovenia i biologii leshcha (*Abramis* brama) v Armenii (Some issues of penetration and biology of bream (*Abramis brama*) in Armenia). Material of regional conference "Investigation and conservation of animals of the South Caucasus", Yerevan, pp 128–130
Ruedi M, Mayer F (2001) Molecular systematics of bats of the genus *Myotis* (Vespertilionidae) suggests deterministic ecomorphological convergences. Mol Phylogenet Evol 21(3):436–448
Salicini I, Ibáñez C, Juste J (2011) Multilocus phylogeny and species delimitation within the Natterer's bat species complex in the Western Palearctic. Mol Phylogenet Evol 61(3):888–898
Satunin KA (1916) Review of faunistic studies of the Caucasian region during the five-year period 1910–1914. Book 29, no. 3
Savvaitova KA, Dorofeeva EA, Markaryan VG, Smoley AI (1989) Foreli ozera Sevan (Trouts of Sevan lake). Rep Zool Inst AS USSR 204:180
Schmidt P (1895) Beitrag zur Kenntnis der Laufspinnen (Araneae Citigradae Thor.) Russlands. Zoologische Jahrbücher, Abtheilung für Systematik, Geographie und Biologie der Thiere 8:439–484
Simmons NB (2005) Order Chiroptera. In: Wilson DE, Reeder DM (eds) Mammal species of the world. John Hopkins University Press, Baltimore, pp 312–529
Smoley AI (1964) O pitanii sigov ozera Sevan (On the nutrition of whitefishes of Sevan lake). Herald AS of ArmSSR, Biol Sci 17(6):49–57
Smoley AI (1979) Dinamika chislennosti lososevyh ryb ozera Sevan v uslovijakh izmenenija ego rezhima (Dynamics of the number of salmon fish of Lake Sevan in the conditions of changing its regime). Reports of Sevan Hydrobiological Station, 17, pp 221–227
Stepanyan I (2021) New data concerning *Hyla orientalis* Bedriaga, 1890 (Anura: Hylidae) in Southern Armenia. Herpetol Notes 14:325–329
Szczerbak NN (2003) Guide to the reptiles of the Eastern Palearctic. Krieger, Malabar
Tailly M, Ananian V, Dumont HJ (2004) Recent dragonfly observations in Armenia, with an updated checklist. Zool Middle East 31:93–102
Ter-Grigorian MA (1973) Chervetsy i shchitovki (Coccoidea). Muchnistye chervetsy (Pseudococcidae). Fauna Armyanskoy SSR (Scale insects. Mealybugs (Pseudococcidae). Fauna of Armenian SSR), Yerevan
Ter-Minassian ME (1947) Opredelitel' Zhukov-dolgonosikov Armenii. (Identification guide of weevils (Coleoptera, Curuculionidae) of Armenia). Zoologicheskiy sbornik Akademii Nauk Armyanskoy SSR 4:1–220
Terterian HE (1968) Moshki (Simuliidae) Fauna Armyanskoy SSR (Blackflies (Simuliidae). Fauna of Armenian SSR), Yerevan
Ter-Voskanyan H, Aghababyan K (2014) Masked Shrike (*Lanius nubicus*) in Armenia. In: XII international conference on electric machines and drive systems, ICEMDS 2014, Barcelona
Treves A, Ullas K (2003) Human-carnivore conflict and perspectives on carnivore management worldwide. Conserv Biol 17(6):1491–1499
Tsytsulina K, Masuda R (2004) Molecular phylogeny of whiskered bats (*Myotis*) in Palaearctic region. In: Mawari SW, Okada H (eds) Neo-science of natural history: integration of geoscience and biodiversity studies. Proceedings of the international symposium Dawn of a new natural history—integration of geoscience and biodiversity studies. Hokkaido University, Sapporo, pp 85–89
Tuniyev BS (1995) On the Mediterranean influence on the formation of herpetofauna of the Caucasian isthmus and its main xerophylous refugia. Russ J Herpetol 2(2):95–119
van Swaay CAM, Warren MS (2006) Prime butterfly areas of Europe: an initial selection of priority sites for conservation. J Insect Conserv 10(1):5–11
van Swaay CAM, Van Strien AJ, Aghababyan K, Åström S, Botham M, Brereton T, Chambers P, Collins S, Domènech FM, Escobés R, Fernández-García JM, Fontaine B, Goloshchapova

S, Gracianteparaluceta A, Harpke A, Heliölä J, Khanamirian G, Julliard R, Kühn E., Lang A, Leopold P, Loos J, Maes D, Mestdagh X, Monasterio Y, Munguira ML, Murray T, Musche M, Õunap E, Pettersson L, Popoff S, Prokofev I, Roth T, Roy D, Settele J, Stefanescu C, Švitra G, Teixeira S. M, Tiitsaar A, Verovnik R, Warren MS (2015) The European Butterfly Indicator for Grassland species 1990–2013. Report VS2015.009, De Vlinderstichting, Wageningen

van Swaay CAM, Van Strien AJ, Aghababyan K, Åström S, Botham M, Brereton T, Carlisle B, Chambers P, Collins S, Dopagne C, Escobés R, Feldmann R, Fernández-García JM, Fontaine B, Goloshchapova S., Gracianteparaluceta A, Harpke A, Heliölä J, Khanamirian G, Komac B, Kühn E, Lang A, Leopold P, Maes D, Mestdagh X, Monasterio Y, Munguira ML, Murray T, Musche M, Õunap E, Pettersson LB, Piqueray J, Popoff S, Prokofev I, Roth T, Roy DB, Schmucki R, Settele J, Stefanescu C, Švitra G, Teixeira SM, Tiitsaar A, Verovnik R, Warren MS (2016) The European Butterfly Indicator for Grassland species 1990–2015. Report VS2016.019, De Vlinderstichting, Wageningen

Vasilyan D (2009) Lower vertebrates of the Late Paleogene and Neogene of Armenia. PhD thesis, Yerevan

Wardikjan SA (1980) Haykakan SSH vnasakar erkarachap titerneri voroshich (Identification guide of Harmful *geometer moths* of Areminan SSR), Yerevan

Wardikjan SA (1985) Atlas genital'nogo apparata pyadenits (Geometridae, Lepidoptera) Armyanskoy SSR (Atlas of the genital apparat of geometr moths (Geometridae, Lepidoptera) of Armenian SSR), Yerevan

Weinberg P (2001) On the status and biology of the wild goat in Daghestan (Russia). J Mount Ecol 6:31

Weinberg P, Jdeidi T, Masseti M, Nader I, de Smet K, Cuzin F (2008) *Capra aegagrus* (Bezoar, Wild Goat)

Wolf C, Ripple WJ (2018) Rewilding the world's large carnivores. R Soc Open Sci 5(3):172235

World Spider Catalog (2023) Version 24. Natural History Museum Bern. https://doi.org/10.24436/2

Zamani A, Marusik YM (2021) A new genus and ten new species of spiders (Arachnida, Araneae) from Iran. ZooKeys 1054:95–126

Zarikian N (2020) A contribution to the checklist of the jumping spiders (Araneae: Salticidae) of Armenia. Bull Iraq Nat Hist Mus 16(2):193–202. https://doi.org/10.26842/binhm.7.2020.16.2.0193

Zarikian NA, Kalashian MY (2021) An annotated checklist of spiders deposited in the Arachnida collection of the Institute of Zoology, Scientific Center of Zoology and Hydroecology of the NAS RA, Yerevan, Armenia. Part I. Arachnologische Mitteilungen 61:11–19

Zazanashvili N, Garforth M, Bitsadze M (eds) (2020) Ecoregional conservation plan for the Caucasus, Tbilisi

Zarikian NA (2021) A Survey of Running Crab Spiders Philodromidae (Araneae) of Armenia Bulletin of the Iraq Natural History Museum, 2021, Volume 16, Issue 4, Pages 495–508

Zarikian NA, Propistsova EA, Marusik YM (2022) On spider families (Arachnida: Araneae) new to Armenia. Israel Journal of Entomology 51: 103–117

Zarikian NA (2022a) New records on Salticidae and Theridiidae (Araneae) spiders from Armenia. Bulletin of the Iraq Natural History Museum 17: 169–185

Zarikian N (2022b) New Records of Araneae from Armenia. Indian Journal of Entomology. 84, 4 (Dec. 2022), 757–760 https://doi.org/10.55446/IJE.2022.688

Zarikian N, Dilbaryan K, Khachatryan A, Harutyunova L (2023) Species composition and diversity of spider (Arachnida: Araneae) in the northern forests of Armenia. Biodiversity, 24 (1-2), 66–75 https://doi.org/10.1080/14888386.2023.2184424

Ziaie H, Gutleb B (1997) New comments on otters in Iran. IUCN Otter Spec Group Bull 14(2):91–92

Chapter 6
Biodiversity Conservation Problems

George Fayvush, Karen Aghababyan, Alla Aleksanyan, Marine Arakelyan, Arsen Gasparyan, Mark Kalashian, Lusine Margaryan, and Siranush Nanagulyan

Forty-four percent of the territory of Armenia is a high mountainous area. The degree of land use is strongly unproportional. The zones under intensive development make 18.2% of the territory of Armenia with concentration of 87.7% of total population. On these areas, the population density exceeds several times the ecological threshold index (200 person/km^2) reaching up to 480–558 person/km^2. The poorly developed zones make 38.0% of the territory, where only 12.3% of total population resides with a very low density of 11–20 person/km^2 (The fifth… 2014).

Due to intensive nature use, the level of anthropogenic changes of natural landscapes in Armenia is high. More than 60% of the territory is under active agriculture; in semidesert and mountainous steppe zones, the figure reaches up to 80–90%. Overexploitation has resulted in reduction and pollution of the territories covered by wild biodiversity, loss of habitats of certain species and changes in the services provided by ecosystems. The natural factors, which are risky for ecosystems and their components, are also mainly conditioned by unconscientious approaches of human towards nature. In the early post-Soviet period, this was connected with the

G. Fayvush (✉) · A. Aleksanyan · A. Gasparyan · L. Margaryan · S. Nanagulyan
Institute of Botany, National Academy of Sciences of Armenia, Yerevan, Armenia
e-mail: g.fayvush@botany.am; a.aleksanyan@botany.am; a.gasparyan@botany.am; lusinemargaryan@ysu.am; snanagulyan@ysu.am

K. Aghababyan
BirdLinks Armenia NGO, Yerevan, Armenia
e-mail: karen.aghababyan@env.am

M. Arakelyan
Yerevan State University, Yerevan, Armenia
e-mail: arakelyanmarine@ysu.am

M. Kalashian
Scientific Center of Zology and Hydroecology, National Academy of Sciences of Armenia, Yerevan, Armenia
e-mail: mark.kalashian@sczhe.sci.am

G. Fayvush (ed.), *Biodiversity of Armenia*,
https://doi.org/10.1007/978-3-031-34332-2_6

hard socio-economic conditions, energy crisis and poverty of wide classes of population in the country. Over the recent years, the negative impact on biodiversity and increased rates of ecosystem degradation have been conditioned by certain intensification of economic and social activities, which is expressed in overuse of biological resources, exploitation of mines, expansion of areas under construction, visible activation of agriculture and tourism development.

As on the entire globe, the biodiversity of Armenia is influenced by natural and anthropogenic factors. Basically almost all of the hierarchical threats suggested by IUCN have some impact on the biodiversity and ecosystems of Armenia. At the same time, nevertheless, all these factors in most cases affect biodiversity indirectly, by influencing ecosystems.

Of the natural factors, climate change has received the most attention in recent years. Climate change is one of the most pressing environment and development challenges confronting humanity today. Over the past few decades, evidence has mounted that planetary-scale changes are occurring rapidly. These are, in turn, changing the patterns of forcing and feedbacks that characterize the internal dynamics of the Earth System. Key indicators, such as the concentration of CO_2 in the atmosphere, are changing dramatically, and in many cases the linkages of these changes to human activities are strong. It is increasingly clear that the Earth System is being subjected to a wide range of new planetary-scale forces that originate in human activities, ranging from the artificial fixation of nitrogen and the emission of greenhouse gases to the conversion and fragmentation of natural vegetation and the loss of biological species. It is these activities and others like them that give rise to the phenomenon of global change (Rizvi et al. 2015).

According to our analyses (Fayvush 2010, 2015; Fayvush and Aleksanyan 2015, 2016; Fayvush et al. 2020), as the result of the climate change an expansion of the arid ecosystems, reduction of the areas covered by forests and subalpine and alpine landscapes and increased vulnerability of forests are expected. Ecologically, most unstable forests on southern slopes will become more xerophilous and arid open woodlands will shift vertically up. In the lower timberline, it is expected to have worsened conditions for seed regeneration of forests along with penetration of semidesert species as well as shift of the lower timberline up. In the lower forest zone of the central and southern Armenia, an impact of the mountainous-steppe vegetation will be observed and the stands of coppiced origin will retreat.

Vulnerability assessment of main natural ecosystems shows that in the period from now until 2100 the following changes are expected (Aleksanyan et al. 2015; Fayvush 2015; Fayvush and Aleksanyan 2016).

Alpine meadows. Prediction of changes of bioclimatic conditions shows that the general direction of condition changes will not be in the direction of subalpine meadows, as expected, but in the direction of subalpine tall grasses and expansion of wetlands.

Subalpine meadows. The transition is predicted to meadow-steppes, possibly extension of forest ecosystems on the territory of current meadows. In forest regions, raising of upper limit of the forest and in non-forest regions transition to meadow-steppe ecosystems probably will occur. It has to be noticed that alpine and subalpine meadows are the most vulnerable natural ecosystems in Armenia.

Forests. In the humid forests of the middle belt probably will begin processes of xerophytization, thinning, and penetration of plants of the steppes, arid woodlands, and shibliak. Some xerophytization of wet forests will move it into the humid forests. Modern forests of subalpine zone with time will be replaced by common humid forests, there will occur rising of upper limit of forest vegetation with a corresponding shift of subalpine crooked forests and park forests.

Meadow-steppes. Mostly is expected to transition of these ecosystems the steppes, in some cases (when the amount of precipitation will be increased), the formation of subalpine tall-grasses, and sometimes will be possible extension on the territory of modern forest ecosystems.

Steppes. The general direction of ecosystem changes is xerophytization. The modern dry steppes can be replaced by phryganoids, the areas of traganth steppes will be expanded. Current relatively mesophile steppe ecosystems can be replaced by drier subtypes.

Semidesert. In most cases, semidesert vegetation is conserved, with an extension of phryganoid zone. Also expansion of areas of desert ecosystems such as solonchaks and saline deserts is expected.

Shibliak and arid woodlands. In general, the conditions of these ecosystems will conserve and even slightly will increase, but natural regeneration of trees and shrubs can worsen, and eventually these ecosystems, especially in the lower mountain belt, can be replaced to phryganoids.

Petrophilous ecosystems and *wetlands* are intrazonal, and their vulnerability depends on their altitudinal and geographical locations.

The climate change, in particular increase of water temperature, can bring the thermal stress of hydrobionts and disruption of physiological processes and change of the behaviour of organisms. The climate change results also in the warming of the bottom cold water layers, due to which during summer months the optimal area for existence of the fish species adapted to cold waters is severely reduced. In Lake Sevan, the most valuable *Salmonidae* fish species prefer cold water. Therefore, the increase in water temperature will create unfavourable conditions for them.

The group of animals most affected by climate change are amphibians and reptiles. Climate change is causing the disappearance of lakes and ponds, leading to the disappearance of many breeding sites and habitats for rare amphibians such as newts (*Ommatotriton ophryticus*). Endemic reptiles of Armenia are also currently changing their ranges. We noted the changes in distribution in such unique species as parthenogenetic rock lizards as *Darevskia unisexualis* and *D. rostombekowi*, which are listed as Vulnerable and Endangered in IUCN, respectively. Many of them are losing their territories. Among the reptile species affected by climate change are both mesophilic species such as *Darevskia practicola*, cat snake *Telescopus fallax*, Armenian steppe viper *Vipera eriwanensis*, and arid species *Phrynocephalus persicus (= horavthi)*, *Eremias pleskei*, which have practically disappeared from their ranges due to the lack of free territories and changing conditions in the present.

From the side of the anthropogenic factor, the greatest threats are posed by human activity, which completely destroys or transforms natural ecosystems. Among them are primarily open mining, which fully destroys ecosystems and landscapes.

6.1 Open Mining

The underground resources of Armenia are very much exploited. It is estimated that in Armenia there are 613 mines with the value of 170 billion US dollars and 60 types of minerals are extracted. Armenia has 5.1% of the total world resources and 7.6% of the confirmed resources of molybdenum (The Fifth… 2014). There are also significant resources of copper, zinc, iron, lead, gold, silver, rhenium, cadmium, tellurium and other metals. Nowadays the mining industry has been declared by the RA Government as a priority sector of the economy. It is intensively developing and continues to have catastrophic consequences. Thousands of hectares of the territory of Armenia are covered by open mines and tailing ponds. The mines in Armenia are mainly concentrated in two regions – in the Alaverdi area of the Lori Region and the Kapan-Qajaran-Agarak area of the Syunik Region (Fig. 6.1). The industry of

Fig. 6.1 Open pit in Teghut (Lori region)

construction materials has also caused damage to the natural environments of Armenia. Quarries of building materials (sand, stone, gravel, etc.) occupy ever larger areas, at the detriment of both natural ecosystems and agricultural lands. At present due to the extraction of construction materials more than 7000 ha of agricultural lands have become unsuitable for use.

6.2 Construction (Urban Development, Road and Reservoir Construction)

Due to the reduction in the total volume of construction works during 2010–2011, at present the risk of their impact on ecosystems is not high. In recent years, the expansion of existing inhabited areas and establishment of new settlements (in the form of summer-house communities) have been slower and almost without occupying new territories, but construction of big greenhouse complexes on large areas has been carried out in Armenia (Fig. 6.2). Artificial fishponds occupy rather large territories in Armenia (especially on Ararat valley), but they are going as refugia for many species of migrating and nestling birds (Fig. 6.3). Works on the road network are mainly aimed at widening and renovating existing roads. On the other hand, the intensification of construction in cities and towns leads to an increase in air pollution and intensification of the extraction of building materials.

Fig. 6.2 Greenhouse complex on Ararat valley

Fig. 6.3 Fishponds on Ararat valley

Urbanization is one of the main reasons for the extinction of many species of reptiles. The construction of a road in the overpopulated area of the Ararat Plain is one of the reasons for the destruction of the habitat of rare reptiles. The dense network of main and secondary roads directly destroys their habitats. Moreover, many reptiles die on our roads and highways every year. Road reconstruction has a strong negative impact on rock lizard species of genus *Darevskia*, including endangered endemic species with a very limited global distribution, as rocks along roads are their main habitat type.

6.3 Agriculture

After the independence of Armenia, the land privatization process has seriously changed the character and status of agriculture in the country. At present, the main land users for agricultural production in the country are farms, which manage more than 82% of the arable lands, 75% of the perennial stands and 50% of the haymaking areas. In the sphere of agriculture, the most serious problems connected with the environment are the losses of water due to ineffective irrigation, as well as

Fig. 6.4 *Cirsium incanum* on abandoned field (Gegharkunik region)

salination of soils, and erosion and pollution by agricultural wastes. For the natural ecosystems being used as pastures, the biggest threat is the disproportionate distribution of the pasture load: distant pastures suffer from undergrazing, which results in changes in ecosystems, in particular the replacement of alpine carpets with alpine meadows as well as the active penetration of subalpine weeds into alpine ecosystems. On the other hand, pastures near settlements and the vicinity of animal watering places suffer from overgrazing. These territories are overused and degraded to various degrees, varying from changes in plant cover to erosion. Abandoned fields are a problem. After the privatization of land in the 1990s, peasants often received very small plots, which were not economically profitable. As a result, many sites remained abandoned and became a reservoir for weeds and other invasive and expanding species. They often form new monodominant communities (*Silybum marianum, Cirsium anatolicum, Cirsium incanum, Onopordum acanthium*, etc.), clogging the surrounding areas with their seeds (Figs. 6.4 and 6.5). A serious threat to biodiversity is the legally prohibited in Armenia, but the ongoing burning of natural pastures and agricultural fields (Fig. 6.6).

Agriculture is the main threat for the four species of reptiles listed as Critically Endangered in Red Book of Armenia and IUCN: three lizards – *Phrynocephalus horvathi, Eremias pleskei, Eremias arguta transcaucasia* and *Testudo greaca armeniaca.* The land which was once ploughed became unsuitable for endangered species of lizards. The transformation of land under gardens and vineyards may fully destroy all living populations. Surveying the habitat of populations of most distinct

Fig. 6.5 Abandoned field in Gegharkunik region

Fig. 6.6 Burning crop residues in the field (Ararat region)

forms of rare subspecies of tortoises *Testudo greaca armeniaca*, an alarming situation has been detected (Arakelyan and Parham 2008). During the last decades, this land has undergone major anthropogenic transformation, resulting in the degradation of the natural environment. Almost all the land is in private hands, and much of these holdings are ploughed. Thus, under habitat transformation, the exceptional biodiversity of semidesert is endangered. The hotspot habitats with many endangered species of reptiles occur on small patches and needs urgent protection. These patches of habitat are situated among agricultural lands. Due to their rocky compositions of this site among flat land, the rocky island is still save their natural composition. Furthermore, the neighbouring plants, fungi and animals take refuge here, in virgin land. However, the small patches of wild land also will be used and their habitat changed in near future. The establishment of mini-reserves to conserve natural populations of plants, fungi and animals on these virgin islands among the land of agriculture use will protect habitats for many endemics and rare species of wildlife.

6.4 Loggings

According to official data, the volume of illegal loggings in Armenia has reduced during the last years. But nevertheless long-term negative changes in the ecological status are observed in areas which have been subject to intensive loggings.

6.5 Hydropower Production

In Armenia, the construction of small hydropower plants (SHPP) is considered essential in development of the renewable energy sector. The construction of SHPPs is done according to the SHPP development scheme approved by the RA Government in 2009. According to this scheme, it is planned to construct 115 SHPPs. As of 1 January 2014, licenses for hydropower production and HPP construction were issued by the RA Committee on Regulation of Public Services to 150 SHPPs. Parallel to the implementation of the scheme on development of small hydropower plants, Armenia has witnessed many problems related to the overuse of water resources and river ecosystems, biodiversity, specially protected areas, landscapes, the social status of the population and life quality. Studies implemented in recent years have analysed the impact of SHPPs on the water regimes of a number of rivers, the loss of biodiversity, natural calamities, tourism development as well as on the socio-economic conditions of communities. The planning and implementation of SHPPs basically do not consider the needs of the water fauna, nor is the impact of the water regime of the mountainous rivers on littoral and aquatic ecosystems and biodiversity assessed or studied.

Fig. 6.7 Solar power stations in Gegharkunik region

6.6 Solar Power Production

Solar power stations are a relatively new, but intensively developing direction of energy in Armenia. Already, new solar power plants have been built, are operating and are planned, occupying significant areas (Fig. 6.7). Unfortunately, data on their long-term impact on natural ecosystems and biodiversity are absent. However, our observations have shown that the fenced areas of solar power plants, which prevent intensive grazing, have a positive effect on the ecosystems of the meadows and steppes of Armenia.

6.7 Recreation and Tourism

The impact of recreation and tourism on ecosystems is mainly connected with recreational trampling of the plant cover, and with anxiety and stress in animals. On the other hand, underuse of forests for recreation has indirectly contributed to the increase of harvested volumes of wood. At the same time, the development of recreation and ecotourism may be more beneficial and advantageous from economic and environmental perspectives, but an appropriate policy and organization to manage these activities are still missing in Armenia.

6.8 Environmental Pollution

The changes in biodiversity and ecosystem services connected with pollution are mainly caused by accumulation of hazardous chemical substances in the soil, pollution of groundwater and rivers, accumulation of industrial wastes and landscape degradation. Illegal and unregulated dumping of household waste is of particular

Fig. 6.8 Illegal dump in Tavush region

concern (Fig. 6.8). It disturbs the conditions for growth, development and reproduction, eliminates valuable, threatened and rare species in natural ecosystems, as well as causes declines in the productivity of agrocenoses and worsens the yield quality.

6.9 Threats to Species

6.9.1 Plants

During analysis of possible impact of climate change on rare and endangered plant species, we have taken into account not only the possible changes in ecosystems and the ecological amplitude of the adaptation of these species, the diversity of habitats in which these species can be conserved, the abundance of their populations, but also other internal and external factors. Herewith it is necessary to consider that climate change can have both negative and positive impacts on populations and distribution of rare and endangered species.

Assessing all these factors, we have concluded that for 239 species included in the Red Book of plants of Armenia (Tamanyan et al. 2010) expected climate change will have no significant impact. These species generally grow in ecosystems that are preserved under any scenario of climate change, or they have fairly wide ecological amplitude, so they can adapt to the changing conditions or easily find new habitats in the case of forced migration.

According to our hypotheses for 139 plant species included in the Red Book of plants of Armenia as a result of climate change conditions will significantly improve. At first, these are thermophilous species, for wider distribution of which now clearly amount of effective temperatures is missing. Herewith frequently does not matter their relation to one or another ecological group of water demand. In particular, among these plants are moisture-loving species (hydro- and hygrophilous), which are growing in water, on the banks of reservoirs, or mesophilic species growing in forests of the lower mountain belt, including early spring ephemera (e.g. *Cyclamen vernum, Sternbergia fischeriana, Pteridium tauricum, Oenanthe silaifolia, Carpesium abrotanoides, Rorippa spaskajae* et al.). Therefore, for all these species expected climate change cannot be considered as a threat for their existence in the territory of Armenia and should pay attention to reducing the negative impact of other factors.

On the other hand, according to the conducted modelling of changes of ecosystems and habitats due to the climate change, for 74 species of vascular plants, included in the Red Book of plants of Armenia, this factor will be the one of threats to determine the possibility of their existence in Armenia. This group of plants includes, first of all, species adapted to mesophilic conditions of subalpine and alpine belts, for which climate change will lead to a dramatic reduction of area and diversity of these ecosystems and habitats. These species are, for example, *Botrychium lunaria, Antennaria caucasica, Eriophorum latifolium, Rhododendron caucasicum, Lomatogonium carinthiacum, Scilla rosenii* and others. This group of plants includes also the inhabitants of freshwater wetland habitats of the lower and middle mountain belts, the areas of this kind of habitats will be clearly reduced due to decreasing of precipitation and increasing of temperature (*Carex pendula, Trigonella capitata, Thelypteris palustris* and others). In addition, the threat of climate change is real for mesophilic species of steppes, meadow-steppes and meadows, populations of which are small in number or are isolated from habitats, which will be available in a result of climate change. This group includes *Acanthus dioscoridys, Sternbergia colchiciflora, Grossheimia caroli-henricii,* and others. As we have noted above, for all of these species climate change will be one of the major threats for existence, but it should be remembered that for most of them there are other no less serious threats also, in particular, that most of them are growing in areas of intensive economic activities and negative impact of anthropogenic factor, which can be decisive.

We analysed the main threats (in particular climate change) for plant species included in the Red Book of Armenia in the Critically Endangered (CR) category. The results of the analysis are shown in Table 6.1.

Table 6.1 Main threats to rare plants species included in the Red Data Book of Plants of Armenia (Tamanyan et al. 2010) in the category CR

Species	Floristic region	Ecosystem	Limiting factors	Impact of climate change	Threats
Lycopodium selago	Idjevan	Moist rocks in forest	Restricted EOO[a] and AOO[b]	None	There is no specific factor that threatens the species
Pteridium tauricum	South Zangezur	Riversides and clearings	Restricted EOO and AOO, loss/degradation of habitat	Positive	Antropogenous factor: grows in the vicinity of village
Ophioglossum vulgatum	Aparan and South Zangezur	Forest	Restricted EOO and AOO, loss/degradation of habitat	Negative	Climate change, forestry activities
Salvinia natans	Lori	Lakes	Restricted EOO and AOO, loss/degradation of habitat	Negative	Natural succession, change in water regime
Thelypteris palustris	Darelegis	Wetland	Restricted EOO and AOO, loss/degradation of habitat	Negative	Change in water regime
Acanthus dioscoridys	Yerevan	Meadow-steppe	Restricted EOO and AOO, degradation of habitat	Negative	Grazing, haymaking, climate change
Sagittaria sagittifolia	Lori, Yerevan	Wetland	Degradation of habitat	Negative	Change in water regime, climate change
Sagittaria trifolia	Lori	Wetland	Degradation of habitat	Negative	Change in water regime, climate change
Allium akaka	Meghri	Screes	Degradation of habitat	Positive	Mining, road constructions
Allium egorovae	Meghri	Subalpine meadows	Degradation of habitat	Negative	Overgrazing, climate change
Allium scabriscapum	Darelegis	Screes	Restricted EOO and AOO	None	There is no specific factor that threatens the species
Allium talyschense	Darelegis	Subalpine belt, screes	Degradation of habitat	Negative	Overgrazing, climate change

(continued)

Table 6.1 (continued)

Species	Floristic region	Ecosystem	Limiting factors	Impact of climate change	Threats
Sternbergia fischeriana	South Zangezur	Forest, forest edges, shibliak	Restricted EOO and AOO	Positive	Forestry activities
Carum komarovii	South Zangezur	Subalpine and alpine belts, srees	Restricted EOO and AOO	Negative	Overgrazing, climate change
Dorema glabrum	Yerevan	Semidesert	Restricted EOO and AOO, degradation of habitat	Positive	Melioration, expansion of arable lands
Falcaria falcarioides	Yerevan	Wetland	Loss/ degradation of habitat	Positive	Drainage of marshes, burning of vegetation, melioration
Froriepia subpinnata	South Zangezur	Forest edges	Restricted EOO and AOO	None	Grazing, haymaking
Oenanthe silaifolia	Yerevan	River banks, saline soils	Restricted EOO and AOO, degradation of habitat	Positive	Melioration, expansion of arable lands, economic constructions
Opopanax persicus	Aparan	Forest edges and glades	Restricted EOO and AOO, loss/ degradation of habitat	None	Intensive recreation, expansion of residential areas
Peucedanum pauciradiatum	Meghri	Semidesert, screes	Restricted EOO and AOO, loss/ degradation of habitat	Positive	Mining, road constructions, expansion of arable lands
Tamamschaniella rubella	Lori	Forest glades	Restricted EOO and AOO	None	Forestry activities
Asphodeline lutea	Idjevan	Forest	Restricted EOO and AOO	None	Forestry activities, expansion of settlements
Amberboa amberboi	Yerevan	Semidesert	Restricted EOO and AOO	Positive	Expansion of arable land, grazing
Centaurea leuzeoides	Darelegis, South Zangezur	Steppe, screes	Restricted EOO and AOO	None	There is no specific factor that threatens the species

(continued)

Table 6.1 (continued)

Species	Floristic region	Ecosystem	Limiting factors	Impact of climate change	Threats
Centaurea schelkovnikovii	Meghri	Subalpine and alpine belts, screes, meadows	Restricted EOO and AOO	Negative	Overgrazing, climate change
Centaurea takhtadzianii	Shirak	Steppe, field edges	Restricted EOO and AOO, loss/ degradation of habitat	Negative	Expansion of arable lands, climate change
Centaurea vavilovii	Yerevan	Meadow-steppe	Restricted EOO and AOO	Positive	Expansion of arable lands, grazing, has invasive potential
Cirsium alatum	Yerevan	Salt marshes	Restricted EOO and AOO, restricted ecological adaptation	Negative	Drainage of marshes, burn of vegetation, climate change
Cousinia qaradaghensis	Meghri	Screes	Restricted EOO and AOO	None	Expansion of arable land
Echinops ritro	Darelegis	Meadow-steppe	Restricted EOO and AOO	Negative	Grazing, land development, climate change
Grossheimia caroli-henricii	Darelegis	Forest edges, road sides	Restricted EOO and AOO, loss/ degradation of habitat	Negative	Road construction, overgrazing, recreational trampling, climate change
Rhaponticoides tamanianae	Shirak, Darelegis	Steppe, steppe shrubs	Restricted EOO and AOO	None	Overgrazing, road construction
Sonchus araraticus	Yerevan	Salt marshes	Restricted EOO and AOO	Negative	Drainage, burn of vegetation, climate change
Steptorhamphus czerepanovii	Yerevan, Darelegis	Screes	Restricted EOO and AOO	None	Land development
Tomanthea carthamoidos	Yerevan	Semidesert	Restricted EOO and AOO, loss/ degradation of habitat	Positive	Land development
Paracaryum laxiflorum	Shirak	Steppe, steppe shrubs	Restricted EOO and AOO	Negative	Overgrazing, climate change

(continued)

Table 6.1 (continued)

Species	Floristic region	Ecosystem	Limiting factors	Impact of climate change	Threats
Arabis laxa	South Zangezur	Forest edges, shibliak	Restricted EOO and AOO	Positive	Grazing, intensity of economic activity
Asperuginoides axillaris	Yerevan	Rocks, screes	Restricted EOO and AOO, loss/ degradation of habitat	None	Construction
Crambe armena	Shirak, Yerevan	Semidesert	Restricted EOO and AOO	Positive	Economic activity, expansion of arable land
Didymophysa aucheri	Aparan, Yerevan	Alpine belt screes	Restricted EOO and AOO	Negative	Climate change
Diptychocarpus strictus	Yerevan	Semidesert	Restricted EOO and AOO, loss/ degradation of habitat	Positive	Arable land expansion, irrigation
Draba hispida	Aparan	Alpine belt, screes	Restricted EOO and AOO	Negative	Climate change
Isatis karjaginii	Yerevan, North Zangezur	Screes, stony slopes	Restricted EOO and AOO	None	Grazing, recreational trampling
Isatis sevangensis	Areguni	Screes, stony slopes	Restricted EOO and AOO, loss/ degradation of habitat	None	Recreational trampling
Pachyphragma macrophyllum	Idjevan	Forest	Restricted EOO and AOO	Positive	Shows invasive properties
Peltariopsis grossheimii	Meghri	Screes, semidesert	Restricted EOO and AOO, loss/ degradation of habitat	Positive	Mining, road construction
Pseudovesicaria digitata	Aparan	Alpine belt screes	Restricted EOO and AOO	Negative	Climate change
Rorippa spaskajae	North Zangezur	Wetland	Restricted EOO and AOO	Positive	Grazing
Sameraria cardiocarpa	Darelegis	Steppe, phryganoids	Restricted EOO and AOO	None	Expansion of arable land

(continued)

Table 6.1 (continued)

Species	Floristic region	Ecosystem	Limiting factors	Impact of climate change	Threats
Sameraria glastifolia	Darelegis, Meghri	Phryganoids, semidesert	Restricted EOO and AOO, loss/ degradation of habitat	Positive	Mining, road construction
Thlaspi umbellatum	South Zangezur	Shibliak	Restricted EOO and AOO, loss/ degradation of habitat	Positive	Construction
Cercis griffithii	Meghri	Semidesert	Restricted EOO and AOO, loss/ degradation of habitat	Positive	Expansion of arable land
Callitriche hermaphroditica	Aparan	Oligotrophe lake, wetland	Restricted EOO and AOO	Positive	Vegetation succession, grazing
Campanula caucasica	Sevan	Meadow-steppe	Restricted EOO and AOO, loss/ degradation of habitat	None	Mining, overgrazing
Campanula massalskyi	Shirak	Rocks	Restricted EOO and AOO	None	Mining
Allochrusa takhtajanii	Yerevan	Semidesert	Restricted EOO and AOO, loss/ degradation of habitat	Positive	Expansion of arable land
Arenaria brachypetala	Yerevan	Semidesert	Restricted EOO and AOO, loss/ degradation of habitat	Positive	Overgrazing, expansion of arable land
Bufonia takhtajanii	Yerevan	Phryganoid	Restricted EOO and AOO	None	Overgrazing, road construction
Cerastium capillatum	Lori	Meadows	Restricted EOO and AOO	Negative	Overgrazing, climate change
Cocciganthe flos-cuculi	Idjevan	Wetland	Restricted EOO and AOO	None	Grazing

(continued)

Table 6.1 (continued)

Species	Floristic region	Ecosystem	Limiting factors	Impact of climate change	Threats
Gypsophila stevenii	Idjevan	Shibliak	Restricted EOO and AOO, loss/ degradation of habitat	None	Road construction, settlement expansion
Silene arenosa	Yerevan	Sandy desert	Restricted EOO, AOO, and habitats	None	Sand excavation
Silene chustupica	South Zangezur	Screes	Restricted EOO and AOO	None	Grazing
Euonymus velutina	South Zangezur	Forest edges, plane grove	Restricted EOO and AOO	None	Economic activity
Anthochlamys polygaloides	Meghri	Sandy riverbeds	Restricted EOO and AOO	None	Change of water regime in Arax river
Beta lomatogona	Shirak, Yerevan	Semidesert, steppe	Restricted EOO and AOO, loss/ degradation of habitat	Positive	Arable land expansion, overgrazing
Bienertia cycloptera	Yerevan	Salt bodies (solonchaks)	Restricted EOO and AOO, loss/ degradation of habitat	Positive	Soil desalinization, melioration, expansion of arable land
Colchicum umbrosum	Idjevan	Forest glades	Restricted EOO and AOO, loss/ degradation of habitat	None	Forestry activities
Merendera greuteri	Shirak	Steppes	Restricted EOO and AOO, loss/ degradation of habitat	None	Arable land expansion, overgrazing
Merendera sobolifera	Yerevan	Salt marshes	Restricted EOO and AOO	Negative	Drainage, burn of vegetation, grazing, climate change
Swida iberica	South Zangezur	Forest edges, plane grove	Restricted EOO and AOO	None	Economic activity
Rosularia persica	Meghri	Rocks	Restricted EOO and AOO	None	Construction

(continued)

Table 6.1 (continued)

Species	Floristic region	Ecosystem	Limiting factors	Impact of climate change	Threats
Citrullus colocynthis	Meghri	Sandy riverbeds	Restricted EOO and AOO	None	Change of water regime in Arax river
Carex cilicica	Darelegis	Wetland	Restricted EOO and AOO	Negative	Grazing, climate change
Carex oligantha	Aparan, South Zangezur	Screes	Restricted EOO and AOO	Negative	Climate change
Carex pyrenaica	Aparan, Sevan	Screes	Restricted EOO and AOO	Negative	Climate change
Kobresia persica	Aparan	Alpine meadows	Restricted EOO and AOO	Negative	Overgrazing, climate change
Cephalaria nachiczevanica	Darelegis	Subalpine meadows	Restricted EOO and AOO	Negative	Overgrazing, haymaking, climate change
Euphorbia aleppica	Yerevan	Semidesert	Restricted EOO and AOO, loss/ degradation of habitat	Positive	Expansion of settlements
Astragalus achundovii	Darelegis	Semidesert	Restricted EOO and AOO, loss/ degradation of habitat	Positive	Expansion of arable land
Astragalus agasii	North Zangezur	Alpine belt, screes	Restricted EOO and AOO	Negative	Grazing, climate change
Astragalus amblolepis	Yerevan	Steppe shrubs	Restricted EOO and AOO, loss/ degradation of habitat	None	Expansion of arable land
Astragalus bylowae	North Zangezur	Steppe	Restricted EOO and AOO, loss/ degradation of habitat	None	Overgrazing, expansion of arable land
Astragalus campylosema	Upper Akhuryan	Meadows	Restricted EOO and AOO, loss/ degradation of habitat	None	Overgrazing, expansion of arable land

(continued)

Table 6.1 (continued)

Species	Floristic region	Ecosystem	Limiting factors	Impact of climate change	Threats
Astragalus commixtus	Yerevan	Phryganoids	Restricted EOO and AOO, loss/ degradation of habitat	None	Overgrazing, road construction
Astragalus corrugatus	Yerevan	Semideserts, saline soils	Restricted EOO and AOO, loss/ degradation of habitat	Positive	Desalinization of soil, irrigation
Astragalus globosus	Upper Akhuryan	Subalpine meadows	Restricted EOO and AOO, loss/ degradation of habitat	Negative	Overgrazing, climate change
Astragalus grammocalyx	Yerevan	Meadow-steppe, juniper open forest	Restricted EOO and AOO, loss/ degradation of habitat	Negative	Expansion of arable land, overgrazing, climate change
Astragalus humilis	North Zangezur	Meadow-steppes	Restricted EOO and AOO, loss/ degradation of habitat	Negative	Expansion of arable land, overgrazing, climate change
Astragalus lunatus	Lori	Steppe	Restricted EOO and AOO, loss/ degradation of habitat	None	Expansion of arable land, overgrazing
Astragalus ordubadensis	Meghri	Phryganoids	Restricted EOO and AOO, loss/ degradation of habitat	Positive	Mining, expansion of arable land
Astragalus refractus	North Zangezur	Meadow-steppes	Restricted EOO and AOO, loss/ degradation of habitat	None	Expansion of arable land, overgrazing
Astragalus schelkovnikovii	Yerevan	Desert, semidesert	Restricted EOO and AOO, loss/ degradation of habitat	Positive	Expansion of arable land

(continued)

Table 6.1 (continued)

Species	Floristic region	Ecosystem	Limiting factors	Impact of climate change	Threats
Astragalus schuschaensis	Sevan	Screes, rocks	Restricted EOO and AOO, loss/ degradation of habitat	Negative	Mining, climate change
Colutea komarovii	Meghri	Phryganoids	Restricted EOO and AOO, loss/ degradation of habitat	None	Expansion of arable land
Onobrychis major	Darelegis	Meadow-steppe	Restricted EOO and AOO, small density of population, loss/ degradation of habitat	None	Expansion of arable land, overgrazing, recreational trampling
Onobrychis meschchetica	North Zangezur	Shibliak	Restricted EOO and AOO, small density of population	Positive	Recreational trampling
Vicia pisiformis	South Zangezur	Forest, steppe shrubs	Restricted EOO and AOO	None	Forestry activities
Frankenia pulverulenta	Yerevan	Salt marshes, saline soils	Restricted EOO and AOO, loss/ degradation of habitat	Positive	Grazing, tillage, drainage of marshes, vegetation burning
Erodium sosnowskyanum	Aparan, Sevan	Alpine meadows	Restricted EOO and AOO	Negative	Grazing, climate change
Ornithogalum gabrielianae	Aparan	Subalpine meadows	Restricted EOO and AOO	None	Overgrazing, recreational trampling
Hypericum armenum	Idjevan, South Zangezur	Rocks	Restricted EOO and AOO	None	There is no specific factor that threatens the species
Hypericum eleonorae	Idjevan, South Zangezur	Rocks	Restricted EOO and AOO	None	There is no specific factor that threatens the species

(continued)

Table 6.1 (continued)

Species	Floristic region	Ecosystem	Limiting factors	Impact of climate change	Threats
Iris iberica	Idjevan	Shibliak	Restricted EOO and AOO, loss/ degradation of habitat	Positive	Grazing, road construction
Luzula forsteri	South Zangezur	Forest	Restricted EOO and AOO	None	Forestry activities
Teucrium canum	Shirak	Steppe	Restricted EOO and AOO, loss/ degradation of habitat	None	Mining (tufa)
Rhinopetalum gibbosum	Yerevan	Desert, semidesert	Restricted EOO and AOO, loss/ degradation of habitat	Positive	Expansion of arable land, sand excavation
Tulipa sylvestris	Idjevan	Steppe, forest edges	Restricted EOO and AOO, loss/ degradation of habitat	None	Tillage
Linum barsegianii	Yerevan	Salt marshes	Restricted EOO and AOO, small density of population, loss/ degradation of habitat	Negative	Grazing, tillage, drainage of marshes, vegetation burning
Nuphar lutea	Upper Akhuryan	Water bodies	Loss/ degradation of habitat	Positive	Change of water regime, drainage of the territory
Corallorhiza trifida	Idjevan, Aparan	Forest	Loss/ degradation of habitat	None	Forestry activities
Listera ovata	Idjevan	Forest	Loss/ degradation of habitat	None	Forestry activities
Ophrys apifera	South Zangezur	Shibliak	Loss/ degradation of habitat	Positive	Expansion of arable land, construction
Paeonia tenuifolia	South Zangezur	Shibliak	Loss/ degradation of habitat	Positive	Mining

(continued)

Table 6.1 (continued)

Species	Floristic region	Ecosystem	Limiting factors	Impact of climate change	Threats
Acantholimon fedorovii	Meghri	Phryganoids	Loss/ degradation of habitat	Positive	Mining, road construction
Acantholimon festucaceum	Meghri	Phryganoids	Loss/ degradation of habitat	Positive	Road construction
Aegilops crassa	Yerevan	Screes, rocks, grinded boulders of rivers	Restricted EOO and AOO, small density of population, loss/ degradation of habitat	None	Construction, recreational trampling
Aeluropus repens	Yerevan	Salt marshes	Restricted EOO and AOO, loss/ degradation of habitat	Positive	Grazing, tillage, drainage of marshes, vegetation burning, expansion of arable land, construction of artificial fishponds
Amblyopyrum muticum	Yerevan	Steppe	Restricted EOO and AOO	Negative	Climate change
Bromopsis gabrielianae	South Zangezur	Alpine screes	Restricted EOO and AOO	Negative	Climate change
Calligonum polygonoides	Yerevan	Desert	Restricted EOO and AOO, loss/ degradation of habitat	None	Sand excavation, grazing, expansion of arable lands
Rheum ribes	Shirak, Yerevan	Steppe	Restricted EOO and AOO, loss/ degradation of habitat	None	Expansion of arable land
Asterolinon linum-stellatum	Meghri	Phryganoids	Restricted EOO and AOO, loss/ degradation of habitat	Positive	Expansion of arable land, road construction
Primula cordifolia	Lori	Meadows	Restricted EOO and AOO, loss/ degradation of habitat	None	Overgrazing

(continued)

Table 6.1 (continued)

Species	Floristic region	Ecosystem	Limiting factors	Impact of climate change	Threats
Adonis wolgensis	Areguni, Aparan	Steppe, steppe shrubs	Restricted EOO and AOO	None	There is no specific factor that threatens the species
Ranunculus villosus	South Zangezur	Forest, meadows	Restricted EOO and AOO	None	Forestry activities, grazing
Reseda globulosa	Meghri	Phryganoids	Restricted EOO and AOO, loss/ degradation of habitat	Positive	Mining, road construction, expansion of arable land
Crataegus szovitsii	Darelegis	Forest edges	Restricted EOO and AOO	None	There is no specific factor that threatens the species
Crataegus ulotricha	South Zangezur	Forest edges	Restricted EOO and AOO	None	There is no specific factor that threatens the species
Potentilla erecta	Lori, Idjevan	Wetlands	Restricted EOO and AOO	Negative	Climate change
Potentilla porphyrantha	Sevan, Darelegis, North Zangezur	Rocks	Restricted EOO and AOO	None	Mining
Pyrus browiczii	Darelegis	Forest edges, open forest	Restricted EOO and AOO	None	There is no specific factor that threatens the species
Cruciata sosnowskyi	Sevan	Alpine screes	Restricted EOO and AOO	Negative	Climate change
Galium valantioides	Lori	Steppe shrubs	Restricted EOO and AOO, loss/ degradation of habitat	None	Grazing, expansion of arable land
Thesium compressum	Yerevan	Salt marshes	Restricted EOO and AOO, loss/ degradation of habitat	Positive	Grazing, tillage, drainage of marshes, vegetation burning, expansion of arable land
Linaria pyramidata	Aparan	Subalpine tall grasses	Restricted EOO and AOO, loss/ degradation of habitat	None	Overgrazing, recreational trampling

(continued)

Table 6.1 (continued)

Species	Floristic region	Ecosystem	Limiting factors	Impact of climate change	Threats
Scrophularia atropatana	Yerevan	Semidesert, steppe, screes	Restricted EOO and AOO, loss/ degradation of habitat	Positive	Grazing, expansion of arable land
Scrophularia takhtajanii	Meghri	Subalpine rocks	Restricted EOO and AOO	Negative	Climate change
Verbascum erivanicum	Meghri	Phryganoids	Restricted EOO and AOO	Positive	Mining, road construction, expansion of arable land
Stelleropsis magakjanii	Darelegis	Meadow-steppe	Restricted EOO and AOO, loss/ degradation of habitat	None	Grazing, expansion of arable land
Centranthus longiflorus	Darelegis	Steppe, screes	Restricted EOO and AOO, loss/ degradation of habitat	None	Grazing, expansion of arable land
Valerianella kotschyi	Darelegis	Steppe, phryganoids	Restricted EOO and AOO, loss/ degradation of habitat	Positive	Grazing, expansion of arable land
Viola caucasica	South Zangezur	Alpine rocks, screes	Restricted EOO and AOO	Negative	Climate change

[a]*EOO* Extent of occurrence
[b]*AOO* Area of occupancy

6.9.2 *Fungi*

One of the protecting measures of fungal genetic pool is to include in the Red Book mostly important and economically valuable species. The new Red Book is an important official document and guidebook for the effective conservation of the exceptional fungi's biota of Armenia.

The creation of the last edition of Red Book of Armenia was implemented during 2007–2009 on the basis of existing data and new field observations of fungi implemented by the specialists of Yerevan State University. Taking into consideration that the conservation of fungi relatively differs from that of the plants, the description of 40 species (macroscopic species of Ascomycetes and Basidiomycetes) to be included in the Red Book has been represented by IUCN's six categories (IUCN

2001). Of the species evaluated, 15 species were classified as Endangered (EN), 12 as Vulnerable (VU), 6 as Critically Endangered (CR), 4 as Data Deficient (DD), 2 as Near Threatened (NT), and 1 as Extinct (EX). The assessment of the species and the short explanation of the assessment are given, as well as the brief description of the species, its distribution, biological, ecological and phytocoenological peculiarities, limiting factors, conservation actions, and dot distribution maps.

6.9.3 Lichens

Conservation strategies for lichens can include various management and legally empowered interventions. One of the effective tools for the conservation of threatened species in the country is the addition to the Red Data Book of Armenia. During the Soviet period, the Red Book of the USSR included three lichen taxa (*Leptogium burnetiae* C.W. Dodge, *Leptogium hildenbrandii* (Garov.) Nyl., and *Usnea florida* (L.) F.H. Wigg.) growing on the territory of independent Armenia. In contrast to plants and macrofungi, lichens were not included in the Red Data Book of Armenia yet. According to the recent assessments of the conservation status of some lichen species, more than 20 of them, such as *Acrocordia cavata* (Ach.) R.C. Harris, *Anaptychia roemeri* Poelt., *Lobaria pulmonaria* (L.) Hoffm. (Fig. 6.9), *Megaspora rimisorediata* (Fig. 6.10), *Megaspora cretacea*, *Ramalina panizzei* De Not., and others can be proposed for inclusion to the upcoming update for the Red Data Book of Armenia (Gasparyan 2016).

On the other hand, 82% of epiphytic lichens grow on the territory of the nature specially protected areas (Gasparyan et al. 2016), while lichen-relevant

Fig. 6.9 *Lobaria pulmonaria* on a bark of trunk of deciduous tree in the Dilijan National Park

Fig. 6.10 *Megaspora rimisorediata* on the bark of trunk of *Juniperus* sp. in the "Khosrov Forest" State Reserve

interventions (e.g., conservation of old-growth forest patches) in the management planning documents for nature specially protected areas and foresties have not been incorporated yet.

6.9.4 Birds

The last assessment of the conservation statuses of Armenian avifauna was conducted in 2009 in frames of the Red Book of Animals of Armenia (Aghasyan and Kalashyan 2010), partly using data of National Bird Monitoring (2003–2009). According to that publication, in Armenia there is one species classified as Regionally Extinct, none of the bird species were classified in the category of Critically Endangered, 18 were assessed as Endangered, 64 as Vulnerable, and 12 species as Data Deficient. The category Near Threatened was not considered for the mentioned edition of the Red Data Book.

The recent publications made in 2019–2022 (specified below) demonstrated a need for thorough revision of the 2010 edition of the Red List of birds of Armenia, as for a number of species the change of their conservation status became obvious. Also, they show that the category Near Threatened should also be considered.

The major threats for the birds and their habitats in Armenia are related to direct persecution, transformation and destruction of habitats, pollution, overharvest of the biological resources, and the climate change.

The direct persecution happens mainly due to poaching on the birds. Poaching has a varied nature, for example, the raptors can be shot for trophy, as the habit of having the mounted specimens at home is still very common among the people. This threat especially affects large eagles, like golden eagle (*Aquila chrysaetos*) and vultures, like bearded vulture (*Gypaetus barbatus*) and griffon vulture (*Gyps fulvus*) (Aghababian 2008; Aghababyan et al. 2019; Aghababyan and Khanamirian 2020). Another way of poaching takes place due to poor knowledge of game birds and red-listed species by hunters. In this case the hunters often shoot protected ducks, like ferruginous pochard (*Aythya nyroca*) or protected waders, like white-tailed lapwing (*Vanellus leucurus*) (Aghababyan 2021). A positive example with the white-headed ducks (*Oxyura leucocephala*) demonstrated how control over the poaching at the restricted area can support the growth of the population of threatened species (Aghababyan 2019). The next reason for poaching is intentional shooting of eatable birds. In such cases, the hunters know the birds very well, as well as are aware about their protected or at least non-huntable status, but nevertheless do shoot those species for meat. Thus, such species, like Caucasian grouse (*Tetrao mlokosiewiczi*) and honey buzzard (*Pernis apivorus*) are often being shot in the country. The last case of poaching is aimed at killing the "pest" species of birds. Thus, the European bee-eaters (*Merops apiaster*) are being shot and trapped, being considered a pest that hunts on bees and affects bee hives, while the northern goshawks (*Accipiter gentilis*) and booted eagle (*Hieraaetus pennatus*) are being shot and trapped as pests which steal pigeons from pigeon breeders (Aghababyan and Stepanyan 2020).

Transformation and destruction of habitats is true for both terrestrial and freshwater ecosystems. The forests of Armenia are suffering from the unsustainable forestry management practices, which lead to habitat fragmentation and removal of snag and hollow trees. It affects the specialists of deep forests, like green warbler (*Phylloscopus nitidus*) and song thrush (*Turdus phylomelos*), hollow nesters like semicollared flycatcher (*Ficedula semitorquata*) and coal tit (*Parus ater*), and the insectivorous species, which feed on the insects that develop in dead trees and snag, such as black woodpecker (*Dryocopus martius*) and middle spotted woodpecker (*Dendrocopos medius*) (Aghababyan et al. 2010, 2017; Aghababyan 2018).

The semideserts are being occupied by horticulture, which completely transforms the habitat under orchards and vineyards. It seriously affects the specialized species, like lesser short-toed lark (*Calandrella rufescens*) and Finsch's wheatear (*Oenanthe finschii*).

The mountain steppes and meadows used for pasture suffer from the overgrazing and uncontrolled mowing. Overgrazing destroys the habitat in several phases. In the earlier stages, it results in change of the plant composition, mainly because the livestock feeds selectively, leaving out the poisonous and thorny plants, which occupy the space of grasses and legumes. Thus, it decreases the plant diversity and therefore the insect diversity of the mountain steppes and meadows, affecting such specialized species as whinchat (*Saxicola rubetra*) and European stonechat (*Saxicola rubicola*). At the later stages, it results in soil erosion, affecting most of the species of the habitat, especially the ground nesters, and even common species such as Eurasian skylark (*Alauda arvensis*) and common quail (*Coturnix coturnix*). The mowing on the flatter areas of meadows is not controlled neither from the point of view of areas for haymaking, nor seasonally. Mowing machinery therefore influences a number of species that nest in the grass, like the common rosefinch (*Carpodacus erythrinus*), or the species which have brood type of the chicks, like corn crake (*Crex crex*) (Aghababyan et al. 2021a).

The wetland ecosystems (according to IUCN 2022 habitat classification scheme) also suffer from transformation. The shorelines of the lakes are being cleaned from the macrophytes, influencing a huge variety of ducks, herons, cormorants, and grebes, as shown in the case of Lake Sevan. The brackish marshes of Ararat Plain are being consistently drained to be used for agriculture, thus destroying habitats for such specialized species, like marbled teal (*Marmaronetta angustirostris*) and black-winged stilt (*Himantopus himantopus*). The drainage of the grass marshes of the mountain plateaus lead to loss of habitats and a critical decline of common crane (*Grus grus*) and a significant decline of northern lapwing (*Vanellus vanellus*) (Aghababyan et al. 2021b). Eventually, the fragmentation of the rivers due to construction of small hydropower plants leads to decline of kingfisher (*Alcedo atthis*) but also influences white-throated dipper (*Cinclus cinclus*).

There are also some anthropogenic activities which are affecting various ecosystems. Those are related to open-pit mining and urban development, which are completely destroying the ecosystems, causing extermination of up to 80% of the bird fauna.

The pollution can be conditionally divided into two types: mechanical and chemical. The mechanical pollution is related to littering of the environment with plastic or other waste, which influences the birds mechanically. Thus, the plastic littering in Lake Sevan causes swallowing of the plastic elements and further mortality of such waterbirds like Armenian gull (*Larus armeniacus*) and great cormorant (*Phalacrocorax carbo*). Another example of mechanical pollution is related to disposal of the waste, which contains plant oil or fish oil. Such oils cover the feathers of the waterbirds breaking their thermoregulation and flight abilities, which has been happening with the white storks (*Ciconia ciconia*) since 2019 (Aghababyan et al. 2022a) (Fig. 6.11).

The chemical pollution is related to the organic pollution of freshwater ecosystems and pollution by heavy metals and Persistent Organic Pollutants (POPs), which include polychlorinated biphenyls, chlorine-organic pesticides and similar agents. The organic pollution of freshwater ecosystems leads to eutrophication of the lakes and ponds (like it is clearly visible on Lake Sevan since 2020), which influences the benthos fauna and ichthyofauna, which in turns affects the special piscivorous birds, like great crested grebe (*Podiceps cristatus*) and black-necked grebe (*Podiceps nigricollis*). The sources of contamination by heavy metals are predominantly related to poor waste disposal and possibly mining. This way, such metals like lead, copper, cadmium, chromium, mercury and others become involved into the natural cycles, and are being accumulated in the organisms of top predators such as peregrine falcon (*Falco peregrinus*) (Aghababyan and Tumanyan 2009) and short-toed snake-eagle (*Circaetus gallicus*) in terrestrial ecosystems and purple heron (*Ardea purpurea*) and great bittern (*Botaurus stellaris*) in the freshwater ecosystems. The

Fig. 6.11 White stork (*Ciconia ciconia*) polluted by oil

same mechanism works with the POPs and the only difference is the source, as the pesticides come from the agricultural lands (orchards, gardens, grain fields, and so on), and the polychlorinated biphenyls mainly come from construction waste. Therefore, the species which suffer from those agents are slightly different, as most belong to suburban areas, like levant sparrowhawk (*Accipiter brevipes*) and Eurasian hobby (*Falco subbuteo*), which breed in the orchards (Ananyan et al. 2011).

The overharvest of biological resources can be directly related to the birds, as well as the objects of their feeding or key species in the ecosystems. Thus, the improper system of decision-making for issuing the permits for harvesting game birds, along with the lack of their monitoring leads to a decrease of game birds' populations, as it was demonstrated on the examples of common pochard (*Aythya ferina*) and chukar (*Alectoris chukar*) (Aghababyan et al. 2021c, 2022c). Such situation is exacerbated by the lack of control over the number of harvested specimens, since the state inspection (the main controlling body) is understaffed and underfinanced.

The examples of overharvest of the birds' food items are mainly related to overfishing. Thus, the poaching on the fishes with electrofishing and chlorine in the small rivers and streams results in critical decline of the kingfisher (*Alcedo atthis*) in such areas. The examples of overharvest of the key elements in ecosystems are already described in the paragraph of deforestation.

The last threat is related to the consequences of the climate change, which are influencing the ecosystems and related bird fauna through several factors, such as warming of the mid temperatures (annual, spring-summer, winter), decrease of precipitations and humidity, increase of paranormal atmospheric events (unexpected freezing or warming, unexpected storms, and so on). Those factors increase aridization of the habitats, inclining the rate of their transformation. Among the various groups of the species, which suffer from these factors, it is worth mentioning that the community of the alpine species, such as white-winged snowfinch (*Montifringilla nivalis*), which is closely related to the melting snow in the brooding season. As the area of the snow patches in May–June declines due to decrease of precipitations and increase of temperature, the range of the species also shrinks accordingly. The increase of temperature affects other species too, for example, the lower border of distribution range of another alpine specialist – wallcreeper (*Tichodroma muraria*) moves up, probably following the lower border of its food items. Similarly, the factors affect most of the specialized bird species in alpine zone and in other habitats.

For conservation of birds in Armenia, 18 Important Bird and Biodiversity Areas (IBA) have been designated by BirdLife International (Heath et al. 2000); however, as the recent revision shows (Aghababyan et al. 2022b), those are not fully sufficient for protection of nationally and internationally important species and their habitat and have to be expanded. Further works on the expansion of the IBAs is preferred to be combined with the other important biodiversity areas, as it has been demonstrated in the case of Teksar Mountain (Aghababyan et al. 2022d).

6.9.5 Amphibians and Reptiles

Two out of seven species of amphibians (Amphibia) and 19 out of 51 species of reptiles (Reptilia) are listed in the latest edition of the Red Book of Armenia (2010). Among amphibians, climate change has affected the Northern banded newt *Ommatotriton ophryticus* due to the drying up of the few ponds found in the northern parts of Armenia in the southern part of their range. Habitat loss and changes in the water regime of water bodies during breeding period have also negatively affected populations of the Eastern spadefoot *Pelobates syriacus*, which are located at the edge of their ranges.

Declines in population densities have been recorded for a number of reptiles inhabiting semidesert areas, notably *Testudo graeca, Phrynocephalus persicus, Trachylepis septemtaeniata, Eumeces schneideri, Eryx jaculus, Malpolon monspessulanus, Eremias pleskei,* etc.

The unique stony semidesert landscape with many endangered replies species is located in the Arax River Valley which is home for 32 species of turtle, snake and lizard species, where 18 of them are listed in IUCN and Red book of Armenia. Currently, the unique semidesert landscape is quickly transforming to vineyards, gardens etc., which fully destroyed the natural habitat of endangered species of amphibian and reptiles. Because of low level of awareness of local people, many “unpleasure” animals such as frogs, lizard, snakes and turtles are killed without any reason. The reptiles are very vulnerable because of the degradation of the natural environment because of uncontrolled use of lands for pasturing, crop, fruit cultivation, poaching, sand open-pit mining, etc. and due to climate change reasons.

Among reptiles inhabiting the semidesert Armenia are especially endangered and need in urgent conservation activity two species of lizards *Phrynocephalus horvathi* and *Eremias pleskei* – globally classified as Critically Endangered and therefore considered to be facing an extremely high risk of extinction in the wild. Each year we are decreasing the chances to save the unique and endangered species of Ararat Valley. *P. p. horvathi (=persicus)* is endemic to the middle part of Araks river valley in E Anatolia, S Armenia, Nakhichevan (Azerbaijan) and around and north of Lake Urmiah in north west Iran. Habitats of the lizard are preserved as small sites of semidesert surrounded by agrarian lands. In the vicinity of Vedi town (Ararat region), the most usual habitat for this species is represented by the few remaining areas of sandy desert covered with scattered xerophytes grass and bushes (typically *Calligonum polygonoides*). Other localities are situated in stone semidesert zones of Armavir region. Our recent survey revealed that from 27 historical localities, currently, the lizards were detected only in three sites with extremely low population density. This means *Ph. horvathi* is on the verge of extinction in Armenia. Among reptiles from semidesert areas with declining population, one the first place is *E. pleskei*. The area of *Eremias pleskei* is restricted to a small area of Ararat plain and north west of Iran. Habitat loss is a particular threat to restricted range species. Now the lizards are detected only in one population with high density in it.

Among threatened animals of Ararat Valley is also a spur-thighed tortoise (*Testudo graeca* Linnaeus 1758). *T. graeca* is listed as Vulnerable in IUCN Red List of Threatened Animals; however, the subspecies *Testudo graeca armeniaca* which Chkhikvadze and Bakradze (1991) described from the Araks valley at Armenian borderland region with Iran and Turkey are globally noted as Critically Endangered. The range-wide studies of *T. graeca* showed that the genetics and morphology of some populations in the Araks region are of particular interest and they are the most morphologically and genetic distinct race of *Testudo* found anywhere (Parham et al. 2006; Arakelyan et al. 2018). The limitation of distribution as mountain chains and river made populations of turtles isolated. For all studied population was recorded a very low density of tortoises (Arakelyan et al. 2018). Among them, the most critical situation for survival was revealed for *T. g. armeniaca* from Armavir and Ararat regions of Armenia where they occupy very restricted territories, which are under strong human press. In additional to destruction, fragmentation of the habitats due to intensive agriculture, overgrazing and urbanization, recently are increasing the number of dealers for markets who take tortoises from local people (mostly cattleman) which are destroying the last surviving populations. Thus, the endemic forms of Ararat tortoise face serious threats to their continued survival, because they tend to have limited geographic distributions.

Among reptiles inhabiting the mountain steppe landscape, the most endangered are steppe racerunner lizards – *Eremias arguta transcaucasica, Ablepharus chernovi, Parvilacerta parva,* three parthenogenetic rock lizards of genus *Darevskia, and Vipera eirwanensis.* The steppe racerunner lizards, *Eremias arguta transcaucasica*, is known from only one locality at south west coast of Lake Sevan. The area of Armenian subspecies located at an altitude of 1920–2000 m above sea level in mountain steppe. This population is an exception habitat for *E. arguta* in high mountains. The severe climate on this high elevation, long-term isolation, and habitat peculiarity has changed the ecology of species and leads to the speciation. It is one of the rarest and most endangered reptiles' species in Armenia, which is on the edge of extinction. For more than 10 years, we have considered this species extinct for Armenia, but in 2006 one previously translocated population was occasionally found again, while maternal populations were disappear. The unique population of steppe runner lizards in Armenia occupies very small area where the core of population occupied near 50 hectares only. Moreover, steppe racerunner lizards are critically endangered because of habitat destruction, fragmentation, agricultural cultivation of land, overgrazing, road mortality, using the territory as dump with associated dogs, cats, rodents, gulls, jackdaws, crows preying up on lizards. Additionally, we recorded the negative competition with other species of lizards. Our monitoring results have shown that the number of steppe runner lizards was three time declined during 2008–2015 while next species of striated green lizards *Lacerta strigata* have increased the density of population. Thus, the in situ and ex situ conservation actions are urgent for saving this species for Armenia. In steppe zone of central Armenia we have discovered the presence of Chernov' skink *Ablepharus chernovi* for fauna of Armenia after more than 40 years of absence of records (Arakelyan et al. 2019). The highly fragmented distribution of dwarf lizard *Parvilacerta parva* in Armenia is located on the north-eastern edge of its geographic range and represented by 31 small and isolated populations (Arakelyan et al. 2018). As result of land development for agriculture, *P. parva* is listed as Critically Endangered in Armenia (Arakelyan et al. 2011) where more than ten locations were destroyed (Arakelyan et al. 2018). Of the seven parthenogenetic species of genus *Darevskia,* the sixth are globally endangered. Three species of parthenogenetic species (*D. unisexualis, D.rostombekowi and D.dahli*) are included in Red Book of Armenia because of habitats of lizard undergoing major anthropogenic transformation. *D. rostombekowi* ("Endangered" in IUCN) has a relatively small range, consisting of several different isolated areas within northern Armenia. It inhabits small alpine relict isolated areas in the forest, mountain meadows, mountain steppes and anthropogenic-transformed habitats. *D. dahli* ("Endangered" in Red Book of Armenia, "Near Threatened" in IUCN) is widespread in north-eastern Armenia and southern part of Georgia. Suitable habitats of the species in north-eastern Armenia are divided into seven vast isolated areas assigned to highland forest, meadow and steppe zones. *D. unisexualis* ("Vulnerable" in IUCN) range covers the territory of northern and central highland Armenia. The species are located at rock outcrops, piles of stones and rocky slopes in the mountain-steppe zone. During last decades, based on fieldwork we have noted the declining of the areas of parthenogenetic lizards due to undergone major anthropogenic transformation of their surroundings and climate change.

Armenian steppe viper, *Vipera eirwanensis*, is one of the most threatened snake species. Vipers of Armenia, as worldwide, face a large number of threats. In the IUCN Red List those species includes the Armenian steppe viper as "Vulnerable." Habitat loss, climate change and persecution by humans are often threats to this species. The construction of new roads, apart from the direct habitat destruction caused, brings more visitors to formerly remote regions.

The rapid decline of endangered species of reptiles has raised concerns about their conservation and the urgent need for action. Therefore, for the management and conservation of endangered species of amphibians and reptiles are important habitat protection and species management.

6.9.6 Invertebrates

The last assessment of the conservation status of Armenian invertebrates was conducted in 2009 in frames of the Red Book of Animals of Armenia (Aghasyan and Kalashyan 2010). According to that publication, in Armenia there are 16 species of mollusks classified as Critically Endangered (13 species) and Endangered (3 species) and 139 species of Arthropods (Insects only) of which one species is assessed as Extinct, 34 as Critically Endangered, 63 as Endangered and 41 as Vulnerable. It must be underlined that representation of Invertebrates in the book is much worse than for Vertebrates, whereas about a 2/5 of the vertebrate species of Armenia are represented in the Red Book; for invertebrates this figure is less than 1% of the estimated total number of species. Due to the absence of respective specialist and initial data, several important groups of invertebrates are not presented at all (e.g. Arachnidan, Crustacean, majority of the insect orders). Besides, new data collected after 2010 show necessity of re-assessment of the status of several species. Finally, biological traits of many species included in the Red Book are understudied which makes it difficult to take conservation measures. Thus, the Invertebrate section of the book needs a thorough revision.

Like all other organisms, invertebrates are sensitive to transformation and destruction of habitats, pollution, overharvest of the biological resources, and the climate change, while the direct persecution in the most cases has quite subordinate importance.

Taking into consideration the close relationship of invertebrates with inhabited ecosystems, it should be expected that the effects of climate change described above for these ecosystems will lead to parallel changes in the conditions of invertebrates' populations. The most threatened species will be those associated with high-mountain – alpine and subalpine ecosystems, namely, representatives of the genus *Dorcadion – D. kasikoporanum, D. bistriatum, D. gorbunovi, D. semilucens, D. cineriferum* and *D. sevliczi*, as well as ground beetle *Trechus infuscatus* and white butterfly *Artogeia bowdeni*, two latter have very restricted range in Aragats mountain. Reduction in their ranges should be expected due to simple geometric considerations.

Intensive agricultural use, especially, overgrazing and uncontrolled mowing are the main threats for several red-listed species inhabiting so-called open landscapes, from semidesert to high-mountain meadows. Destruction of vegetation cover and soil compaction due to grazing leads to the worsening of breading condition for most of herpetobiont species including *Dorcadion* species and *Trechus infuscatus* above mentioned, and some other ground beetles – *Deltomerus khnzoriani, Chylotomus alexandri, Poecilus festivus*, longhorn *Conizonia kalashiani*, as well as orthopterans *Scotodrymadusa satunini, Bicolorana roeseli, Poecilimon geoktshaicus* and *Poecilimonella armeniaca*. The same reasons may have led to the disappearance of the grasshopper *Saga pedo* which was known only from steppe near Sevan and was not collected here since 1957. Parallel with grazing mowing can lead to the worsening of development conditions of several hortobiont species which primarily applies to numerous species of butterflies, for example, blues *Plebejus transcaucasicus, Tomares romanovi, Maculinea alcon monticola, M. nausithous*, as well as Hymenopterans including all seven species of bumblebees of the genus *Bombus* (*Bombus armeniacus, B. portshinskii, B. alagesianus, B. pratorum, B. terrestris, B.niveatus, B. daghestanicus*).

Probably, the greatest concern is the state of species inhabiting semideserts of Ararat Plane. These ecosystems are the subject of intensive agricultural use being nearly completely transformed to orchards, vineyards and arable lands, and, in recent years, to greenhouses. As a result, the areas of such species are highly fragmented and their habitats are situated under intensive anthropogenic pressure. In particular, these habitats can suffer from uncontrolled use of insecticides and fertilizers, which penetrate neighbouring agricultural lands. An example is the numerous red-listed species inhabiting Goravan Sands near Vedi town. Majority of these species have very limited distribution with coinciding Extent of Occurrence and Area of Occupancy and restricted to a small territory of very peculiar community of sandy desert. These are endemic of Goravan Sands click beetle *Cardiophorus araxicola* and some other species in Armenia known mostly from here – chafers *Pharaonus caucasicus, Anisoplia reitteriana* and *Tanyproctus araxidis*, darkling beetle *Cyphostethe semenovi*, jewel-beetle *Sphenoptera khnzoriani*, wingless grasshopper *Nocarodes nodosus*, assassin fly *Machimus erevanensis*. To protect this ecosystem, a special state sanctuary with an area of about 80 hectares has been organized, but a significant part of the areas of Red Book species has remained outside the sanctuary, and in recent years, these territories have been almost completely ploughed under orchards and vineyards – without any assessment. In fact, the very existence of the habitat inside sanctuary is under threat as well, since irrigation of these lands can lead to a change in the hydrological regime of the soil and worsening of the conditions in the sanctuary.

The wetland ecosystems also suffer from transformation. For freshwater molluscs, most influent threats are the changes of hydrological regime of the water bodies inhabited. In particular, among the most threatened species classified as Critically Endangered or Endangered, *Acroloxus lacustris, Gyraulus albus, G. laevis, Planorbis carinatus*, very likely disappeared in Lake Sevan and small nearby reservoirs due to the decrease in the lake level. Violation of the hydrological regime and water pollution in small reservoirs inhabited by *Musculium strictum* and *Odhneripisidium annandalei* are also considered major threats to these species. For sure, the same threats can be considered the major for all 70 dragonfly species red-listed. Abrupt changes in the shoreline of Lake Sevan together with overexploitation for recreational purposes pose a threat to the endemic of these habitats, the ground beetle *Dyschirius sevanensis*. Changes of hydrological regime of salt marshes inhabited by Ararat cochineal *Porphyrophora hammelii* which is preserved in very fragmented plots in Ararat Plane surrounded by irrigated agricultural lands is a major threat for this endemic species as well. Cleaning of vegetation along shorelines of some rivers in Ararat plane may lead to worsening of habitat conditions of comb-clawed beetle *Ctenicera pectinicornis*.

The state of forest species included in the Red Book causes relatively least concern. The most important threat for some xylophagous species is a common practice of forestry which provides removal of old and overgrown trees from forest ecosystems which leads to a reduction of microhabitats suitable for the development of such species as longhorns *Cerambyx cerdo acuminatus* and *Rosalia alpina* (both included in the Annex II of Bern Convention), click-beetle *Ctenicera pectinicornis*, Armenian endemic comb-clawed beetle *Isomira armena*. Unsustainable forestry and, probably, climate change can lead to changes in microhabitat conditions for endemic of Syunik Region snail *Euxina akramowskii*. Though irregular but also important threat for forest insects is aviation pesticide treatment against outbreaks of phyllophagous caterpillars which affects the butterfly fauna as a whole. For example, according to our observations aviation pesticide treatment carried out in 1990s in some forests of Central Armenia led to significant decrease of population number of Apollo butterfly *Parnassius apollo kashtshenkoi* included in the Annex II of Bern Convention.

Urbanization and development of infrastructure lead to deep transformation and destruction of habitats. In this sense, most threatened species can be considered some narrowly distributed species inhabiting landscapes in the city limits of Yerevan. These are endemics of Armenia darkling beetle *Ectromopsis bogatchevi* and weevils *Cyclobaris richteri* and *Baris mirifica* distributed mainly or even exclusively in the neighbouring of Nubarashen settlement.

It must be emphasized that for the most of threatened species it is hard to identify specific threats which usually influence the populations condition synergistically. This must be taken into consideration when elaborating conservation measures. Protection of invertebrates is only possible in the frame of habitats' protection, which is best within the system of protected areas. But in the existing protected areas, biodiversity of Armenia presented insufficiently. From 155 red-listed species only 95 are presented in the existing PAs of Armenia (Fayvush et al. 2013). Also among 30 prime butterfly areas allocated by the Butterflies of Armenia project, only 12 are covered by existing PAs (https://www.butterfly-conservation-armenia.org/prime-butterfly-areas.html). Thus, conservation of invertebrate fauna of Armenia needs serious improvement of the system of protected areas of the country. Outside of protected areas, more strict and effective regulation of economic activity is quite necessary. These include regulation of grazing and mowing, some changes in forestry practice, regulation of irrigation and reducing water pollution, etc. Agricultural transformation of virgin lands should be carried out only if the EIA procedure is carried out.

It is well known that the best method of conservation of individual species of plants and animals is conservation of their habitats with the whole ecosystem. The best method of ecosystem's conservation is to conserve them in the specially protected areas (SPAs). The network of protected areas in Armenia began to take shape back in the Soviet times; at the moment there are 3 state reserves, 4 national parks, 27 sanctuaries, and 232 natural monuments. The total area of protected areas in Armenia covers 13.1% of the territory of the republic. However, despite a very large area, the network of protected areas does not cover all the diversity of the most important habitats. So, in recent decades, some new SPAs were proposed but not approved by the Government yet.

References

Aghababian K (2008) The distribution and number of golden eagles in Armenia. In: Research and conservation of the raptors in Northern Eurasia: materials of the 5th conference on raptors of Northern Eurasia. Ivanovo, 4–7 February 2008, pp 169–170

Aghababian K, Tumanyan S (2009) On the breeding peregrines *Falco peregrinus brookei* in some regions of Armenia. In: Sielicki J, Mizera T (eds) Peregrine falcon populations – status and perspectives in the 21st century. Turul, Warsaw, pp 99–108

Aghababyan K (2018) Habitat requirements of the Semicollared Flycatcher *Ficedula semitorquata* in Armenia. A BOU-funded project report. BOU, Peterborough, UK

Aghababyan K (2019) Recent counts of White-headed Ducks *Oxyura leucocephala* in Armash wetlands, Armenia. Sandgrouse 41:5–6

Aghababyan K (2021) Status and conservation of White-tailed Lapwing *Vanellus leucurus* in Armenia. Wader Study 128(1):87–92

Aghababyan K, Khanamirian G (2020) The state of Bearded Vulture, *Gypaetus barbatus* (Linnaeus, 1758) in Armenia. Bird Census News 32(1–2):11–16

Aghababyan K, Stepanyan H (2020) Booted Eagle *Hieraaetus pennatus* J. F. Gmelin, 1788 in Armenia: update on conservation status. J Life Sci 14:14–21

Aghababyan K, Ananyan V, Kalashyan M, Khanamiryan G (2010) Analysis of forest pests and pestholes exacerbated by climate change and climate variability in Syunik Marz of Armenia and to identification of the most applicable prevention measures. Report under the "Adaptation to Climate Change Impacts in Mountain Forest Ecosystems of Armenia" UNDP/GEF/00051202 Project

Aghababyan K, Ter-Voskanyan H, Khachatryan A, Gevorgyan V, Khanamirian G (2017) The state of semi-collared Flycatcher (*Ficedula semitorquata* Homeyer, 1885) in Armenia. In: Proceedings of the III International Scientific Conference "Biological diversity and conservation problems of the fauna – 3", Yerevan

Aghababyan K, Khanamirian G, Gevorgyan V (2019) An update of current situation of Griffon Vultures *Gyps fulvus* (Hablitz, 1783) in Armenia. Tichodroma J Slovak Ornithol Soc 31:3–10

Aghababyan K, Khachatryan A, Ghazaryan A, Gevorgyan V (2021a) About the state of Corn Crake *Crex crex* Bechstein 1803 in Armenia. Bird Census News 34(1):9–17

Aghababyan K, Khanamirian G, Ghazaryan A, Gevorgyan V (2021b) About conservation status of Northern Lapwing *Vanellus vanellus* in Armenia. J Ecol Nat Resour 5(3):1–9

Aghababyan K, Khachatryan A, Baloyan S, Ghazaryan A, Gevorgyan V (2021c) Assessing the current status of the Common Pochard *Aythya ferina* in Armenia. Wild 71:147–166

Aghababyan K, Khanamirian G, Manukyan M, Ghazaryan H, Pahutyan N, Khechoyan A, Hambardzumyan K, Sharimanyan M, Avetisyan K, Tsaturyan A, Arabyan A, Sahakyan K, Sayadyan A, Khachatryan A, Martirosyan B, Gevorgyan V, Ghazaryan A, Sundar G (2022a) White Storks *Ciconia ciconia* L. became victims of environmental pollution of Ararat Plain of Armenia. J Ecol Nat Resour 6(1):1–10

Aghababyan K, Khanamirian G, Khachatryan A, Grigoryan V, Tamazyan T, Baloyan S (2022b) Revision of important bird and biodiversity areas of Armenia. Int J Zool Anim Biol 5(1):1–27

Aghababyan K, Aebischer N, Baloyan S (2022c) The modern state of Chukar *Alectoris chukar* J. E. Gray, 1830 in Armenia. Ornis Hungarica 30(1):80–96

Aghababyan K, Khanamirian G, Khachatryan A, Martirosyan B, Grigoryan V, Zuerker T, Baloyan S (2022d) Evaluation of importance of Teksar Mountain of Armenia for bird and butterfly protection. Int J Zool Anim Biol 5(5):1–14

Aghasyan A, Kalashyan M (eds) (2010) Red Data Book of the Republic of Armenia. Yerevan

Aleksanyan A, Khudaverdyan S, Vaseashta A (2015) Modeling river ecosystems vulnerability assessment from climate change – case study of Armenia. Pol J Environ Stud 24(2):871–877

Ananyan V, Tumanyan S, Janoyan G, Aghababyan K, Bildstein K (2011) On breeding Levant Sparrowhawk in selected regions of Armenia. Sandgrouse 33(2):114–119

Arakelyan M, Parham JF (2008) The geographic distribution of turtles in Armenia and Nagorno-Karabakh Republic (Artsakh). Chelonian Conserv Biol 7(1):70–77

Arakelyan M, Danielyan F, Corti C, Sindaco R, Leviton AE (2011) Herpetofauna of Armenia and Nagorno-Karabakh. Salt Lake City SSAR

Arakelyan M, Türkozan O, Hezaveh N, Parham J (2018) Ecomorphology of tortoises (Testudo graeca complex) from the Araks river valley. Russ J Herpetol 25(4):245–252

Arakelyan M, Ananian V, Petrosyan R (2019) Rediscovery of Chernov's skink (*Ablepharus chernovi* Darevsky, 1953) in Armenia. Herpetol Notes 12:475–477

Chkhikvadze VM, Bakradze MA (1991) On the systematic position of the recent land turtle of from the Arax valley. Trudy Tbilisskogo Gosudartswenogo Univiversiteta [The works of Tbilisi State University] 305:59–63

Fayvush G (coord) (2010) Climate change impacts, vulnerability assessment and adaptation. In: Second national communication on climate change under the United Nations Framework Convention on Climate Change, Yerevan, pp 47–77

Fayvush G (coord) (2015) Climate change impacts: vulnerability assessment and adaptation. In: Third national communication on climate change under the United Nations Framework Convention on Climate Change. Yerevan, pp 51–87

Fayvush GM, Aleksanyan AS (2015) Some evidences of climate change impact on the flora and vegetation of alpine belt of Armenia and forecasts of main ecosystems change. Electron J Nat Sci 1:19–22

Fayvush GM, Aleksanyan AS (2016) Climate change as threat to plant diversity of Armenia. Takhtadjania 3:112–126

Fayvush G, Tamanyan K, Kalashyan M, Vitek E (2013) "Biodiversity hotspots" in Armenia. Ann Naturhist Mus Wien B 115:11–20

Fayvush G, Ghazaryan H, Jenterejyan K, Aleksanyan A (2020) Natural ecosystems and biodiversity. In: Fourth national communication on climate change. Yerevan, pp 101–117

Gasparyan A (2016) The list of epiphytic lichens proposed for registration in the Red Book of the Republic of Armenia, Yerevan

Gasparyan A, Aptroot A, Burgaz AR, Otte V, Zakeri Z, Rico VJ, Araujo E, Crespo A, Divakar PK, Lumbsch HT (2016) Additions to the lichenized and lichenicolous mycobiota of Armenia. Herzogia 29(2):692–705

Heath MF, Evans MI, Hoccom DG, Payne AJ, Peet NB (eds) (2000) Important bird areas in Europe: priority sites for conservation. Cambridge

IUCN (2001) IUCN Red List categories and criteria. Version 3.1. Gland, Switzerland

IUCN (2022) Habitats classification scheme (Version 3.1). https://www.iucnredlist.org/resources/habitat-classification-scheme

Parham JF, Türkozan O, Stuart BL, Arakelyan M, Shafei S, Macey JR, Werner YL & Papenfuss TJ (2006) Genetic Evidence for Premature Taxonomic Inflation in Middle Eastern Tortoises. Proc Cal Acad Sci 57(3):955–964
Rizvi AR, Baig S, Verdone M (2015) Ecosystem based adaptation: knowledge gaps in making an economic case for investing in nature based solutions for climate change. IUCN
Tamanyan K, Fayvush G, Nanagjulyan S, Danielyan T (eds) (2010) The Red Book of plants of the Republic of Armenia. Yerevan
The fifth national report to convention on biological diversity (2014) Yerevan

Chapter 7
Problems of Invasive Plants and Animals

George Fayvush, Alla Aleksanyan, Marine Arakelyan, Hripsime Hovhannisyan, and Mark Kalashian

Alien invasive species are considered the most important threat to biodiversity and ecosystem services in many countries of the world. At the same time, they have a very significant impact on the standard of living of the population (IPBES 2018). Very often, they are the main reason for the extinction of local species (Bellard et al. 2015) and cause major changes in natural ecosystems (Cacabelos et al. 2020; Liu et al. 2020), influencing the services provided by nature. Over the past hundred years, the number of cases of intentional and unintentional introductions of alien species has sharply increased worldwide (Seebens et al. 2017). Naturally, a very large number of alien species were introduced intentionally for economic benefit or to improve the living conditions of the population, possibly for cultural or educational purposes (Liu et al. 2012; Pipek et al. 2020; Pyšek et al. 2020). Many other alien species have been introduced unintentionally, for example, with ballast waters, transport, soil, and forage material or food (Saul et al. 2017). It has been shown that the pathways by which alien species have been transported to new regions have changed over time (Hulme 2009; Essl et al. 2015). In addition, it is clear that many of the most problematic species found their way to their new home in several ways and, most likely, more than once (Wilson et al. 2009; Essl et al. 2015; Saul et al. 2017). As a result of many years of research on alien invasive plant species in

G. Fayvush (✉) · A. Aleksanyan · H. Hovhannisyan
Institute of Botany after A. Takhtadjan NAS RA, Yerevan, Armenia
e-mail: gfayvush@yahoo.com; alla.alexanyan@gmail.com; ripi91@mail.ru

M. Arakelyan
Yerevan State University, Yerevan, Armenia
e-mail: arakelyanmarine@ysu.am

M. Kalashian
Scientific Center of Zoology and Hydroecology, National Academy of Sciences of Armenia, Yerevan, Armenia
e-mail: mark.kalashian@sczhe.sci.am

G. Fayvush (ed.), *Biodiversity of Armenia*,
https://doi.org/10.1007/978-3-031-34332-2_7

Armenia, we have accumulated diverse material on their distribution throughout the territory of the republic and the impact on natural ecosystems.

Here, we consider only the most dangerous and potentially dangerous invasive plant species. Invasive animals and fungi will be the object of special studies in the near future.

7.1 Invasive Plants

From the list of plant species (Fayvush et al. 2020) that were imported or brought to Armenia at one time or another (more than 400 names), we have selected 14 species. These species already now pose a threat to natural ecosystems, biodiversity, public health, agricultural or forestry activities; or, according to our forecasts, they can become so in the near future due to climate change, growing economic activity, or due to their own invasive potential. The data obtained during our research are combined in Table 7.1.

Acer negundo L., a North American species introduced to Europe in the seventeenth century, is currently considered an invasive species in the majority of European countries, as well as in Japan and New Zealand (CABI 2022a). It appeared in Armenia probably at the beginning of the twentieth century, since the first herbarium collection in the herbarium of the Takhtajan Institute of Botany of the National Academy of Sciences of the Republic of Armenia (ERE) dates back to 1924. Starting from the 1930s, it was widely used in landscaping settlements and creating protective forest zones along roads and railways. At present, having escaped from plantings owing to good seed and vegetative propagation, it penetrates into natural ecosystems, mainly in coastal habitats along streams and rivers in the lower and middle mountain zones up to 1500 m a.s.l. At the same time, populations of this species were found on the territory of the Shikahogh Reserve and the Khosrov Forest Reserve, as well as on the territory of the Dilijan, Sevan, and Arevik National Parks (Fig 7.1). The degree of threat to natural ecosystems is assessed by us as "high."

Ailanthus altissima (Mill.) Swingle. The homeland of the species is China. It was introduced to many countries of the world as an ornamental plant, has now spread almost all over the world, and is considered a dangerous invasive species in many countries. It was deliberately introduced into Armenia in the 1930s for landscaping settlements (Favush and Tamanyan 2014). Excellent seed and vegetative reproduction allowed it to escape from plantings and spread widely in Armenia, mainly in fairly humid habitats of the lower and middle mountain zones. Owing to vegetative propagation, it often forms dense thickets and penetrates into natural ecosystems, forests and light forests, forming monodominant communities, changing natural ecosystems. It grows on the territory of a number of specially protected natural areas (Erebuni, Shikakhokh, and Khosrov Forest Reserves, Dilijan and Arevik National Parks) (Fig. 7.2). The degree of threat to natural ecosystems is assessed by us as "very high."

Table 7.1 The most dangerous invasive and potentially invasive plant species in Armenia

Species	Pathway	Method of reproduction and distribution	Time of appearance in Armenia	Modern distribution in Armenia (by floristic regions)
Acer negundo	Intentional introduction	Seeds and vegetatively	1920s	Lori, Idjevan, Aparan, Yerevan, Sevan, Areguni, North and South Zangezur, Meghri
Ailanthus altissima	Intentional introduction	Seeds and vegetatively	1930s	Lori, Idjevan, Aparan, Yerevan, Darelegis, North and South Zangezur, Meghri
Ambrosia artemisiifolia	Self-penetration and distribution	Seeds	1983	Lori, Idjevan, Aparan, Yerevan
Buddleja davidii	Intentional introduction	Seeds and vegetatively	Mid-twentieth century	Idjevan, Meghri
Cirsium incanum	Self-penetration and distribution	Seeds and vegetatively	No later than the nineteenth century	All floristic regions
Clematis vitalba	Intentional introduction	Seeds	1940	Idjevan
Conyza canadensis	Unintentional introduction	Seeds	No later than the nineteenth century	Lori, Idjevan, Aparan, Yerevan, Darelegis, North and South Zangezur, Meghri
Grindelia squarrosa	Unintentional introduction	Seeds	2015	Shirak
Helianthus tuberosus	Intentional introduction	Seeds and vegetatively	Mid-twentieth century	Lori, Idjevan, Aparan, Yerevan
Hippophae rhamnoides	Intentional introduction	Vegetatively	Mid-twentieth century	Sevan, Areguni
Leucanthemum vulgare	Self-penetration and distribution	Seeds and vegetatively	1952 – Lori	Upper Akhuryan, Shirak, Lori, Idjevan, Aparan, Sevan, Areguni, Yerevan, since 2003 North Zangezur
Robinia pseudoacacia	Intentional introduction	Seeds and vegetatively	1930s	Lori, Idjevan, Yerevan, South Zangezur
Silybum marianum	Self-penetration and distribution	Seeds	1967 and 1980	Lori, Idjevan, North and South Zangezur, Meghri
Solidago canadensis	Intentional introduction	Seeds and vegetatively	Mid-twentieth century	Lori, Aparan, Sevan, Areguni, Yerevan

Ambrosia artemisiifolia L. The species is native to North and Central America. Currently, the species is found on all continents except Antarctica. Owing to high seed productivity (seeds are small, easily carried by wind with seeds of grain crops and by road and rail transport), it spreads intensively on its own and is considered a dangerous invasive species in most countries.

Fig. 7.1 *Acer negundo*

Fig. 7.2 *Ailanthus altissima*

In Armenia, it was first discovered in 1983 at the mouth of the Aghstev River (Gabrielyan and Tamanyan 1985), where it most likely penetrated from the territory of Azerbaijan (Fig. 7.3). In 1997, it was discovered at the mouth of the Debed River at a distance of about 50 km from the first locality, where, most likely, it entered from the territory of Georgia, since no intermediate subpopulations between these two points have been found so far. Until approximately 2010, populations occupied insignificant areas in disturbed habitats in the lower reaches of the Debed and Aghstev rivers. Then the intense distribution of the species across the territory of Armenia began: along the valleys of these rivers, the species rose to a height of 1700–1800 m a.s.l. and reached the cities of Spitak and Dilijan (most likely the seeds were transferred by road and rail transport), growing in disturbed habitats along roads and railways. In addition, one population was found in cultivated fields in the village of Dsegh and also much to the south: near the Yerevan–Sevan highway. It should be noted that the species was found on lawns in Yerevan and in flower beds on the Ararat Plain. The species was found on the territory of the Dilijan National Park. We assess the degree of threat to natural ecosystems as "high" and the threat to public health as "very high" (the pollen of the species is considered a very strong allergen).

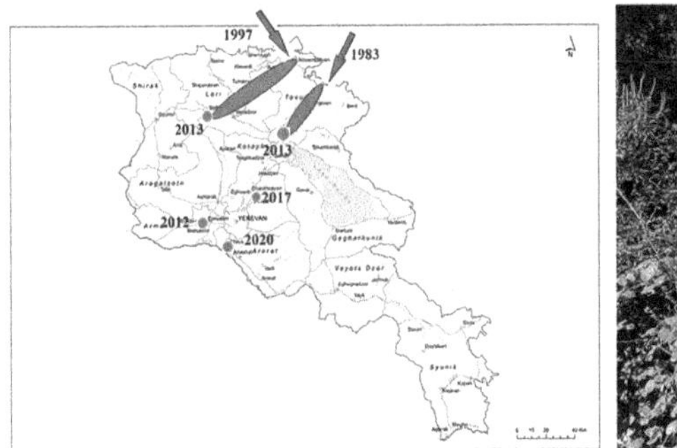

Fig. 7.3 *Ambrosia artemisiifolia*

Fig. 7.4 *Buddleja davidii*

Buddleja davidii Franch. The homeland of the species is China. It was widely used as an ornamental shrub in many countries of the world since the eighteenth century, and it is currently considered a dangerous invasive species in the United States, Canada, Great Britain, New Zealand, and New Guinea (CABI 2022b). In Armenia, it has been grown as an ornamental plant since the second half of the twentieth century in Yerevan (a few plantings in parks) and in settlements in the north and south of the republic. At present, we have found individual specimens that "escaped" from the culture and grow in disturbed habitats in the Ijevan and Meghri floristic regions (Fig. 7.4). It is also found on the territory of the Arevik National Park. So far, we have assessed this species as "potentially invasive."

Cirsium incanum (S.G. Gmel.) Fisch. ex M. Bieb. Southeastern Europe and the countries of the Eastern Mediterranean are considered to be the homeland of the species. It is often considered as a synonym for *Cirsium arvense* (L.) Scop., but we follow the authors of the *Flora of Armenia* and consider it an independent species (Arevshatyan 1995). Back in the seventeenth century, the species was discovered in the United States, and now it has spread across Europe, Western Asia, South and North America, Africa, Australia, and New Zealand owing to its high seed productivity (CABI 2022c). It is considered a dangerous invasive species. It came to Armenia no later than the nineteenth century (possibly earlier); the first herbarium collections in the ERE herbarium date back to 1920. That is, the species is among the very first collections. It can be assumed that the seeds of the species came to Armenia with forage of the troops of the Russian army during the Russo-Turkish war. Currently, the species is distributed throughout the territory of Armenia; it is found both in disturbed habitats and abandoned agricultural lands and in natural ecosystems of meadows and steppes. It is distributed on the territory of all state reserves (Erebuni, Shikakhokh, and Khosrov Forest) and national parks (Arpi Lich, Dilijan, Sevan, and Arevik) of Armenia (Fig. 7.5). We assess the degree of threat to natural ecosystems as "very high."

Clematis vitalba L. The birthplace of the species is Europe and the Mediterranean region. The species naturalized in North America and Australia in the nineteenth and twentieth centuries, where it is considered invasive (CABI 2022d). The species was most likely brought to Armenia as an ornamental plant in the first half of the twentieth century, and it was grown on a personal plot in the city of Noyemberyan, since the first collection of this species stored in the ERE herbarium is dated 1940. The species was considered very rare (until recently, only two localities were known) and was included in the Red Book of Plants of Armenia (Tamanyan et al. 2010) as disappearing. Our observations have shown that at present the species is intensely spreading in the Ijevan floristic region, where on the territory between the villages of Dzhujevan and Koghb along the edges of the forest it forms dense thickets, penetrating into forest ecosystems. Its distribution has already reached the city of Ijevan (Fig. 7.6). Although the distribution of this species in Armenia is still limited to one floristic region, we assess the degree of threat to natural ecosystems as "high."

Conyza canadensis (L.) Cronquist. The homeland of the species is North America; it entered Europe in the nineteenth century and quickly spread owing to high seed productivity (seeds are small, easily carried by the wind over long distances). Currently, it is widely distributed around the globe (CABI 2022e). It appeared in Armenia probably at the end of the nineteenth century, most likely from the territory of Georgia or Azerbaijan, since the first collections in the ERE

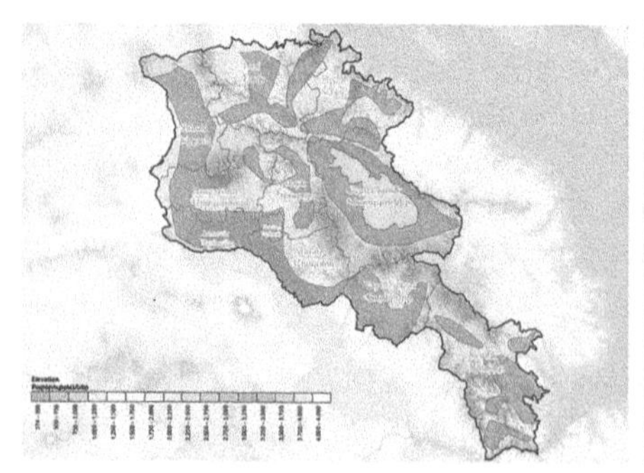

Fig. 7.5 *Cirsium incanum*

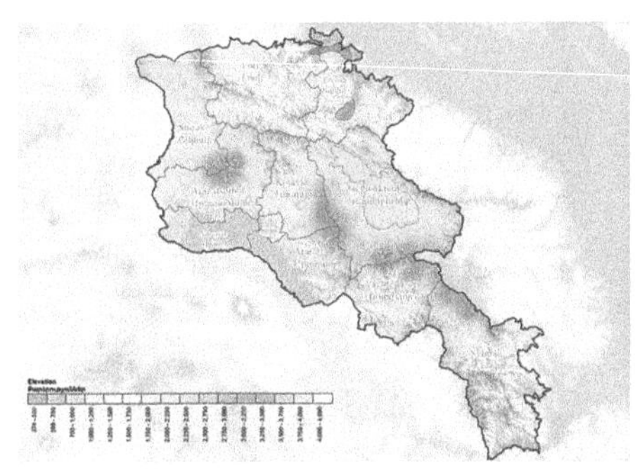

Fig. 7.6 *Clematis vitalba*

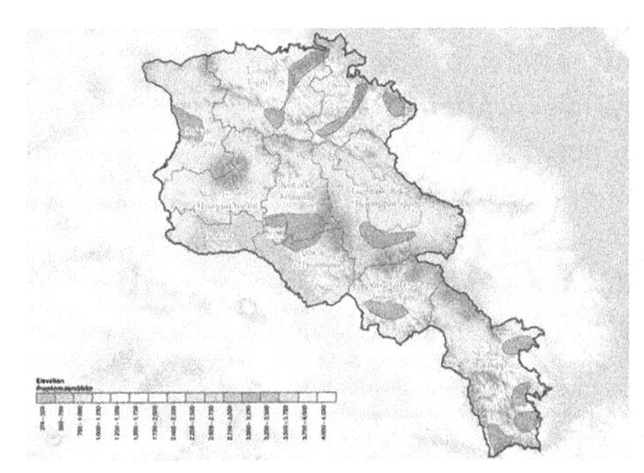

Fig. 7.7 *Conyza canadensis*

herbarium date back to 1925 and are confined to Northern Armenia. At present, the species is widely distributed throughout the territory of the republic (from north to extreme south), growing mainly in disturbed habitats. However, it also penetrates and naturalizes in some steppe and forest ecosystems in the lower and middle mountain zones up to an elevation of 1600 m a.s.l. The species is registered on the territory of state reserves (Erebuni, Shikakhokh, Khosrov Forest) and national parks (Dilijan and Arevik). The degree of threat to natural ecosystems is assessed by us as "high" (Fig. 7.7).

Grindelia squarrosa (Pursch) Dunal. The homeland of the species is North America. The species was introduced to Europe in the twentieth century (to Ireland in 1915), where it spread widely throughout the countries of Central and Eastern Europe (CABI 2022f). It was first recorded in Russia in 1976. In 2013, it was found in Georgia (Jinjolia and Shakarishvili 2014) near the village of Karsani (near Mtskheta). In 2015, the species was found on the territory of Armenia (Fig. 7.8) (Gabrielyan et al. 2016). The authors of this work suggest that the species entered Armenia "after the devastating Spitak earthquake of 1988, which engulfed all of Northern Armenia, with humanitarian aid coming from many countries, including America" (p. 131). We doubt the correctness of this assumption, since our research

Fig. 7.8 *Grindelia squarrosa*

in 2020 found new localities of this species at a distance of more than 10 km, that is, the species has spread to this distance over the past 5 years. Thus, it should have captured a much larger area over the past 30 years after the earthquake and humanitarian aid supplies. In our opinion, the appearance of this species in Georgia and Armenia is associated with the development of tourism in recent decades (both locations in Georgia and Armenia are located near major tourist sites) and the intensification of car traffic. Despite the fact that the species spreads quite quickly over disturbed habitats, it has not yet been registered in natural ecosystems, and we assess it as "potentially invasive."

Helianthus tuberosus L. The homeland of the species is North America. It was introduced into Europe at the beginning of the seventeenth century; later it was widely distributed throughout the globe (Europe, South America, Australia, New Zealand, China, Japan, India, and South Africa), growing mainly as a valuable food plant and as an ornamental species (CABI 2022g). In Armenia, it began to be grown in the middle of the twentieth century, mainly in the north of the republic and in the Ararat Valley. The species reproduces by seeds, but the main mode of reproduction is vegetative, owing to which plants quickly capture territories, forming dense monodominant thickets. At present, the species, having "escaped" from cultivation, forms dense thickets along the banks of rivers and in moist habitats up to an elevation of 1400–1500 m a.s.l. in the Lori, Ijevan, and Yerevan floristic regions of Armenia (Fig. 7.9). The degree of threat to natural ecosystems is assessed by us as "high."

Hippophae rhamnoides L. This species cannot be considered as an alien invasive species in the literal sense, since its natural populations exist in Armenia. However, a large amount of planting material, which was used on a large scale in the lake basin, was brought to the republic from Russia in the mid-1950s during the intensive afforestation of the released soils of Lake Sevan. At present, the species has formed dense impenetrable thickets owing to intense vegetative reproduction, expanding its range and capturing new territories both along the lake shore and at a distance, penetrating into meadow-steppe and meadow communities in the Sevan and Areguni floristic regions. The degree of threat to natural ecosystems is assessed by us as "high" (Fig. 7.10).

Fig. 7.9 *Helianthus tuberosus*

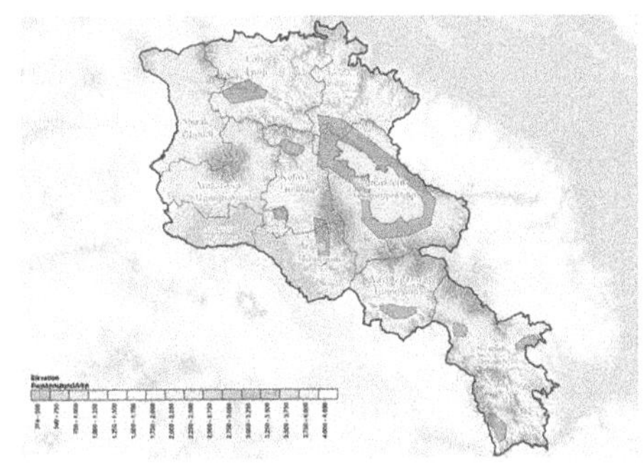

Fig. 7.10 *Hippophae rhamnoides*

Leucanthemum vulgare Lam. Eurasia is considered the birthplace of the species, where it is very widespread in Europe and Siberia. It was introduced to North America in the middle of the eighteenth century as an ornamental species, and it is currently found in South America, Australia, New Zealand, and South and East Africa (CABI 2022h). In Armenia, it was first discovered in 1952 in the Lori floristic region on the border with Georgia, from where it began its rapid spread, being encountered in the Shirak, Ijevan, Aparan, and Yerevan regions; in 2003, it was discovered in Northern Zangezur (Fig. 7.11). Distribution proceeds mainly by seeds, and the species first captures large areas in disturbed habitats and abandoned fields, from where it easily penetrates into pastures and meadows at elevations of 1300–2200 m a.s.l. owing to intensive vegetative reproduction. Overgrazing significantly contributes to the resettlement and consolidation of the species on pastures. We assess the degree of threat to natural ecosystems as "very high."

Robinia pseudoacacia L. The species is native to North America. It was introduced to Europe as an ornamental plant at the beginning of the seventeenth century, where it began to be widely used in green spaces. It is currently considered a dangerous invasive species in the majority of European countries and South Africa

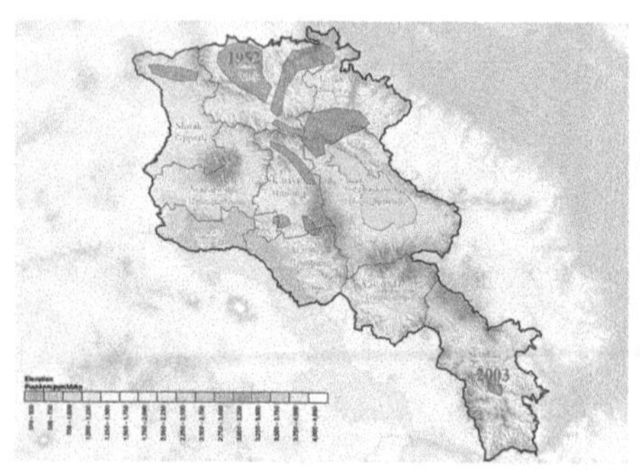

Fig. 7.11 *Leucanthemum vulgare*

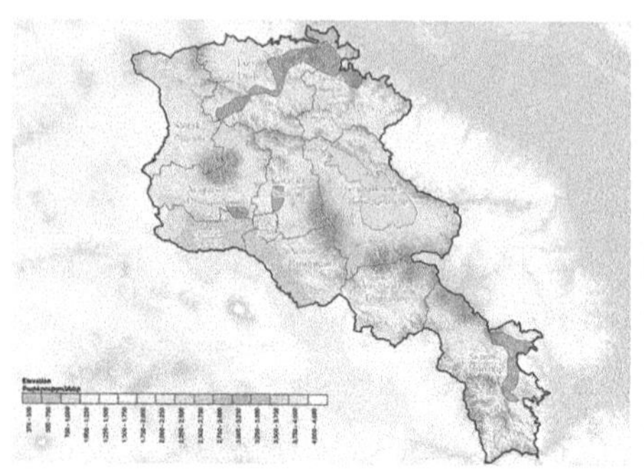

Fig. 7.12 *Robinia pseudoacacia*

(CABI 2022i). It began to be widely used in landscaping settlements and for creating forest protection zones along roads and railways in Armenia in the 1930s. It did not show invasive properties almost throughout all of the twentieth century; however, its intense vegetative reproduction began in plantings, where dense coppice thickets formed, as a result of mass felling of trees due to the energy and economic crisis of the 1990s. At the same time, individual specimens began to appear far from plantations (probably due to seed reproduction), penetrating into natural ecosystems. At present, this phenomenon is most pronounced in the Lori, Ijevan, and South Zangezur floristic regions Fig. 7.12. The degree of threat to natural ecosystems is assessed by us as "high."

Silybum marianum (L.) Gaertn. The Mediterranean is considered the birthplace of the species, but it has long been known from the Arabian Peninsula and India. It is currently distributed in North and South America, Japan, Australia, New Zealand, and sub-Saharan Africa (CABI 2022j). It is considered a dangerous invasive species in the majority of the countries. In Armenia, it was first discovered in 1967 in the South Zangezur floristic region in a "plane grove" near the border with Azerbaijan. It was also found in 1980 in the north of the republic in the Ijevan floristic region on the border with Georgia (near the main highway linking Armenia and Georgia). Over the past years, the species has intensely spread around these localities; new places of penetration into Armenia from the side of Azerbaijan in the North Zangezur

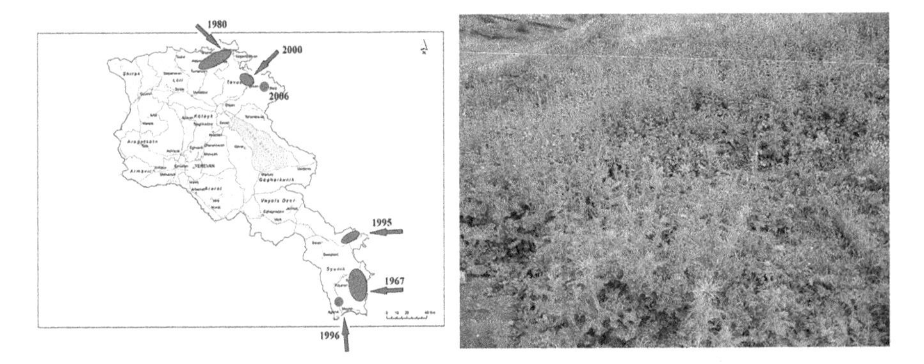

Fig. 7.13 *Silybum marianum*

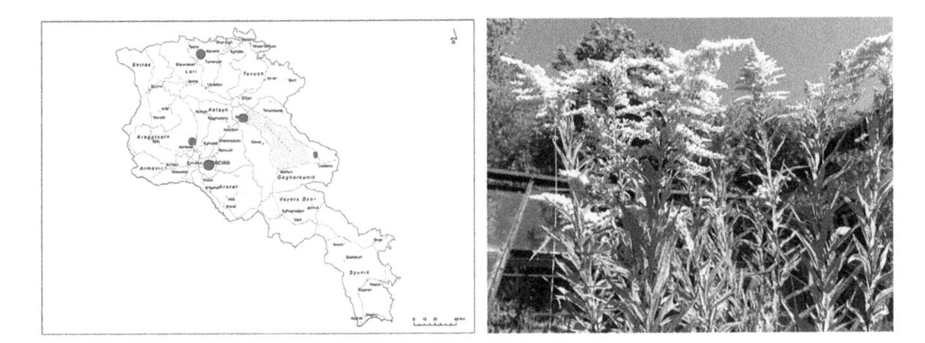

Fig. 7.14 *Solidago canadensis*

floristic region and from Iran in the Meghri region were found (Fig. 7.13). Very often, it forms dense monodominant thickets owing to high seed productivity, especially in disturbed habitats; it penetrates into natural ecosystems and reproduces in forests, open woodlands, steppes, and meadows. We assess the degree of threat to natural ecosystems as "very high."

Solidago canadensis L. The species is native to North America. In the seventeenth century, it was introduced to Europe as an ornamental plant, and already in the nineteenth century, it was noted as invasive, penetrating into natural ecosystems. At present, the species is considered invasive throughout almost all of Europe, as well as in Japan, China, and Australia (CABI 2022k). It is propagated by seeds and vegetatively, often forming dense monodominant thickets. In Armenia, the exact time of the beginning of cultivation on personal plots could not be established, but it has been grown in the Yerevan Botanical Garden in the floriculture department since the 1970s, and individual "runaway" specimens are still found throughout the garden (Fig. 7.14). In addition, the species is still cultivated as an ornamental plant in some gardens and parks and spreads beyond them by self-seeding. We evaluate the species as "potentially invasive."

In addition to those mentioned above, a number of alien species deserve special attention (Table 7.2). They were discovered in Armenia relatively recently, and are

Table 7.2 Potentially invasive plants in Armenia

Species	Pathway	Method of reproduction and distribution	Time of appearance in Armenia	Modern distribution in Armenia (by floristic regions)
Acalypha australis	Self-penetration	Seeds	1955	Idjevan
Aegopodium podagraria	Unintentional introduction	Seeds	1967 2007	Lori – Vanadzor Botanical Garden, Shirak
Aira elegantissima	Self-penetration	Seeds	1978	Idjevan, Meghri
Amorpha fruticosa	Intentional introduction	Seeds and vegetatively	Mid-twentieth century	Aparan, Yerevan
Asterolinon linum-stellatum	Self-penetration and distribution	Seeds	1979	Meghri. Very rare species, is included in the Red Data Book of Armenia
Bellis perennis	Unintentional introduction	Seeds and vegetatively	1984	Lori, Idjevan, South Zangezur
Caragana arborescens	Intentional introduction	Seeds and vegetatively	Mid-twentieth century	Almost in all regions in protective forest lines
Centaurea diffusa	Self-penetration and distribution	Seeds	1965	Yerevan Botanical Garden, Shirak, Aparan, Yerevan, Darelegis, South Zangezur
Cirsium congestum	Self-penetration and distribution	Seeds	1974 1978 1985	Yerevan Darelegis Meghri
Cuscuta campestris	Self-penetration and distribution	Seeds	1947	Idjevan, Shirak, Aparan, Yerevan, Darelegis, North and South Zangezur, Meghri
Ecballium elaterium	Self-penetration and distribution	Seeds	1959	Idjevan, South Zangezur, Meghri
Erigeron annuus	Self-penetration and distribution	Seeds	1970	Aparan (Arzni), Idjevan
Impatiens glandulifera	Unintentional introduction	Seeds	1996	Lori (Pushkino)
Imperata cylindrica	Self-penetration	Seeds	1939	Meghri
Iva xanthifolia	Self-penetration	Seeds	1994	Shirak
Legousia falcata	Self-penetration and distribution	Seeds	1978 2005	Meghri South Zangezur
Oenothera biennis	Unintentional introduction	Seeds	1958	Idjevan

(continued)

Table 7.2 (continued)

Species	Pathway	Method of reproduction and distribution	Time of appearance in Armenia	Modern distribution in Armenia (by floristic regions)
Paspalum dilatatum	Self-penetration	Seeds	1990	Aparan (Ararat valley)
Paspalum paspaloides	Self-penetration and distribution	Seeds	1947 1990	Idjevan Aparan (Ararat valley)
Paulownia tomentosa	Intentional introduction	Seeds and vegetatively	2007	Introduced in different regions of Armenia
Picris echioides	Self-penetration	Seeds	2016	South Zangezur
Polygonum orientale	Intentional introduction	Seeds	2012	Yerevan (Masis)
Psilurus incurvus	Self-penetration	Seeds	1956	Meghri. Very rare species, is included in the Red Data Book of Armenia
Salix babylonica	Intentional introduction	Seeds and vegetatively	Mid-twentieth century	Lori, Idjevan, Aparan, Yerevan, Darelegis
Siegesbeckia orientalis	Self-penetration and distribution	Seeds	1956	Idjevan, Yerevan, South Zangezur
Tagetes minima	Self-penetration and distribution	Seeds	2014	Yerevan, Aparan

invasive in many countries of the world. Although they are found in nature, mainly in disturbed habitats, they have an invasive potential, but have not yet shown their invasive properties here. These are predominantly thermophilous species, which, especially in connection with the forecasted climate change, can significantly expand their range in our country and turn out to be dangerous invasive species. First of all, it is absolutely necessary to organize monitoring of these species in order to detect the features of their distribution and penetration into natural ecosystems.

7.2 Invasive Animals

Despite the importance of studying the invasive animal species, there is little information about the invasive animals of Armenia in the literature. In the first approach of estimating the number of invasive animal species in Armenia (Arakelyan et al, in litt.), we counted a total of 34 species of alien animals, which may have negative impact of ecosystems of Armenia. Most of them are terrestrial invertebrates (21 insects, 2 nematodes and 1 mollusk), three freshwater invertebrates (2 leeches and river prawn), four species of fish and two mammals. Unfortunately, studies of their harmful effects of animal invasion have not been conducted in the country. The most dangerous invasive animals are not currently monitored or controlled. However, it should be noted, that infection with invasions is increasing, since among the listed animal species, ten of the species were registered in Armenia for the first time in the last 5 years. Some data of recently discovered and potentially most harmful species of insects are presented below.

Leptoglossus occidentalis Heidemann, 1910 – Western conifer seed bug. Natively occupied west of the North American continent, from Mexico to southern Canada. Now range expanded to Midwest and the Northeast states of the United States, all of Mexico; the species found in Central and South Americas. Introduced to Europe, due to active expansion, the species has been recorded in almost all European countries, also penetrated into Ukraine, the south of the European Russia, found in eastern Kazakhstan and in the Caucasus: in Georgia and recently in Armenia entering, most likely, from neighbouring territories of Georgia (Kalashian et al. 2021 and works cited there). Known from Lori and Tavush provinces, somewhere registered large indoor overwintering clusters. The bug is tropically associated mainly with conifers; it is indicated for about 40 species of pine (Pinaceae) and cypress (Cupressaceae); it is also recorded on pistachios (*Pistacia* spp.) and almonds *Amygdalus communis* L. The bugs feed mainly on the generative organs of forage plants, reducing seed production. In addition, the bugs can serve as a carrier of fungus *Sphaeropsis sapinea* (Fr.) = *Diplodia pinea* (Desm.), causing pine diplodiosis. Finally, large indoor overwintering clusters characteristic of the bug can be uncomfortable for people (Fig. 7.15).

Halyomorpha halys (Stål, 1855) – Brown marmorated stink bug. Natively occupied southeast of Asia, including Russian Far East, China, Korean peninsula, Taiwan, Japan and North Vietnam. Penetrated into the USA in 1996 and now occupies most of the states and reached South provinces of Canada. In Europe, the species was first found in Switzerland in 2004 and spread in most countries of South and Central Europe, to the East reached Asian part of Turkey, South European Russia, Georgia, Abkhazia and Kazakhstan. Very recently penetrated into Armenia; was registered Yerevan in 2020 and in Lori province in 2022 (Kalashian et al. 2022, and works cited there). *H. halys* harms various fruit and berry, vegetable, grain and leguminous crops, ornamental plantings, forest species, etc., causing significant economic damage. In Armenia, harmful impact not registered so far but it can be assumed that the bug is established in the country; its further distribution and increase of impact can be predicted as well, especially in the conditions of forecasted climate change (Fig. 7.16).

Nezara viridula (Linnaeus, 1758) – Southern green stink bug. The bug is considered a cosmopolitan species occurring in tropical, subtropical, and warm temperate regions of all continents between latitudes 45°N and 45°S (CABI 20221). Armenia remained a "white spot" in distribution maps of the species, but in 2021 the bug was found in Lori province (near Teghut vill.) and in 2022 near Alaverdi town of the same province where numerous feeding bugs of different stages were presented, and

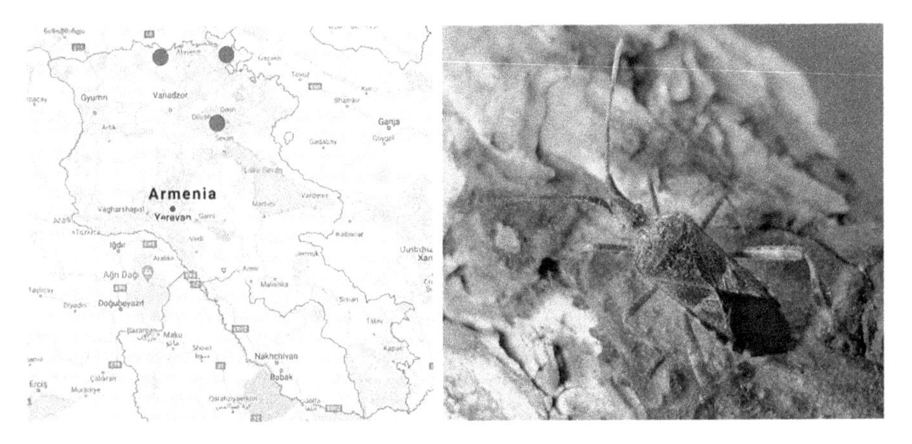

Fig. 7.15 *Leptoglossus occidentalis*

Fig. 7.16 *Halyomorpha halys*

in Ijevan town of Tavush province, most likely penetrating here from neighbouring territories of Georgia and/or Azerbaijan. This highly polyphagous species is considered as an important pest of a wide range of grain, fruit and vegetable crops, although preferring leguminous plants. *N. viridula* is a thermophilic species and currently can survive only in warmest territories of Armenia, but increase of the range and harmful impact can be predicted, especially in the conditions of forecasted climate change. Consequently, the bug can become a serious pest of common bean *Phaseolus vulgaris* widely cultivated in Armenia, as well as of other leguminous plants, fruit trees and ornamentals (Fig. 7.17).

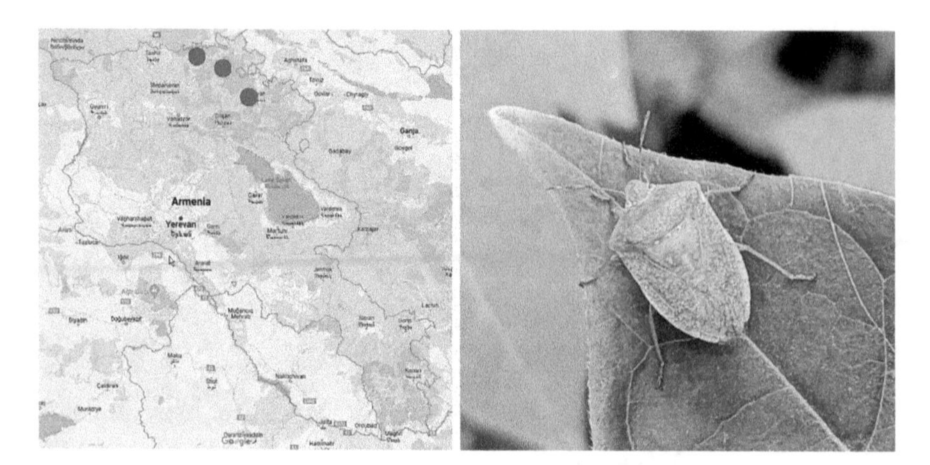

Fig. 7.17 *Nezara viridula*

Harmonia axyridis (Pallas, 1773) – Asian ladybird. Native range occupies vast territories in Asia from Northern Kazakhstan and Southern Siberia to the Russian Far East, Sakhalin, the Kuril Islands, Japan, the Korean Peninsula, China (including Taiwan), and North Vietnam. During twentieth century, the species was imported to the United States and several European countries as an effective tool in biocontrol of aphids and coccids and then successfully naturalized and penetrated into natural ecosystems, beginning large-scale expansion from secondary centres. Thus, it expanded to almost all of Europe, reaching South and Central European Russia. Its sources of outbreaks are found in North, East, and South Africa and Central and South America. Now *H. axyridis* is considered as one of 100 most dangerous invasive species in Europe (Hulme and DAISIE 2009). Its expansion from native range to the west to the Central Asia is continuing (Kalashian et al. 2017, 2019, and works cited there). In Armenia, it was first registered in 2016, and then actively distributed through the country, now it is known from Lori, Gegharkinik, Tavush, Aragatsotn, Ararat, Vayots Dzor and Syunik provinces reaching to Meghri; inhabits territories at altitudes interval from about 500 to 3000 m. Through its invasive range, the species is considered as a competitor of local guilds of aphido- and coccidophages, pest of fruits (in particular, grape) on agricultural lands and even in the vineries. During overwintering period, it can provoke allergic reactions in humans. In Armenia the harmful influence has not been studied (Fig. 7.18).

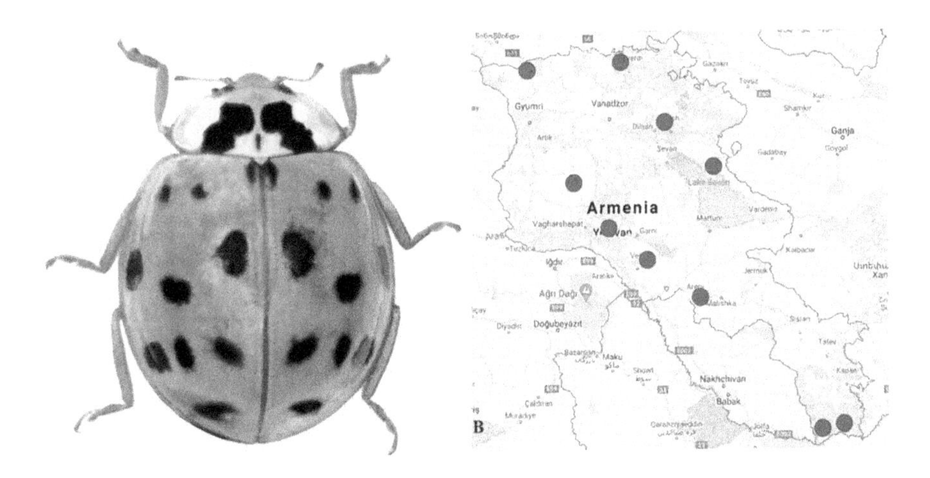

Fig. 7.18 *Harmonia axyridis*

References

Arevshatyan IG (1995) *Cirsium* mill. In: Flora Armenii. Koeltz Scientific Books, Czech Republic, pp 286–305

Bellard C, Cassey P, Blackburn TM (2015) Alien species as a driver of recent extinctions. Biol Lett 12(2):0623. https://doi.org/10.1098/rsbl.2015.0623

CABI (2022a) Invasive species compendium. www.cabi.org/isc/datasheet/2862. Cited April 29

CABI (2022b) Invasive species compendium. www.cabi.org/isc/datasheet/10314. Cited April 29

CABI (2022c) Invasive species compendium. www.cabi.org/isc/datasheet/13628. Cited April 29

CABI (2022d) Invasive species compendium. www.cabi.org/isc/datasheet/14280. Cited April 29

CABI (2022e) Invasive species compendium. www.cabi.org/isc/datasheet/15251. Cited April 29

CABI (2022f) Invasive species compendium. www.cabi.org/isc/datasheet/114634. Cited April 29

CABI. (2022g) *Invasive species compendium.* www.cabi.org/isc/datasheet/26716. Cited April 29

CABI (2022h) Invasive species compendium. www.cabi.org/isc/datasheet/13357. Cited April 29

CABI (2022i) Invasive species compendium. www.cabi.org/isc/datasheet/47698. Cited April 29

CABI (2022j) Invasive species compendium. www.cabi.org/isc/datasheet/50304. Cited April 29

CABI (2022k) Invasive species compendium. www.cabi.org/isc/datasheet/50599. Cited April 29

CABI (2022l) Invasive species compendium. www.cabidigitallibrary. https://doi.org/10.1079/cabicompendium.36282

Cacabelos E, Martins GM, Faria J, Prestes ACL, Costa T, Moreu I, Neto AI (2020) Limited effects of marine protected areas on the distribution of invasive species, despite positive effects on diversity in shallow water marine communities. Biol Invasions 22(3):1169–1179. https://doi.org/10.1007/s10530-019-02171-x

Essl F, Bacher S, Blackburn TM, Booy O, Brundu G, Brunel S, Cardoso AC, Eschen R, Gallardo B, Galil B, Garcia-Berthou E, Genovesi P, Groom Q, Harrower C, Hulme PE, Katsanevakis S, Kenis M, Kuhn I, Kumschick S, Jeschke JM (2015) Crossing frontiers in tackling pathways of biological invasions. Bioscience 65(8):769–782. https://doi.org/10.1093/biosci/biv082

Fayvush GM, Tamanyan KG (2014) Invazivnye i ekspansivnye vidy rastenii Armenii (invasive and expansive plant species of Armenia), Yerevan

Fayvush G, Aleksanyan A, Wong LJ, Pagad S (2020) Global register of introduced and invasive species – Armenia. Version 1.4. Invasive species specialist group ISSG. Checklist dataset. https://doi.org/10.15468/w96h3j. Accessed via GBIF.org on 2021-01-16

Gabrielyan ETS, Tamanyan KG (1985) New genus and rare species from the flora of Armenia. Biol Zh Arm 38(2):164–166

Gabrielyan ETS, Timukhin IN, Tuniev BS, Agababyan MV (2016) New invasive genus *Grindelia* (Asteraceae) from Armenia and new localities from Northwestern and Western Transcaucasia. Takhtadjania 3:130–132

Hulme PE (2009) Trade, transport and trouble: managing invasive species pathways in an era of globalization. J Appl Ecol 46(1):10–18. https://doi.org/10.1111/j.1365-2664.01600.x

Hulme PE, DAISIE (eds) (2009) Handbook of alien species in Europe. Springer, Dordrecht

IPBES (2018) The IPBES regional assessment report on biodiversity and ecosystem Services for Europe and Central Asia. Secretariat of the Intergovernmental Science-Policy Platform on Biodiversity and Ecosystem Services. https://doi.org/10.5281/ZENODO.3237429

Jinjolia L, Shakarishvili N (2014) *Grindelia squarrosa* (Rursh) Dunal—a new alien genus and species for flora of Georgia. Bull Georgian Natl Acad Sci 8(3):64–68

Kalashian MYu, Ghrejyan TL & ad Karagyan GH (2022) Brown Marmorated Stink Bug *Halyomorpha halys* (Stål, 1855) (Heteroptera: Pentatomidae) penetrated into Armenia. Russ J Biol Invasions, 13(3), pp 305–308. https://doi.org/10.1134/S2075111722030080

Kalashian MY, Ghrejyan TL, Karagyan GH (2017) Harlequin ladybird *Harmonia axyridis* pall. (Coleoptera, Coccinellidae) in Armenia. Russ J Biol Invasions 8(4):313–315. https://doi.org/10.1134/S207511171704004X

Kalashian MY, Ghrejyan TL, Karagyan GH (2019) Expansion of harlequin ladybird *Harmonia axyridis* pall. (Coleoptera, Coccinellidae) in Armenia. Russ J Biol Invasions 10(2):153–156. https://doi.org/10.1134/S2075111719020073

Kalashian MY, Ghrejyan TL, Karagyan GH (2021) First finding of Western conifer seed bug *Leptoglossus occidentalis* Heid. (Heteroptera, Coreidae) in Armenia. Russ J Biol Invasions 12(3):274–276. https://doi.org/10.1134/S2075111721030073

Liu X, McGarrity ME, Li Y (2012) The influence of traditional Buddhist wildlife release on biological invasions: religious release promotes species invasions. Conserv Lett 5(2):107–114. https://doi.org/10.1111/j.1755-263X.2011.00215.x

Liu X, Blackburn TM, Song T, Wang X, Huang C, Li Y (2020) Animal invaders threaten protected areas worldwide. Nat Commun 11(1):2892. https://doi.org/10.1038/s41467-020-16719-2

Pipek P, Blackburn TM, Delean S, Cassey P, Şekercioğlu ÇH, Pyšek P (2020) Lasting the distance: the survival of alien birds shipped to New Zealand in the 19th century. Ecol Evol 10(9):3944–3953. https://doi.org/10.1002/ece3.6143

Pyšek P, Hulme PE, Simberloff D, Bacher S, Blackburn TM, Carlton JT, Dawson W, Essl F, Foxcroft LC, Genovesi P, Jeschke JM, Kühn I, Liebhold AM, Mandrak NE, Meyerson LA, Pauchard A, Pergl J, Roy HE, Seebens H, Richardson DM (2020) Scientists' warning on invasive alien species. Biol Rev 95(6):1511–1534. https://doi.org/10.1111/brv.12627

Saul WC, Roy HE, Booy O, Carnevali L, Chen HJ, Genovesi P, Harrower CA, Hulme PE, Pagad S, Pergl J, Jeschke JM (2017) Assessing patterns in introduction pathways of alien species by linking major invasion data bases. J Appl Ecol 54(2):657–669. https://doi.org/10.1111/1365-2664.12819

Seebens H, Blackburn TM, Dyer EE, Genovesi P, Hulme PE, Jeschke JM, Pagad S, Pyšek P, Winter M, Arianoutsou M, Bacher S, Blasius B, Brundu G, Capinha C, Celesti-Grapow L, Dawson W, Dullinger S, Fuentes N, Jager H, Essl F (2017) No saturation in the accumulation of alien species worldwide. Nat Commun 8(1):14435. https://doi.org/10.1038/ncomms14435

Tamanyan KG, Fayvush GM, Nanagjulyan SG, Danielyan TS (eds) (2010) Red book of plants of Armenia. Yerevan

Wilson JRU, Dormontt EE, Prentis PJ, Lowe AJ, Richardson DM (2009) Something in the way you move: dispersal pathways affect invasion success. Trends Ecol Evol 24(3):136–144. https://doi.org/10.1016/j.tree.2008.10.007

Annexes

G. Fayvush (ed.), *Biodiversity of Armenia*,
https://doi.org/10.1007/978-3-031-34332-2

Annex 1: List of Species of the Flora of Armenia Investigated Karyologically

Species	2n	Ploidity[a]	Geoelement[b]	Belt[c]	Vegetation	Habitats[d]
Alliaceae						
Allium affine Ledeb.	16	di-	Arm-Ir	l-m	Steppe	E1.2E
Allium albidum Fisch. ex M.Bieb.	16	di-	E.Med	l-sb	Steppe, meadows	E1.2E, E2.32
Allium atroviolaceum Boiss.	16, 32	di-, tetra	A.Med	l-u	Semideserts, steppe, open forest	E1.33, E1.45, E1.2E, F5.34
Allium aucheri Boiss.	16	di-	Arm-Atr	u-sb	Meadow, forest	E2.32, G1.A1
Allium callidictyon C.A.Mey. ex Kunth	16	di-	Arm-Atr	l-m	Semideserts, steppe	E1.33, E1.45, E1.2E
Allium cardiostemon Fisch. & C.A.Mey.	16, 16+B	di-	Arm-Atr	m-sb	Steppe, meadows, forest, open forest	E1.2E, E2.32, G1.A1, F5.34
Allium derderianum Regel	16, 16+B	di-	Atr	a	Screes	H2.3
Allium dictyoprasum C.A.Mey. ex Kunth	16	di-	Arm-Ir	l-sb	Semidesert, steppe, open forest, screes	E1.33, E1.45, E1.2E, F5.34, H2.6
Allium flavum L. subsp. *tauricum* (Besser ex Rchb.) Stearn	16	di-	Cauc-Arm-Ir	l-u	Semidesert, steppe, open forest	E1.33, E1.45, E1.2E, F5.34
Allium fuscoviolaceum Fomin	16	di-	Cauc-Arm	l-u	Steppe, open forest	E1.2E, F5.34
Allium gramineum K.Koch	16, 16+B	di-	Arm	m-sb	Steppes, meadows	E1.2E, E2.32

Allium jajlae Vved.	16	di-	Cauc-Arm	l-u	Steppe, meadows, forest, open forest	E1.2E, E2.32, G1.A1, F5.34
Allium karsianum Fomin	16	di-	Arm	m-sb	Steppe, meadows	E1.2E, E2.32
Allium kunthianum Vved.	16, 32	di-, tetra-	Cauc-Arm-Atr	u-a	Steppe, meadows, forest	E1.2E, E2.32, G1.A1
Allium leucanthum K.Koch	16, 32	di-, tetra-	Arm-Ir	l-m	Screes, deserts, forest	H2.6, H5.32, G1.A1
Allium materculae Bordz.	16	di-	Arm-Atr	l-m	Semideserts	E1.33, E1.45
Allium moschatum L.	16	di-	E.Med.	n-sb	Semideserts, steppe, meadows	E1.33, E1.45, E1.2E, E2.32
Allium paradoxum (M.Bieb.) G.Don	16	di-	Arm-Atr	l-sb	Forest	G1.6H, G1.A1
Allium pseudoampeloprasum Miscz. ex Grossh.	16	di-	Arm	s-u	Steppe, meadows, rocks	E1.2E, E2.32, H3.1
Allium pseudoflavum Vved.	18	di-	Arm	l-sb	Semideserts, steppe, open forest	E1.33, E1.45, E1.2E, F5.34
Allium rotundum L.	16	di-	E.A.Med	l-u	Steppe, meadows, forest	E1.2E, E2.32, G1.A1
Allium rubellum M. Bieb.	16, 16+B	di-	Ir-Tur	l-m	Semideserts, steppe, open forest, rocks	E1.33, E1.45, E1.2E, F5.34, H3.2
Allium rupestre Steven	16	di-	Cauc-Arm	m	Screes	H2.6
Allium saxatile M.Bieb.	16	di-	Palearct	l-u	Steppes, screes, meadows	E1.2E, H2.3, E2.32
Allium schoenoprasum L.	16, 16+B, 32, 48	di-, tetra-, hexa-	Holarct	sb-a	Meadows, rocks, screes	E4.3A, H3.1, H2.3

Species	2n	Ploidity[a]	Geoelement[b]	Belt[c]	Vegetation	Habitats[d]
Allium struzlianum Ogan.	16	di-	Arm (endemic)	m-u	Screes	H2.4
Allium szovitsii Regel	48	hexa-	Arm	sb	Meadows, screes	E2.32, H2.3
Allium victorialis L.	16	di-	Palearct	m-sb	Meadows, forest edges	E2.32, E5.4
Allium vineale L.	32	tetra-	Eur-A.Med.	m-sb	Meadows, wetlands	E2.32, D4.1
Allium woronowii Miscz. ex Grossh.	16	di-	Arm	m	Steppe	E1.2E
Nectaroscordum tripedale (Trautv.) Grossh.	18	di-	Arm-Ir	m-sb	Meadows, forest	E2.32, G1.6H, G1.A1
Amaryllidaceae						
Galanthus alpinus Sosn.	24	di-	Arm	l-m	Forest, open forest	G1.6H, G1.A1, F5.34
Galanthus artjuschenkoae Gabrielian	72	hexa-	Arm	l-m	Forest	G1.6H, G1.A1
Galanthus transcaucasicus Fomin	24	di-	Arm	l	Forest	G1.6H, G1.A1
Apiaceae						
Astrodaucus orientalis (L.) Drude	20	di-	E.A.Med	l-m	Semideserts, steppe	E1.33, E1.45, E1.2E
Bupleurum rotundifolium L.	16	di-	Palearct	l-u	Semideserts, steppe, meadows	E1.33, E1.45, E1.2E, E2.32
Carum carvi L.	20	di-	Eur-A.Med	m-sb	Steppe, meadows	E1.2E, E2.32
Carum caucasicum (M.Bieb.) Boiss.	22	di-	Cauc-Arm-Ir	a	Meadows	E4.3A, E4.4
Chaerophyllum aureum L.	22	di-	Eur-A.Med	m	Forest	G1.A1

Chamaesciadium acaule (M.Bieb.) Boiss.	20	di-	Cauc-Arm-Ir	a	Meadows	E4.3A, E4.4
Daucus carota L.	18	di-	Palearct	l-m	Steppe, open forest	E1.2E, F5.34
Elaeosticta glaucescens (DC.) Boiss.	22	di-	Arm-Atr	l	Semideserts	E1.33, E1.45
Eryngium campestre L.	14	di-	Eur-A.Med	l-m	Semideserts, steppe	E1.33, E1.45, E1.2E
Falcaria falcarioides (Bornm. & H.Wolff) H.Wolff	22	di-	Arm-Ir	l-u	Wetlands	D4.1
Falcaria vulgaris Bernh.	22	di-	Palearct	l-u	Semideserts, steppe, meadows	E1.33, E1.45, E1.2E, E2.32
Ferula orientalis L.	22	di-	Arm-Ir	m-u	Steppe, meadows	E1.2E, E2.32
Ferula rigidula DC.	22	di-	Arm-Ir	m	Steppe	E1.2E
Ferulago setifolia K.Koch	22	di-	Arm	m-u	Steppe, meadows, steppe shrubs	E1.2E, E2.32, F2.33
Grammosciadium daucoides DC.	20	di-	Arm	u-a	Forest, meadows	G1.6H, G1.A1, E2.32
Grammosciadium platycarpum Boiss. & Hausskn.	20	di-	Arm-Ir	m-u	Forest, meadows	G1.6H, G1.A1, E2.32
Heracleum pastinacifolium K.Koch	22	di-	Arm	m	Forest, forest edges, wetlands	G1.A1, E5.4, D4.1
Malabila dasyantha (K.Koch) Grossh.	22	di-	Cauc-Arm-Ir	l-u	Semideserts, steppe, meadows	E1.33, E1.45, E1.2E, E2.32
Pimpinella aurea DC.	20	di-	Arm-Ir	l-m	Semideserts, steppe	E1.33, E1.45, E1.2E

Species	2n	Ploidity[a]	Geoelement[b]	Belt[c]	Vegetation	Habitats[d]
Prangos lophoptera Boiss.	22	di-	Arm-Atr	l-m	Steppe, screes, rocks	E1.2E, H2.5, H2.6, H3.1, H3.2
Seseli libanotis (L.) K.Koch	22	di-	Eur-Med	l-sb	Steppe, meadows, forest	E1.2E, E2.32, G1.A1
Tordylium maximum L.	22	di-	Eur-Med	l-m	Semideserts, steppe	E1.33, E1.45, E1.2E
Torilis arvensis (Huds.) Link	12	di-	Eur-A.Med	l-u	Steppe, steppe shrubs	E1.2E, F2.33
Trinia leiogona (C.A.Mey.) B.Fedtsch.	18	di-	Arm	m-u	Steppe, meadows	E1.2E, E2.32
Xanthogallum purpurascens Ave-Lall.	22	di-	Cauc-Arm-Ir	m-u	Forest, meadows	G1.6H, G1.A1, E2.32
Zosima orientalis Hoffm.	12	di-	Ir-Tur	l-m	Steppe	E1.2E
Acoraceae						
Acorus calamus L.	34	aneu-	Holarct	l	Water ecosystems	C3.24
Asparagaceae						
Asparagus officinalis L.	40	tetra-	Eur-A.Med	l-u	Steppe, meadows, forest edges	E1.2E, E2.32, E5.4
Asparagus persicus Baker	60	hexa-	E.A.Med	l-m	Semideserts, steppe	E1.33, E1.45, E1.2E
Asparagus veriticillatus L.	20	di-	A.Med	l-sb	Forest edges, steppe, steppe shrubs	E5.4, E1.2E, F2.33
Asphodelaceae						

Asphodeline dendroides (Hoffm.) Woronow ex Grossh.	28	tetra-	Arm-Atr	l-m	Semideserts, steppe	E1.33, E1.45, E1.2E
Asphodeline taurica (Pall.) Kunth	28	tetra-	E.Med	m-u	Steppe shrubs, steppe	E1.2E, F2.33
Eremurus spectabilis M.Bieb.	14	di-	Cauc-E.A.Med	m-sb	Steppe, steppe shrubs	E1.2E, F2.33
Asteraceae						
Acroptilon repens (L.) DC.	26	di-	E.A.Med	l-m	Semideserts, steppe	E1.33, E1.45, E1.2E
Aetheopappus pulcherrimus (Willd.) Cass.	36	hexa-	Cauc	u-a	Meadow, screes	E2.32, H2.3, H2.4
Amberboa moschata (L.) DC.	32	di-	Arm	l-m	Semideserts, steppe	E1.33, E1.45, E1.2E
Amberboa sosnowskyi Iljin	32	di-	Atr	l-m	Semideserts, steppe	E1.33, E1.45, E1.2E
Anthemis cretica L. subsp. *iberica* (M.Bieb.) Grierson	36	tetra-	Cauc	u-a	Screes	H2.3, H2.4
Arctium lappa L.	36	tetra-	Palearct	l-m	Ruderal habitats, wetlands	I1.53, D4.1
Artemisia absinthium L.	18	di-	Palearct	l-a	Ruderal habitats	I1.53
Artemisia fragrans Willd.	18	di-	Cauc-Ir-Tur	l-m	Semideserts	E1.45
Artemisia splendens Willd.	18	di-	Cauc-Arm-Ir	m-a	Steppe, meadows	E1.2E, E2.32

Species	2n	Ploidity[a]	Geoelement[b]	Belt[c]	Vegetation	Habitats[d]
Artemisia vulgaris L.	16	di-	polychor	l-m	Forest edges, ruderal habitats	E5.4, I1.53
Callicephalus nitens (M.Bieb. ex Willd.) C.A.Mey.	32	tetra-	Arm-Ir	l-m	Juniper forest, open forest	F5.13, G3.93, F5.34
Carduus crispus L.	16	di-	Eur-Med	l-u	Steppe, wetlands	E1.2E, D4.1
Carduus nutans L.	16	di-	Eur-A.Med	l-a	Steppe, meadows	E1.2E, E2.32
Carduus onopordioides Fisch. ex M.Bieb.	30	tri-	Arm-Atr	l-m	Semideserts, steppe	E1.33, E1.45, E1.2E
Carthamus lanatus L.	64	octa-	Eur-A.Med	l-m	Steppe, open forest	E1.2E, F5.34
Carthamus oxyacanthus M.Bieb.	24	di-	E.A.Med	l-m	Semideserts, steppe	E1.33, E1.45, E1.2E
Centaurea aggregata Fisch. & C.A.Mey.	18	di-	Arm-Ir	l-u	Steppe, open forest, screes	E1.2E, F5.34, H2.6
Centaurea alexandri Bordz.	18	di-	Arm (endemic)	l-u	Semideserts, open forest	E1.33, E1.45, F5.34
Centaurea behen L.	36, 36+3B	tetra-	E.A.Med	l-u	Semideserts, steppe, juniper forest	E1.33, E1.45, E1.2E, F5.13, G3.93
Centaurea brugeriana (DC.) Hand.-Mazz. ssp. *belangeriana* (DC.) Bornm.	22	di-	Ir-Tur	l	Ruderal habitats	I1.53
Centaurea carduiformis DC.	20, 20+B	di-	Arm-Atr	l-m	Semideserts, steppe, juniper forest	E1.33, E1.45, E1.2E, F5.13, G3.93
Centaurea cheiranthifolia Willd.	40	tetra-	Cauc-Arm-Atr	sb-a	Meadows, tall grasses	E4.3A, E4.4, E5.5A
Centaurea cyanus L.	24	di-	Eur-A.Med	m	Forest edges, fields	E5.4, I1.53

Centaurea depressa M.Bieb.	16	di-	E.A.Med	l-u	Ruderal habitats, fields	I1.53
Centaurea diffusa Lam.	18	di-	Eur-Med	l-m	Steppe	E1.2E
Centaurea erivanensis (Lipsky) Bordz.	32	tetra-	Arm	l-m	Semideserts, steppe	E1.33, E1.45, E1.2E
Centaurea gulissaschvilii Dumbadze	18	di-	Arm	l-u	Rocks	H3.1
Centaurea hajastana Tzvel.	30	di-	Arm (endemic)	m	Steppe	E1.2E
Centaurea iberica Trev. ex Spreng.	16	di-	E.A.Med	l-u	Semideserts, steppe, forest edges	E1.33, E1.45, E1.2E, E5.4
Centaurea ovina Pall. ex Willd.	18	di-	Atr	l-sb	Semideserts, steppe, forest edges	E1.33, E1.45, E1.2E, E5.4
Centaurea phaeopappoides Bordz.	26, 26+3B	di-	Atr	m-u	Open forest, rocks, screes	F5.34, H3.2, H2.6
Centaurea polypodiifolia Boiss.	16	di-	Arm-Ir	l-m	Semideserts, steppe, open forest	E1.33, E1.45, E1.2E, F5.34
Centaurea pseudoscabiosa Boiss. & Buhse ssp. *glehnii* (Trautv.) Wagenitz	18	di-	Arm	m-sb	Steppe, meadows, open forest	E1.2E, E2.32, F5.34
Centaurea ruthenica Lam.	30	di-	Palearct	m-u	Steppe	E1.2E
Centaurea salicifolia M.Bieb. ex Willd.	22	di-	Cauc-Arm-Atr	l-sb	Forest, steppe, meadows	E1.2E, G1.A1, E2.32

Species	2n	Ploidity[a]	Geoelement[b]	Belt[c]	Vegetation	Habitats[d]
Centaurea solstitialis L.	16	di-	Eur-A.Med	l-sb	Semideserts, steppe, open forest	E1.33, E1.45, E1.2E, F5.34
Centaurea sosnowskyi Grossh.	20	di-	Arm-Ir	l-u	Rocks, open forest	H3.1, F5.34
Centaurea szovitsiana Boiss.	16	di-	Arm-Ir	m-u	Steppe	E1.2E
Centaurea tamanianae Agababian	30	di-	Arm (endemic)	m-u	Steppe, steppe shrubs	E1.2E, F2.33
Centaurea triumfettii All.	40	tetra-	Med	m-a	Steppe, meadows, forest edge	E1.2E, E2.32, E5.4
Centaurea vavilovii Takht. & Gabrielian	30	di-	Arm (endemic)	u	Steppe	E1.2E
Centaurea virgata Lam. ssp. *squarrosa* (Willd.) Gugler	36	tetra-	Cauc-E.A.Med	l-u	Steppe, semideserts, ruderal habitats	E1.33, E1.45, E1.2E, I1.53
Cephalorhynchus kirpicznikovii Grossh.	18	di-	Arm	l-m	Steppe, steppe shrubs	E1.2E, F2.33
Cephalorhynchus tuberosus (Steven) Schchian	18	di-	E.Med	l-m	Forest, steppe shrubs	G1.A1, F2.33
Chartolepis biebersteinii Jaub. & Spach	18	di-	Arm-Atr	m-u	Steppe	E1.2E
Chartolepis glastifolia (L.) Cass.	36	tetra-	Arm-Ir	m-sb	Open forest, meadows, steppe	F2.33, E2.32, E1.2E
Cheirolepis persica Boiss.	44	tetra-	Arm-Ir	m-u	Semideserts, steppe, rocks	E1.33, E1.45, E1.2E, H3.1
Chondrilla juncea L.	16	di-	Eur-A.Med	l-m	Semideserts, steppe, forest, meadows	E1.33, E1.45, E1.2E, G1.A1, E2.32

Chondrilla latifolia M.Bieb.	15	tri-	A.Med	m	Steppe, open forest	E1.2E, F2.33
Cicerbita macrophylla (Willd.) Wallr.	32	tetra-	adventive	m-sb	Forest, steppe shrubs, meadows	G1.6H, G1.A1, F2.33, E2.32
Cicerbita racemosa (Willd.) Beauverd.	16	di-	Cauc	m-a	Meadows, steppe, wetlands	E2.32, E1.2E, D4.1
Cichorium glandulosum Boiss. & Huet	18	di-	Arm-Ir	l-m	Steppe, screes	E1.2E, H2.6
Cichorium intybus L.	18	di-	Palearct	l-m	Steppe, meadows, forest, semidesert	E1.33, E1.45, E1.2E, E2.32, G1.6H, G1.A1
Cirsium aduncum Fisch. & C.A.Mey. ex DC.	34	di-	Arm-Atr	m-sb	Meadows, steppe, wetlands	E2.32, E1.2E, D4.1
Cirsium arvense (L.) Scop.	34	di-	Holarct	l-u	Steppe, wetlands	E1.2E, D4.1
Cirsium ciliatum (Murray) Moench.	68	tetra-	W.Palearct	l-u	Steppe, forest edges, semideserts	E1.2E, E5.4, E1.33, E1.45
Cirsium congestum Fisch. & C.A.Mey. ex DC.	34	di-	Iran	l-u	Semideserts, steppe, open forest	E1.33, E1.45, E1.2E, F2.33
Cirsium cosmelii (Adam) Fisch. ex Hohen.	20	di-	Arm	m-a	Steppe, forest, meadows	E1.2E, G1.A1, E2.32, E4.3A
Cirsium echinus (M.Bieb.) Hand.-Mazz.	34	di-	Cauc-Arm-Atr	l-a	Steppe, meadows, forest	E1.2E, E2.32, G1.A1
Cirsium elodes M.Bieb.	34	di-	Arm-Ir	l-m	Wetlands, marshes, forest	D2, D4.1, G1.A1

Species	2n	Ploidity[a]	Geoelement[b]	Belt[c]	Vegetation	Habitats[d]
Cirsium esculentum (Siev.) C.A.Mey.	34	di-	Palearct	m-a	Meadows	E2.32
Cirsium incanum (S.G.Gmel.) Fisch. ex M.Bieb.	34	di-	Holarct	l-sb	Semidesert, meadows, steppes, fields	E1.33, E1.45, E2.32, E1.2E, I1.3
Cirsium obvallatum (M.Bieb.) Fisch.	34	di-	Cauc-Arm-Atr	m-sb	Forest, meadows, tall grasses, wetlands	G1.A1, E2.32, E5.5A, D4.1
Cirsium osseticum (Adam) Petrak.	34	di-	Cauc-Arm-Atr	m-u	Forest	G1.A1
Cirsium vulgare (Savi) Ten.	68	tetra-	Holarct	l-u	Ruderal habitats, fields	I1.53, I1.3
Cnicus benedictus L.	22	di-	A.Med	l-m	Semideserts, steppe, juniper forest	E1.33, E1.45, E1.2E, F5.13, G3.93
Conyza canadensis (L.) Cronquist	44	tetra-	Holarct	l-m	Ruderal habitats	I1.53
Cousinia fedorovii Takht.	24	di-	Arm (endemic)	u	Steppe	E1.2E
Cousinia gabrielianae Takht. & Tamanian	24	di-	Arm	l-m	Open forest	F2.33
Cousinia gigantolepis Rech. fil.	20	di-	Arm-Atr	m-u	Meadows, steppe	E2.32, E1.2E
Cousinia macrocephala C.A.Mey.	18	di-	Arm-Atr	m-u	Steppe	E1.2E
Cousinia qaradaghensis Rech. fil.	16	di-	Arm-Atr	m-u	Screes	H2.3
Crepis alpina L.	10	di-	E.A.Med	l-m	Semideserts, steppe, forest edges	E1.33, E1.45, E1.2E, E5.4

Crepis ciliata K.Koch	40, 40+2B	octa-	Arm-Atr	m-sb	Meadows, forest edges	E2.32, E5.4
Crepis foetida L.	10	di-	Eur-A.Med	l-m	Steppe, semideserts, steppe shrubs	E1.33, E1.45, E1.2E, F2.33
Crepis marschallii (C.A.Mey.) F.W.Schultz	8, 8+B	di-	Cauc-Arm-Atr	l-m	Steppe, steppe shrubs	E1.2E, F2.33
Crepis micrantha Czer.	8	di-	E.Med	l-m	Steppe, steppe shrubs, forest edges	E1.2E, F2.33, E5.4
Crepis pannonica (Jacq.) K.Koch	8, 8+1-6B, 9, 9+1-2B, 10, 10+1-3B, 11, 11+1-3B, 12, 12+1-3B, 13+3B, 14+B, 15+1-4B, 16	di-, tri-, aneu-	Pan-Pont	l-m	Steppe, steppe shrubs, forest	E1.2E, F2.33, G1.6H, G1.A1
Crepis pulchra L.	8	di-	A.Med	l-m	Steppe, forest	E1.2E, G1.6H, G1.A1
Crepis sahendi Boiss. & Buhse	10	di-	Arm-Ir	u-a	Meadows	E4.3A, E4.4
Crepis sancta (L.) Babc.	10, 10+1-2B	di-	Eur-A.Med	l-m	Semideserts, steppe, steppe shrubs	E1.33, E1.45, E1.2E, F2.33
Crepis sibirica L.	10	di-	Palearct	m-a	Forest, meadows	G1.6H, E4.4
Crepis sonchifolia (M.Bieb.) C.A.Mey.	8	di-	Cauc	m-sb	Rocks	H3.2
Crepis wildenowii Czer.	10	di-	Arm	a	Screes, rocks	H2.5, H2.6, H3.1, H3.2
Crupina vlgaris Cass.	30	hexa-	A.Med	l-u	Steppe, steppe shrubs, screes	E1.2E, F2.33, H2.5
Doronicum oblongifolium DC.	60	high ploidy	E.Palearct	v-a	Meadows, rocks, forest edges, tall grasses	E4.3A, E4.4, H3.1, E5.4, E5.5A

Species	2n	Ploidity[a]	Geoelement[b]	Belt[c]	Vegetation	Habitats[d]
Echinops transcaucasicus Iljin	36	hexa-	Arm-Atr	l-m	Steppe, meadows	E1.2E, E2.32
Erigeron uniflorus L.	18	di-	A.Med	a	Meadows, rocks, screes	E2.32, H3.2, H2.6
Erigeron venustus Botsch.	18	di-	Arm-Atr	sb-a	Meadows	E2.32
Garhadiolus angulosus Jaub. & Spach	10	di-	E.A.Med	l-m	Semideserts, open forest, steppe	E1.33, E1.45, F5.34, E1.2E
Garhadiolus papposus Boiss. & Buhse	10	di-	E.A.Med	l	Semideserts	E1.33, E1.45
Gnaphalium supinum L.	28	tetra-	Palearct	sb-a	Meadows, screes	E4.3A, H2.3
Grossheimia macrocephala (Muss.-Puschk. ex Willd.) Sosn. & Takht.	18	di-	Arm-Atr	m-sb	Meadows, steppe, forest, tall grasses	E2.32, E1.2E, G1.A1, E5.5A
Gundelia aragatsi Vitek, Fayvush, Tamanyan & Gemeinholzer ssp. *steineri* Vitek, Fayvush, Tamanyan & Gemeinholzer	18	di-	Arm	l-u	Open forest	F5.34
Gundelia armeniaca Nersesian	18	di-	Arm (endemic)	m	Steppe	E1.2E
Hieracium cincinnatum Fries	36	tetra-	Arm-Atr	m-sb	Steppe, rocks	E1.2E, H3.2

Hieracium cymosum L.	36	tetra-	Palearct	l-a	Steppe, meadows, forest, screes, semideserts	E1.2E, E2.32, G1.A1, H2.5, E1.33, E1.45
Hieracium murorum L.	36	tetra-	Palearct	m-a	Rocks, forest	H3.1, H3.2, G1.A1
Hieracium pilosella L.	36	tetra-	Holarct	m-a	Steppe, meadows, forest	E1.2E, E2.32, G1.A1
Hieracium piloselloides Vill.	54	hexa-	Eur-Med	m-sb	Steppe, forest, meadows	E1.2E, E2.32, G1.A1
Hieracium prenanthoides Vill.	27	tri-	Palearct	m-a	Forest, meadows, tall grasses	E2.32, G1.A1, E5.5A
Hieracium umbellatum L.	18	di-	Holarct	m-sb	Meadows, steppe, screes	E2.32, E1.2E, H2.6
Hieracium verruculatum Link	18	di-	E.Med	m-sb	Steppe, meadows, forest	E2.32, E1.2E, G1.A
Hieracium virosum Pall.	27	tri-	A.Med	m-sb	Steppe, meadows	E2.32, E1.2E
Hyalea pulchella (Ledeb.) K.Koch	18	di-	E.A.Med	l-u	Semideserts, steppe, juniper forest	E1.33, E1.45, E1.2E, F5.13, G3.93
Jurinea blanda (M.Bieb.) C.A.Mey.	36	tetra-	Cauc	m-u	Steppe, juniper forest	E1.2E, F5.13, G3.93
Jurinea moschus (Habl.) Bobr.	34	di-	Cauc-Arm-Ir	u-a	Screes, rocks, meadows	H2.3, H3.1, E4.3A
Jurinea squarrosa Fisch. & C.A.Mey.	36	tetra-	Arm	u-a	Rocks, screes	H3.1, H2.5
Koelpinia linearis Pall.	42, 56	hexa-, octa-	A.Med	l-m	Semideserts, steppe, open forest	E1.33, E1.45, E1.2E, F5.34
Lactuca chaixii Vill.	18	di-	Med	m	Forest, steppe shrubs	G1.6H, G1.A1, F2.33

Species	2n	Ploidity[a]	Geoelement[b]	Belt[c]	Vegetation	Habitats[d]
Lactuca georgica Grossh.	18	di-	Arm-Ir	m-u	Screes, forest	H2.6, G1.A1
Lactuca serriola L.	18	di-	Palearct	l-m	Ruderal habitats, meadows, steppe	I1.53, E2.32, E1.2E
Lactuca takhtadzhianii Sosn.	18	di-	Arm-Atr	l-m	Semideserts, steppe	E1.33, E1.45, E1.2E
Lactuca tatarica (L.) C.A.Mey.	18	di-	Palearct	l-m	Semideserts, steppe	E1.33, E1.45, E1.2E
Lactuca wilhelmsiana Fisch. & C.A.Mey. ex DC.	18	di-	Iran	l-m	Forest, steppe shrubs	G1.6H, G1.A1, F2.33
Lactucella undulata (Ledeb.) Nazarova	18	di-	E.A.Med	l-m	Screes, deserts	H2.6, H5.32
Lapsana grandiflora M.Bieb.	14	di-	Cauc-Arm-Ir	l-sb	Forest, meadows	G1.6H, G1.A1, E2.32
Leontodon asperrimus (Willd.) Boiss. ex Bal.	8	di-	Arm-Ir	l-m	Steppe, semideserts, meadows	E1.33, E1.45, E1.2E, E2.32
Leontodon crispus Vill.	8	di-	Med	l-u	Steppe, meadows	E1.2E, E2.32
Leontodon hispidus L.	14	di-	Eur-A.Med	sb-a	Meadows	E2.32
Leucanthemum vulgare Lam.	18, 18+B	di-	Palearct	m-sb	Meadows, steppe, forest	E1.2E, E2.32, G1.6H, G1.A1
Mycelis muralis (L.) Dumort.	18	di-	Eur-Med	l-m	Forest, wetlands	G1.6H, G1.A1, D4.1
Oligochaeta divaricata (Fisch. & C.A.Mey.) K.Koch	24	tetra-	Arm-Ir	l	Semideserts, deserts	E1.33, E1.45, H5.32
Onopordum acanthium L.	34	di-	Holarct	l-u	Ruderal habitats	I1.53

Onopordum armenum Grossh.	34	di-	Arm-Atr	l-m	Steppe	E1.2E
Picris hieracioides L.	10	di-	Palearct	l-sb	Steppe, meadows, forest	E1.2E, E2.32, G1.6H, G1.A1
Picris pauciflora Willd.	10	di-	Med	l-m	Semideserts, steppe	E1.33, E1.45, E1.2E
Picris strigosa M.Bieb.	10	di-	Cauc-Arm-Ir	l-m	Steppe, meadows, semideserts	E1.33, E1.45, E1.2E, E2.32
Podospermum armeniacum Boiss. & Huet	14	di-	Arm-Ir	l-m	Semideserts, steppe	E1.33, E1.45, E1.2E
Podospermum canum C.A.Mey.	14	di-	Cauc-Arm-Ir	l	Semideserts	E1.33, E1.45
Podospermum laciniatum (L.) DC.	14	di-	A.Med	l-m	Steppe, semideserts	E1.33, E1.45, E1.2E
Podospermum meyeri K.Koch	14	di-	Cauc-Arm-Ir	sb-a	Meadows, screes	E2.32, H2.5
Psephellus somcheticus Sosn.	28	di	Arm	m-sb	Steppe, meadows, screes, forest edges	E1.2E, E2.32, H2.6, E5.4
Psephellus taochius Sosn.	30, 30+B	di-	Arm	l-u	Steppe, steppe shrubs	E1.2E, F2.33
Reichardia glauca Mattews	18	di-	Cauc-Arm-Ir	l-m	Semideserts, steppe	E1.33, E1.45, E1.2E
Scariola orientalis (Boiss.) Sojak	18, 36	di-, hexa-	E.A.Med	l-m	Screes	H2.5, H2.6
Scariola viminea (L.) F.W.Schmidt	18	di-	Eur-A.Med	l-m	Screes	H2.5, H2.6
Scorzonera aragatzi Kuthath.	14	di-	Arm (endemic)	m-u	Steppe, meadows, screes	E1.2E, E4.3A, H2.3

Species	2n	Ploidity[a]	Geoelement[b]	Belt[c]	Vegetation	Habitats[d]
Scorzonera bicolor Freyn & Sint.	14, 28	di-, tetra-	Arm-Atr	l-m	Screes, meadows	H2.3, E4.3A
Scorzonera biebersteinii Lipsch.	12	di-	Cauc-Arm-Ir	l-m	Desert, steppe, open forest	H5.32, E1.2E, F5.34
Scorzonera gorovanica Nazarova	14	di-	Arm (endemic)	l-m	Screes, desert	H2.6, H5.32
Scorzonera latifolia (Fisch. & C.A.Mey.) DC.	12, 12+2B	di-	Arm-Ir	m-u	Meadows, steppe shrubs, screes	E2.32, F2.33, H2.6
Scorzonera leptophylla (DC.) Grossh.	14	di-	Iran	m-u	Steppe, meadows, screes	E1.2E, E2.32, H2.6
Scorzonera papposa DC.	14	di-	Arm-Ir	l-m	Screes	H2.6
Scorzonera pseudolanata Grossh.	12, 12+B	di-	E.A.Med	l	Semideserts, desert	E1.33, E1.45, H5.32
Scorzonera rigida Auch. ex DC.	12	di-	Arm-Ir	m-u	Screes, steppe, open forest	H2.6, E1.2E, F5.34
Scorzonera safievii Grossh.	12	di-	Arm	m	Steppe, meadows	E1.2E, E2.32
Scorzonera seidlitzii Boiss.	12	di-	Arm-Atr	m-sb	Meadows	E2.32
Scorzonera semicana DC.	28	tetra-	Arm	m	Steppe, screes	E1.2E, H2.6
Scorzonera suberosa K.Koch	14	di-	Arm	m-u	Steppe, meadows, screes	E1.2E, E2.32, H2.6
Scorzonera turkeviczii Krasch. & Lipsch.	14	di-, tetra	Arm-Atr	m	Steppe, steppe shrubs, screes	E1.2E, F2.33, H2.6

Senecio integrifolius (L.) Clairv.	48	hexa-	Palearct	u-a	Forest, meadows	G1.6H, G1.A1, E2.32
Senecio taraxacifolius (M.Bieb.) DC.	34	di-	Cauc-Arm-Atr	u-a	Rocks	H3.1
Serratula coriacea Fisch. & C.A.Mey. ex DC.	26	di-	Arm-Atr	l-u	Semideserts, steppe, screes	E1.33, E1.45, E1.2E, H2.6
Serratula coronata L.	22	di-	Palearct	m	Forest edges, meadows, wetlands	E5.4, E2.32, D4.1
Serratula radiata (Waldst. & Kit.) M. Bieb.	60	hexa-	Eur	l-sb	Steppe, semideserts, open forest	E1.33, E1.45, E1.2E, F5.34
Serratula serratuloides (Fisch. & C.A.Mey. ex DC.) Takht.	28	tetra-	Arm-Atr	l-a	Screes, juniper forest, semideserts	H2.6, F5.13, G3.93, E1.33, E1.45
Solidago virgaurea L.	18	di-	Holarct	l-a	Meadows, forest	E2.32, G1.6H, G1.A1
Sonchus araraticus Nazarova et Barsegian	18	di-	Arm (endemic)	l	Salt marshes	D6.2
Sonchus arvensis L.	36	hexa-	polychor	l-m	Steppe, steppe shrubs, meadows	E1.2E, F2.33, E2.32
Sonchus asper (L.) Hill.	18	di-	Holarct	l-m	Wetlands, ruderal habitats	D4.1, I1.53
Sonchus oleraceus L.	32	tetra-	polychor	l-m	Wetlands, ruderal habitats	D4.1, I1.53
Sonchus palustris L.	18	di-	Palearct	l-m	Wetlands	D4.1
Sonchus sosnowskyi Schchian	18	di-	Arm (endemic)	u	Wetlands	D4.1

Species	2n	Ploidity[a]	Geoelement[b]	Belt[c]	Vegetation	Habitats[d]
Steptorhamphus czerepanovii Kirp.	16	di-	Arm	l-m	Steppe, screes	E1.2E, H2.6
Steptorhamphus persicus (Boiss.) O. Fedtsch. & B.Fedtsch.	16	di-	E.A.Med	l-m	Rocks, screes	H3.2, H2.6
Steptorhamphus tuberosus (Jacq.) Grossh.	16	di-	E.A.Med	l-m	Open forest, steppe shrubs, screes	F5.34, F2.33, H2.6
Stizolophus balsamita (Lam.) Cass. ex Takht.	26	di-	E.A.Med	l-m	Semideserts, steppe	E1.33, E1.45, E1.2E
Tanacetum abrotanifolium (L.) Druce	36	hexa-	Arm-Ir	l-m	Meadows, wetlands, steppe	E2.32, D4.1, E1.2E
Tanacetum argyrophyllum (K.Koch) Tzvel.	18	di-	Arm-Ir	l-m	Screes	H2.5, H2.6
Tanacetum balsamitoides (Nabel.) Chandjian	18	di-	Arm-Ir	l-sb	Meadows, tall grasses	E2.32, E5.5A
Tanacetum chiliophyllum (Fisch. & C.A.Mey. ex DC.) Sch.Bip.	36	hexa-	Arm-Ir	l-a	Screes	H2.3, H2.4, H2.5, H2.6
Tanacetum parthenium (L.) Sch. Bip.	18	di-	A.Med	l-a	Rocks, forest, wetlands	H3.2, G1.6H, G1.A1, D4.1
Tanacetum pinnatum Boiss.	18	di-	Arm-Ir	l	Semideserts	E1.33, E1.45
Tanacetum vulgare L.	18	di-	Palearct	l-u	Steppe, meadows	E1.2E, E2.32

Taraxacum bessarabicum (Hornem.) Hand.-Mazz.	16, 32	di-, tetra-	polychor	l-sb	Meadows, forest, wetlands	E2.32, G1.6H, G1.A1, D4.1
Taraxacum ceratophorum (Ledeb.) DC.	80	high ploidy	Holarct	m-a	Meadows, forest	E2.32, G1.6H, G1.A1
Taraxacum montanum (C.A.Mey.) DC.	32, 40, 56	tetra-, aneu-	Iran	m-u	Open forest, steppe, screes	F5.34, E1.2E, H2.6
Taraxacum officinale Wigg.	24, 32	tetra-	polychor	l-a	Semideserts, steppe, meadows, forest	E1.33, E1.45, E1.2E, E2.32, G1.6H, G1.A1
Taraxacum serotinum (Waldst. & Kit.) Poir.	16	di-	Pan-Pont-Sarm	m	Steppe	E1.2E
Taraxacum stevenii DC.	16, 32	di-, tetra	Cauc-Ir-Tur	sb-a	Meadows	E4.3A, E4.4
Tomanthea aucheri DC.	18	di-	Arm-Atr	l-u	Semideserts, steppe, juniper forest	E1.33, E1.45, E1.2E, F5.13, G3.93
Tomanthea daralaghezica (Fomin) Takht.	18	di-	Arm	m-u	Semideserts, steppe, juniper forest	E1.33, E1.45, E1.2E, F5.13, G3.93
Tomanthea spectabilis (Fisch. & C.A.Mey.) Takht.	18	di-	Arm-Atr	m-u	Steppe, meadows, forest edges	E1.2E, E2.32, E5.4
Tragopogon armeniacus Kuthath.	12	di-	Arm (endemic)	m-u	Steppe, steppe shrubs, screes	E1.2E, F2.33, H2.6
Tragopogon buphtalmoides (DC.) Boiss.	24, 24+B, 36	tetra-, hexa-	Arm-Ir	m-u	Steppe, steppe shrubs	E1.2E, F2.33
Tragopogon collinus DC.	12	di-	Iran	l	Semideserts	E1.33, E1.45

Species	2n	Ploidity[a]	Geoelement[b]	Belt[c]	Vegetation	Habitats[d]
Tragopogon coloratus C.A.Mey.	12, 24	di-, tetra-	Arm-Ir	l-m	Steppe, semideserts	E1.2E, E1.33, E1.45
Tragopogon dubius Scop.	12	di-	Eur	m	Steppe, steppe shrubs, forest edges	E1.2E, F2.33, E5.4
Tragopogon graminifolius DC.	12, 24, 36	di-, tetra-, hexa-	Cauc-Ir-Tur	l-m	Steppe, steppe shrubs, wetlands	E1.2E, F2.33, D4.1
Tragopogon kemulariae Kuthath.	12	di-	Arm-Ir	m-u	Steppe, meadows	E1.2E, E2.32
Tragopogon krascheninnikovii S.Nikit.	12	di-	Iran	l-m	Steppe, semideserts	E1.2E, E1.33, E1.45
Tragopogon latifolius Boiss.	12, 24	di-, tetra	Arm	m-sb	Meadows, steppe	E1.2E, E2.32
Tragopogon marginatus Boiss. & Buhse	12	di-	Arm-Atr	l-m	Semideserts, steppe	E1.2E, E1.33, E1.45
Tragopogon pterocarpus DC.	12	di-	Arm-Ir	l-m	Semideserts, steppe	E1.2E, E1.33, E1.45
Tragopogon pusillus M.Bieb.	12	di-	Iran	l	Semideserts	E1.33, E1.45
Tragopogon reticulatus Boiss. & Huet	12	di-, tetra-, hexa-	Arm-Ir	sb-a	Meadows	E2.32
Tragopogon segetus Kuthath.	12	di-	Arm	l-u	Steppe, juniper forest	E1.2E, F5.13, G3.93
Tragopogon serotinus Sosn.	12	di-	Arm	l-m	Semideserts, deserts	E1.33, E1.45, H5.32
Tragopogon sosnowskyi Kuthath.	12	di-	Arm	m-u	Steppe, meadows, steppe shrubs	E1.2E, E2.32, F2.33
Tragopogon tuberosus K.Koch	24	tetra-	Cauc	m	Steppe, meadows	E1.2E, E2.32

Tripleurospermum caucasicum (Willd.) Hayek	18, 36	di-, tetra	E.Med	a	Rocks, screes	H3.1, H2.3
Tripleurospermum disciforme (C.A.Mey.) Sch.Bip.	36	tetra-	E.A.Med	l-m	Wetlands	D4.1
Tripleurospermum parviflorum (Willd.) Pobed.	18, 18+B	di-	E.A.Med	l	Semideserts	E1.33, E1.45
Tripleurospermum sevanense (Manden.) Pobed.	18	di-	Arm-Ir	m-u	Wetlands, meadows	D4.1, E2.32
Tripleurospermum tenuifolium (Kit.) Freyn	36	tetra-	Eur-Cauc	l	Ruderal habitats	I1.53
Tripleurospermum transcaucasicum (Manden.) Pobed.	18	di-	Arm	l-a	Wetlands, meadows	D4.1, E2.32
Urospermum picroides (L.) Scop. ex F.W.Schmidt	10	di-	A.Med	l	Semideserts, open forest	E1.33, E1.45, F5.34
Berberidaceae						
Leontice armeniaca Belanger	16	di-	Atr	l	Semideserts	E1.33, E1.45
Boraginaceae						
Buglossoides arvensis (L.) Johnst.	28	tetra-	Eur-A.Med	l-m	Steppe, semideserts, steppe shrubs	E1.2E, E1.33, E1.45, F2.33
Cerinthe minor L.	18	di-	A.Med	l-sb	Steppe, meadows, steppe shrubs	E1.2E, E2.32, F2.33

Species	2n	Ploidity[a]	Geoelement[b]	Belt[c]	Vegetation	Habitats[d]
Cynoglossum officinale L.	24	tetra-	Eur-A.Med	l-m	Semideserts, steppe, ruderal habitats	E1.2E, E1.33, E1.45, I1.53
Echium russicum J.F.Gmel.	18	di-	Eur-Med	m-sb	Steppe, meadows	E1.2E, E2.32
Myosotis alpestris F.W.Schmidt	24	tetra-	Eur-Med	sb-a	Meadows, screes	E1.2E, E4.3A, E4.4
Myosotis sylvatica Ehrh. ex Hoffm.	48	hexa-	Arm-Ir	m-u	Forest, steppe shrubs	G1.6H, G1.A1, F2.33
Onosma microcarpa Stev. ex DC.	16	di-	Arm-Ir	l-m	Steppe, semideserts, open forest	E1.2E, E1.33, E1.45, F5.34
Rochelia disperma (L.fil.) K.Koch	20	di-	Eur-A.Med	l-m	Semideserts, steppe, screes	E1.2E, E1.33, E1.45, H2.6
Brassicaceae						
Arabis carduchorum Boiss.	16	di-	Arm	sb-a	Meadows, screes	E4.3A, E4.4, H2.3, H2.4
Barbarea minor K.Koch	16	di-	Arm-Ir	u-sb	Wetlands	D4.1
Cardamine uliginosa M.Bieb.	16	di-	Cauc-Arm-Ir	l-sb	Wetlands, river banks	D4.1, C3.24
Coluteocarpus vesicaria (L.) Holmboe	14	di-	Cauc-Arm-Ir	m-a	Steppe, meadows, open forest	E1.2E, E4.3A, E4.4, F5.34
Conringia orientalis (L.) Dumort.	14	di-	A.Med	n-u	Ruderal habitats, fields	I1.53, I1.3
Crambe armena N.Busch	60	high ploidy	Arm	l	Semideserts	E1.33, E1.45
Didymophysa aucheri Boiss.	16	di-	Cauc-Arm-Ir	a	Screes	H2.3

Diptychocarpus strictus (Fisch. ex M.Bieb.) Trautv.	14	di-	E.A.Med	l	Semideserts	E1.33, E1.45
Draba araratica Rupr.	16	di-	Arm	a	Rocks	H3.1
Draba bruniifolia Steven	32	tetra-	Cauc-Arm-Ir	sb-a	Rocks, screes	H3.1, H2.3
Draba siliquosa M.Bieb.	16	di-	Cauc	sb-a	Rocks, screes	H3.1, H2.3
Erysimum gelidum Bunge	14, 16	di-	Arm	sb-a	Screes	H2.3
Fibigia suffruticosa (Vent.) Sweet	16	di-	Arm-Ir	m	Steppe, screes	E1.2E, H2.6
Hesperis matronalis L.	14	di-	Palearct	l-sb	Steppe, meadows, forest, steppe shrubs	E1.2E, E2.32, G1.6H, G1.A1, F2.33
Isatis buschiana Schischk.	14	di-	Arm	l-m	Semideserts, steppe	E1.33, E1.45, E1.2E
Isatis steveniana Trautv.	14, 28	di-, tetra	Arm-Ir	m-u	Steppe, screes, open forest	E1.2E, H2.5, H2.6, F5.34
Meniocus linifolius (Steph.) DC.	42	hexa-	Eur-A.Med	m-u	Steppe, ruderal habitats	E1.2E, I1.53
Murbeckiella huetii (Boiss.) Rothm.	24	tetra-	Cauc-Arm-Ir	sb-a	Screes	H2.3
Thlaspi perfoliatum L.	36	hexa-	Eur-A.Med	l-m	Meadows, steppe	E1.2E, E2.32
Turritis glabra L.	12	di-	Palearct	m-u	Meadows, forest edges, steppe shrubs	E2.32, E5.4, F2.33

Species	2n	Ploidity[a]	Geoelement[b]	Belt[c]	Vegetation	Habitats[d]
Campanulaceae						
Campanula collina Sims	68	tetra-	Cauc	m-a	Meadows	E2.32, E4.3A, E4.4
Campanula coriacea P.H.Davis	34	di-	Arm-Ir	l-sb	Rocks	H3.1, H3.2
Campanula saxifraga M.Bieb. subsp. *aucheri* (A.DC.) Ogan.	34	di-	Arm-Atr	u-a	Meadows	E4.3A, E4.4
Campanula stevenii M.Bieb.	16, 32	di-, tetra-	Eur-Cauc	m-a	Steppe, meadows	E1.2E, E2.32
Campanula tridentata Schreb.	34	di-	Arm	sb-a	Meadows	E4.3A, E4.4
Michauxia laevigata Vent.	34	di-	Iran	l-u	Semideserts, open forest, screes	E1.33, E1.45, F5.34, H2.6
Caryophyllaceae						
Arenaria dianthoides Smith	22	di-	Arm-Ir	m-sb	Meadows, steppe	E2.32, E1.2E
Arenaria gypsophylloides L.	22	di-	Arm-Ir	m-sb	Steppe, meadows	E2.32, E1.2E
Arenaria rotundifolia M.Bieb.	44	tetra-	Cauc	sb-a	Wetlands	D4.1
Arenaria serpyllifolia L.	22	di-	Holarct	m-sb	Steppe, meadows, screes	E2.32, E1.2E, H2.5
Cerastium cerastoides (L.) Britt.	38	di-	Palearct	sb-a	Screes	H2.3
Cerastium dichotomum L.	34	di-	A.Med	l-u	Screes, ruderal habitats	H2.5, H2.6, I1.53

Cerastium longifolium Willd.	38	di-	Arm-Ir	l-m	Steppe, open forest	E1.2E, F5.34
Cerastium szovitsii Boiss.	36	tetra-	Atr	sb-a	Screes	H2.3
Dianthus calocephalus Boiss.	30	di-	Arm-Ir	m-sb	Meadows, forest edges	E2.32, E5.4
Dianthus canescens K.Koch	90	hexa-	Arm	l-u	Semideserts, steppe	E1.2E, E1.33, E1.45
Dianthus floribundus Boiss.	30	di-	Arm-Ir	l-m	Semideserts, steppe	E1.2E, E1.33, E1.45
Herniaria incana Lam.	72	octa-	Eur-A.Med	l-sb	Semideserts, steppe, meadows	E1.2E, E1.33, E1.45, E2.32,
Holosteum marginatum Fisch. & C.A.Mey.	20	di-	Cauc-Iran	l-m	Semideserts, steppe	E1.2E, E1.33, E1.45
Melandrium album (Mill.) Garcke	24	di-	W.A.Med	l-u	Semideserts, steppe	E1.2E, E1.33, E1.45
Melandrium latifolium (Poir.) Maire	24	di-	Palearct	m-sb	Meadows, forest edges	E2.32, E5.4
Minuartia aizoides (Boiss.) Bornm.	26	di-	Cauc-Arm-Ir	sb-a	Rocks	H3.1, H3.2
Minuartia oreina (Mattf.) Schischk.	30	di-	Arm-Ir	a	Rocks, screes	H3.1, H3.2, H2.3, H2.4
Paronychia kurdica Boiss.	14	di-	Arm-Ir	l-m	Semideserts, steppe	E1.2E, E1.33, E1.45
Sagina procumbens L.	22	di-	Eur-Med	l-u	Wetlands	D2, D4.1

Species	2n	Ploidity[a]	Geoelement[b]	Belt[c]	Vegetation	Habitats[d]
Scleranthus uncinnatus Schur	22	di-	A.Med	l-u	Steppe, semideserts	E1.2E, E1.33, E1.45
Silene bupleuroides L.	24	di-	Arm-Ir	l-u	Steppe, semideserts	E1.2E, E1.33, E1.45
Silene chlorifolia Smith	24	di-	Cauc-Arm-Ir	m-u	Steppe, meadows	E2.32, E1.2E
Silene compacta Fisch. ex Hornem.	24	di-	Cauc-Arm-Ir	m-sb	Meadows, forest edges	E2.32, E5.4
Silene conoidea L.	20	di-	A.Med	l-m	Steppe	E1.2E
Silene depressa M.Bieb.	24	di-	Arm	u-sb	Rocks	H3.1, H3.2
Silene iberica M.Bieb.	24, 48	di-, tetra-	Arm	l-u	Steppe, forest edges	E1.2E, E5.4
Silene italica (L.) Pers.	24, 48	di-, tetra-	A.Med	m-sb	Forest, forest edges, meadows	G1.6H, G1.A1, E5.4, E2.32
Silene lasiantha K.Koch	24, 48	di-, tetra-	Arm-Ir	m-a	Steppe, meadows	E2.32, E1.2E
Silene marschallii C.A.Mey.	24, 48	di-, tetra-	Arm-Ir	l-m	Semideserts, steppe	E1.2E, E1.33, E1.45
Silene meyeri Fenzl ex Boiss. & Buhse	24	di-	Atr	sb-a	Rocks	H3.1, H3.2
Silene noctiflora L.	24	di-	Eur	m-sb	Forest, forest edges, steppe shrubs	G1.6H, G1.A1, E5.4, F2.33
Silene propinqua Schischk.	24	di-	Arm-Atr	m-u	Steppe, meadows	E2.32, E1.2E
Silene ruprechtii Schischk.	24	di-	Cauc-Arm-Ir	u-a	Meadows, screes	E4.3A, E4.4, H2.3, H2.4
Silene sisianica Boiss. & Buhse	48	tetra-	Arm-Atr	m-u	Rocks, steppe, screes	H3.1, E1.2E, H2.5
Silene spergulifolia (Willd.) M.Bieb.	24	di-	Cauc-Arm-Ir	l-u	Screes, steppe, meadows	H2.5, E1.2E, E2.32
Silene subconica Friv.	24	di-	Eur-A.Med	l-u	Semideserts, steppe	E1.2E, E1.33, E1.45

Silene viscosa (L.) Pers.	24	di-	Pan-Pont-Sarm	m-u	Steppe	E1.2E
Silene wallichiana Klotzsch.	24	di-	Iran	m-u	Steppe, medwows	E1.2E, E2.32
Spergularia maritima (All.) Chiov.	18	di-	Holarct	m-u	Semideserts, steppe	E1.2E, E1.33, E1.45
Chenopodiaceae						
Atriplex sagittata Borkh.	18	di-	Eur-A.Med	l	Semideserts	E1.33, E1.45
Beta corolliflora Zosimovic ex Buttler	36	tetra-	Eur-Med	m-u	Meadows	E2.32
Beta lomatogona Fisch. & C.A.Mey.	18	di-	Arm-Ir	l	Semideserts	E1.33, E1.45
Chenopodium foliosum Aschers.	18	di-	Eur-A.Med	l-m	Semideserts, steppe	E1.2E, E1.33, E1.45
Chenopodium hybridum L.	18	di-	Holarct	l-m	Semideserts, steppe, ruderal habitats	E1.2E, E1.33, E1.45, I1.53
Eurotia ceratoides (L.) C.A.Mey.	36	tetra-	Eur-A.Med	l-m	Semidesert, steppe	E1.2E, E1.33, E1.45
Microcnemum coralloides (Loscos & Pardo) Buen	18	di-	A.Med	l	Salt marshes	D6.2
Spinacia tetrandra Steven	12	di-	Iran	l	Semideserts, wetlands	E1.33, E1.45, D4.1
Suaeda altissima (L.) Pall.	18	di-	Palearctic	l	Semideserts, wetlands	E1.33, E1.45, D4.1
Suaeda heterophylla (Kar. & Kir.) Bunge	18	di-	A.Med.	l	Semideserts, wetlands	E1.33, E1.45, D4.1

Species	2n	Ploidity[a]	Geoelement[b]	Belt[c]	Vegetation	Habitats[d]
Suaeda salsa (L.) Pall.	36	tetra-	Palearctic		Semideserts, wetlands	E1.33, E1.45, D4.1
Colchicaceae						
Colchicum szovitsii Fisch. & C.A.Mey.	18	di-	Cauc-Arm-Ir	l-u	Steppe, meadows, juniper forest, wetlands	E1.2E, E2.32, F5.13, G3.93, D4.1
Merendera mirzoevae Gabrielian	18	di-	Arm (endemic)	l-u	Meadow, forest	E2.32, G1.6H, G1.A1
Merendera sobolifera Fisch. & C.A.Mey.	42	hexa-	E.A.Med	l	Salt marshes	D6.2
Merendera trigyna (Steven ex Adam) Stapf	20	di-	Cauc-Arm-Atr	l-u	Semideserts, steppe	E1.2E, E1.33, E1.45
Convallariaceae						
Polygonatum glaberrimum K.Koch	20	di-	Cauc-Arm-Atr	m	Forest	G1.6H, G1.A1
Polygonatum orientale Desf.	18	di-	E.A.Med	l-m	Forest, steppe shrubs	G1.6H, G1.A1, F2.33
Crassulaceae						
Sedum annuum L.	22	di-	Eur-Med	m-a	Meadows, rocks, screes	E2.32, E4.3A, E4.4, H3.1, H3.2, H2.3, H2.4, H2.5, H2.6
Sedum oppositifolium Sims	28	tetra-	Cauc-Arm-Atr	m-a	Rocks, screes	H3.1, H3.2, H2.3, H2.4, H2.5, H2.6
Sedum tenellum M.Bieb.	14	di-	Cauc-Arm-Atr	sb-a	Rocks, screes	H3.1, H3.2, H2.3, H2.4, H2.5, H2.6

Cucurbitaceae						
Bryonia alba L.	20	di-	Eur-A.Med	l-u	Rocks, forest, ruderal habitats	H3.1, H3.2, G1.6H, G1.A1, I2.22
Cyperaceae						
Carex aterrima Hoppe subsp. *medwedewii* (Lescov) T.V.Egorova	52	di-	Cauc-Arm-Ir	a	Meadows	E4.3A, E4.4
Carex cilicica Boiss.	54	di-	Arm-Ir	sb	Wetlands	D4.1
Carex decasyi (C.B.Clarke) O. Yano & S.R.Zhang	32	tetra-	Cauc-Arm-Ir	a	Meadows	E4.3A, E4.4
Carex diluta M.Bieb.	74	di-	Cauc-Ir-Tur	sb	Wetlands	D4.1
Carex hartmaniorum A.Cajander	52	di-	Eur	sb	Meadows	E4.3A, E4.4
Carex hordeistichos Vill.	58	di-	Eur-A.Med	m	Steppe	E1.2E
Carex muricata L. subsp. *aschokae* Molina Gonz., Acedo & Llamas	58	di-	Cauc-Ir-Tur	u	Meadows	E2.32
Carex nigra (L.) Reichard subsp. *transcaucasica* (T.V.Egorova) Jim. Mejias, G.E.Rodr.-Pa., Amini Rad & Martin Bravo	84	di-	Cauc	sb	Meadows	E2.32

Species	2n	Ploidity[a]	Geoelement[b]	Belt[c]	Vegetation	Habitats[d]
Carex phyllostachys C.A.Mey.	56	di-	Med	l	Forest	G1.A1
Carex secalina Willd. ex Wahlenb.	50	di-	Palearct	u	Meadows	E2.32
Carex songorica Kar. & Kir.	82	di-	E.A.Med	u	Wetlands	D4.1
Carex tomentosa L.	48	di-	Palearct	sb	Meadows	E2.32
Carex tristis M.Bieb.	38	di-	Cauc-Arm-Atr	u-a	Meadows	E4.3A
Dipsacaceae						
Scabiosa bipinnata K.Koch	36	tetra-	Cauc	l-sb	Steppe, meadows, forest	E1.2E, E2.32, G1.A1
Elaeagnaceae						
Elaeagnus angustifolia L.	28	tetra-	A.Med	l-u	Wetlands, semideserts, steppe	E1.2E, E1.33, E1.45, D4.1
Hippophae rhamnoides L.	24	di-	Eur-A.Med	l-u	Wetlands	D4.1
Ephedraceae						
Ephedra procera Fisch. & C.A.Mey.	14	di-	E.A.Med	l-u	Semideserts, steppe, screes	E1.2E, E1.33, E1.45, H2.5, H2.6
Fabaceae						
Astragalus arguricus Bunge	32	tetra-	Cauc-Arm	m	Steppe, screes	E1.2E, H2.5
Astragalus asterias Steven ex Ledeb.	16	di-	Med	m	Steppe,semideserts, screes	E1.2E, E1.33, E1.45, H2.5
Astragalus aureus Willd.	16	di-	Cauc-Arm-Ir	m-u	Steppe	E1.2E
Astragalus falcatus Lam.	16	di-	Cauc	m-u	Forest edges, steppe shrubs	E5.4, F2.33

Astragalus galegiformis L.	16	di-	Cauc	m-u	Steppe	E1.2E
Astragalus gezeldarensis Grossh.	48	hexa-	Arm	a	Screes	H2.3, H2.4
Astragalus gjunaicus Grossh.	64	octa-	Arm	m-u	Steppe	E1.2E
Astragalus glycyphyllos L.	16	di-	Palearct	l-m	Forest edges, steppe shrubs	E5.4, F2.33
Astragalus goktschaicus Grossh.	40	penta-	Arm (endemic)	m-u	Meadows, wetlands	E2.32, D4.1
Astragalus hajastanus Grossh.	16	di-	Arm	l-m	Semideserts, steppe	E1.2E, E1.33, E1.45
Astragalus incertus Ledeb.	16	di-	Arm	m-a	Steppe, meadows	E1.2E, E2.32
Astragalus polygala Pall.	14	di-	Cauc	u-a	Meadows	E2.32
Astragalus sevangensis Grossh.	32, 64	tetra-, octa-	Arm	m-u	Steppe	E1.2E
Astragalus shagalensis Grossh.	16+2B	di-	Arm	m	Forest edges, meadows	E5.4, E2.32
Cicer anatolicum Alef.	16	di-	Arm-Ir	m	Steppe, steppe shrubs	E1.2E, F2.33
Colutea komarovii Takht.	16	di-	Atr	l	Open forest	F5.34
Glycyrrhiza glabra L.	16	di-	Holarct	l	Wetlands	D4.1
Goebelia alopecuroides (L.) Bunge	36	hexa-	Palearct	l-m	Wetlands	D4.1

Species	2n	Ploidity[a]	Geoelement[b]	Belt[c]	Vegetation	Habitats[d]
Halimodendron halodendron (Pall.) Voss	16	di-	E.A.Med	l	Semideserts	E1.33, E1.45
Hedysarum caucasicum M.Bieb.	14	di-	Cauc	sb-a	Meadows	E4.3A, E4.4
Lathyrus aphaca L.	14	di-	Eur-A.Med	l-m	Gardens, fields, wetlands	I1.3, I2.22, D4.1
Lathyrus chloranthus Boiss.	14	di-	Arm-Ir	l-m	Wetlands	D4.1
Lathyrus cicera L.	14	di-	A.Med	l-m	Gardens, fields	I1.3, I2.22
Lathyrus hirsutus L.	14	di-	Eur-A.Med	l-m	Forest edges, steppe shrubs	E5.4, F2.33
Lathyrus incurvus (Roth) Roth	14	di-	Eur-A.Med	l-m	Wetlands	D4.1
Lathyrus miniatus M.Bieb. ex Steven	14	di-	Cauc	l-u	Forest edges, steppe shrubs	E5.4, F2.33
Lathyrus nissolia L.	14	di-	Eur-A.Med	m-u	Steppe, meadows	E1.2E, E2.32
Lathyrus pratensis L.	14	di-	Eur-A.Med	m-u	Meadows, forest edges	E5.4, E2.32
Lathyrus sphaericus Retz	14	di-	Eur-A.Med	l-m	Gardens, fields	I1.3, I2.22
Lathyrus tuberosus L.	14	di-	Palearct	l-m	Steppe, ruderal habitats	E1.2E, I1.53
Lathyrus vinealis Boiss. & Noe	14	di-	E.A.Med	m	Steppe, ruderal habitats	E1.2E, I1.53
Lens ervoides (Brign.) Grande	14	di-	A.Med	l-m	Steppe, semideserts	E1.2E, E1.33, E1.45
Lens orientalis (Boiss.) Schmalh.	14	di-	A.Med	l	Semideserts	E1.33, E1.45

Lotus caucasicus Kuprian. ex Juz.	24	tetra-	Cauc	l-u	Steppe, meadows, forest edges	E1.2E, E2.32, E5.4
Lotus corniculatus L.	24	tetra-	Eur-A.Med	l-m	Meadows, wetlands	E2.32, D4.1
Medicago hemicycla Grossh.	16, 32	di-, tetra-	Arm-Atr	m-u	Steppe, meadows, forest edges	E1.2E, E2.32, E5.4
Medicago minima (L.) Bartalini	16	di-	Eur-A.Med	l-m	Open forest, ruderal habitats, steppe	F5.34, I1.53, E1.2E
Medicago sativa L.	16, 32	di-, tetra-	A.Med	l-a	Steppe, meadows, semideserts, steppe shrubs	E1.2E, E2.32, E1.33, E1.45, F2.33
Ononis arvensis L.	32	tetra-	Palearct	m-u	Wetlands, ruderal habitats	D4.1, I1.53
Oxytropis cyanea M.Bieb.	16	di-	Cauc	sb-a	Screes	H2.3, H2.4
Pisum elatius M.Bieb.	14	di-	Med	l-m	Fields, steppe shrubs, wetlands	I1.3, F2.33, D4.1
Sphaerophysa salsula (Pall.) DC.	16	di-	E.A.Med	l	Salt marshes	D6.2
Trifolium alpestre L.	16	di-	Eur	m-sb	Meadows, steppe shrubs, forest	E2.32, F2.33, G1.6H, G1.A1
Trifolium ambiguum M.Bieb.	48	hexa-	Eur-Cauc	n-a	Meadows, steppe	E1.2E, E2.32, E4.3A, E4.4
Trifolium arvense L.	14	di-	A.Med	l-u	Steppe, forest edges, meadows	E1.2E, E2.32, E5.4
Trifolium campestre Schreb.	14	di-	Eur-A.Med	m-a	Meadows, forest edges, wetlands	E2.32, E4.3A, E4.4, E5.4, D4.1
Trifolium repens L.	32	tetra-	Eur-A.Med	l-u	Meadows, steppe	E1.2E, E2.32

Species	2n	Ploidity[a]	Geoelement[b]	Belt[c]	Vegetation	Habitats[d]
Trifolium trichocephalum M.Bieb.	48	hexa-	Cauc	u-a	Meadows, forest edges	E2.32, E5.4
Vavilovia formosa (Steven) Fed.	14	di-	Cauc-Arm-Ir	a	Screes	H2.3
Vicia akhmaganica Kazar.	10	di-	Arm (endemic)	a	Screes	H2.3
Vicia alpestris Steven	14	di-	Cauc	sb-a	Meadows, screes	E4.3A, E4.4, H2.3, H2.4
Vicia anatolica Turrill	10	di-	E.A.Med	l-m	Semideserts, steppe	E1.2E, E1.33, E1.45
Vicia ciliatula Lipsky	10	di-	Cauc	l-m	Meadows, forest edges	E2.32, E5.4
Vicia ervilia (L.) Willd.	14	di-	Eur	l-m	Steppe, meadows	E1.2E, E2.32
Vicia grandiflora Scop.	14	di-	Eur-Med	l-m	Meadows, steppe shrubs	E2.32, F2.33
Vicia hirsuta (L.) S.F.Gray	14	di-	Palearct	l-m	Forest, steppe shrubs	G1.A1, F2.33
Vicia hyrcanica Fisch. & C.A.Mey.	12	di-	E.A.Med	m	Steppe, fields	E1.2E, I1.3
Vicia lutea L.	14	di-	Eur	l	Fields	I1.3
Vicia narbonensis L.	14	di-	A.Med	l-m	Wetlands	D4.1
Vicia nissoliana L.	10	di-	A. Med	sb-a	Meadows, screes	E4.3A, E4.4, H2.3, H2.4
Vicia pannonica Crantz	12	di-	Pan-Pont	m-l	Steppe, fields	E1.2E, I1.3
Vicia peregrina L.	14	di-	A.Med	l-m	Meadows, fields	E2.32, I1.3
Vicia sativa L.	12, 14	di-	Palearct	l-m	Meadows, fields	E2.32, I1.3

Vicia tetrasperma (L.) Schreb.	14	di-	Palearct	l-m	Steppe shrubs, forest edges	F2.33, E5.4
Vicia variabilis Freyn & Sint.	24	tetra-	Cauc-Arm-Ir	l-m	Meadows, forest	E2.32, G1.A1
Vicia villosa Roth	14	di-	E.A.Med	l	Meadows, fields	E2.32, I1.3
Frankeniaceae						
Frankenia hirsuta L.	10	di-	A.Med	l-m	Salt marshes, semideserts	D6.2, E1.33, E1.45
Fumariaceae						
Corydalis alpestris C.A.Mey.	16	di-	Cauc	sb-a	Meadows	E4.3A, E4.4
Gentianaceae						
Gentiana dshimilensis K.Koch	26	di-	Med	sb-a	Meadows	E4.3A, E4.4
Geraniaceae						
Geranium albanum M.Bieb.	28	tetra-	Cauc-Arm-Atr	l-m	Forest edges, meadows	E2.32, E5.4
Geranium collinum Steph.	28	tetra-	E.A.Med	m	Meadows, forest edges	E2.32, E5.4
Geranium columbinum L.	18	di-	Eur-Med	l	Forest edges, steppe shrubs	F2.33, E5.4
Geranium dissectum L.	22	di-	Eur-A.Med	l	Open forest, gardens, ruderal habitats	F5.34, I2.22, I1.53
Geranium divaricatum Ehrh.	26	di-	Eur-A.Med	m-u	Open forest, gardens, ruderal habitats	F5.34, I2.22, I1.53

Species	2n	Ploidity[a]	Geoelement[b]	Belt[c]	Vegetation	Habitats[d]
Geranium ibericum Cav.	28	di-	Cauc	m-u	Steppe, meadows	E1.2E, E2.32
Geranium lucidum L.	40	tetra-	Eur-A.Med	l-m	Meadows, steppe, wetlands	E1.2E, E2.32, D4.1
Geranium molle L.	26	di-	Eur-A.Med	l-sb	Meadows, steppe shrubs	E2.32, F2.33
Geranium platypetalum Fisch. & C.A.Mey.	42	hexa-	Cauc-Arm-Ir	sb-a	Meadows	E4.3A, E4.4
Geranium pusillum L.	26	di-	Eur-A.Med	l-u	Forest edges, meadows	E2.32, E5.4
Geranium pyrenaicum Burm.fil.	26	di-	Eur-Med	m-u	Forest, meadows	E2.32, G1.6H, G1.A1
Geranium rotundifolium L.	26	di-	Eur-A.Med	l-u	Open forests, steppe shrubs, ruderal habitats	F5.34, E5.4, I1.53
Geranium sylvaticum L.	28	di-	Palearct	m-a	Meadows, forest edges	E2.32, E5.4
Geranium tuberosum L.	28	tetra-	A.Med	l-u	Steppe, fields	E1.2E, I1.3
Globulariaceae						
Globularia trichosantha Fisch. & C.A.Mey.	16	di-	E.A.Med	l-u	Steppe, meadows	E1.2E, E2.32
Hyacinthaceae						
Bellevalia fominii Woronow	8	di-	Arm-Atr	l-u	Steppe, forest	E1.2E, G1.A1
Bellevalia glauca (Lindl.) Kunth	8, 32	di-, octa-	Cauc-Arm-Ir	l-u	Semideserts, steppe	E1.2E, E1.33, E1.45

Bellevalia longistyla (Miscz.) Grossh.	8, 32	di-, octa-	Arm-Atr	l-u	Semideserts, steppe	E1.2E, E1.33, E1.45
Bellevalia paradoxa (Fisch. & C.A.Mey.) Boiss.	8	di-	Arm-Atr	m-a	Meadows, steppe	E1.2E, E2.32
Bellevalia pycnantha (K.Koch) Losinsk.	16	tetra-	Arm-Ir	l-m	Steppe	E1.2E
Bellevalia sarmatica (Pall. ex Georgi) Woronow	8	di-	Pan-Pont-Sarm	l-m	Steppe	E1.2E
Hyacinthella atropatana (Grossh.) Mordak & Zakhar.	22	di-	Atr	l	Semideserts	E1.33, E1.45
Muscari caucasicum (Griseb.) Baker	18	di-	Cauc-Arm-Ir	l-sb	Semideserts, steppe, open forest	E1.2E, E1.33, E1.45, F5.34
Muscari longipes Boiss.	18	di-	Arm-Ir	l-u	Semideserts, steppe	E1.2E, E1.33, E1.45
Muscari neglectum Guss.	36, 45, 54	penta-, hexa-	Eur-A.Med	l-m	Fields	I1.3
Muscari pallens (M.Bieb.) Fisch.	36	tetra-	Cauc	l-u	Forest, meadows	E2.32, G1.6H, G1.A1
Muscari sosnowskyi Schchian	36	tetra-	Arm	m-a	Meadows, steppe, forest	E1.2E, E2.32, G1.6H, G1.A1
Muscari szovitsianum Baker	36, 45	tetra-, penta-	Cauc	l-sb	Meadows, steppe	E1.2E, E2.32
Muscari tenuiflorum Tausch	18	di-	Eur-A.Med	l-sb	Steppe, open forest	E1.2E, F5.34

Species	2n	Ploidity[a]	Geoelement[b]	Belt[c]	Vegetation	Habitats[d]
Ornithogalum balansae Boiss.	24, 25, 26	di-, aneu-	Cauc	u-a	Meadows	E4.3A, E4.4
Ornithogalum brachystachys K.Koch	24, 24+1-12B, 28-36	di-, aneu-	Arm	l-u	Steppe, open forest	E1.2E, F5.34
Ornithogalum gabrielianae Agapova	58, 58+2B	aneu-	Arm (endemic)	u-a	Meadows	E4.3A
Ornithogalum hajastanum Agapova	14, 14+1-8B, 16	di-	Arm	l-sb	Forest, steppe, meadows, juniper forest	G1.A1, E1.2E, E2.32, F5.13, G3.93
Ornithogalum montanum Cirillo	20	tetra-	A.Med	l-sb	Screes, steppe, juniper forest	H2.5, H2.6, E1.2E, F5.13, G3.93
Ornithogalum navaschinii Agapova	32	tetra-	Cauc-Arm-Atr	l-m	Steppe, semideserts	E1.2E, E1.33, E1.45
Ornithogalum schelkovnikovii Grossh.	26	di-	Arm-Ir	m-sb	Steppe, meadows, forest	G1.A1, E1.2E, E2.32
Ornithogalum sigmoideum Freyn & Sint.	24	di-	Arm-Atr	m-sb	Steppe, meadows	E1.2E, E2.32
Ornithogalum tempskyanum Freyn & Sint.	18	di-	Arm	m-sb	Open forest, meadows	F5.34, E2.32
Ornithogalum transcaucasicum Miscz. ex Grossh.	30	hexa-	Arm	u-a	Meadows, forest	E2.32, G1.A1
Puschkinia scilloides Adam	10	di-	Cauc	m-a	Meadows, steppe	E1.2E, E2.32
Scilla armena Grossh.	14, 30	di-, hexa-	Arm	m-sb	Forest, meadows, steppe	G1.A1, E1.2E, E2.32

Scilla rosenii K.Koch	12	di-	Arm	u-sb	Meadows, wetlands	E2.32, D4.1
Scilla siberica Haw.	12	di-	Eur-Cauc	l-m	Forest, steppe shrubs	G1.A1, F5.34
Hypericaceae						
Hypericum eleonorae A.Jelen.	18	di-	Arm (endemic)	m-u	Rocks	H3.1, H3.2
Iridaceae						
Crocus adamii J.Gay	10	di-	Cauc-E.Med	l-a	Steppe, meadows, open forest	E1.2E, E2.32, F5.34
Gladiolus atroviolaceus Boiss.	90	high ploidy	E.A.Med	l-m	Semideserts, steppe, open forest	E1.2E, E1.33, E1.45, F5.34
Iris atropatana Grossh.	18	di-	Atr	l-u	Semideserts, open forest	E1.33, E1.45, F5.34
Iris caucasica Hoffm.	22	di-	Cauc-Arm-Ir	l-sb	Semideserts, steppe, open forest	E1.2E, E1.33, E1.45, F5.34
Iris demetrii Achv. & Mirzoeva	34, 38	aneu-	Arm	m-sb	Steppe, open forest, meadows, wetlands	E1.2E, F5.34, E2.32, D4.1
Iris elegantissima Sosn.	20	di-	Arm-Atr	l-u	Semideserts, steppe	E1.2E, E1.33, E1.45
Iris furcata M.Bieb.	24	di-	Cauc	m-sb	Steppe, meadows	E1.2E, E2.32
Iris grossheimii Woronow ex Grossh.	20	di-	Atr	m-sb	Steppe, open forest, meadows	E1.2E, F5.34, E2.32
Iris iberica Hoffm.	20	di-	Arm	l	Open forest	F5.34
Iris imbricata Lindl.	24	di-	Atr	m-a	Steppe, open forest	E1.2E, F5.34
Iris lineolata (Trautv.) Grossh.	20	di-	Arm-Ir	l-a	Semideserts, steppe, open forest	E1.2E, E1.33, E1.45, F5.34
Iris lycotis Woronow	20	di-	Arm-Ir	l-a	Semideserts, steppe	E1.2E, E1.33, E1.45
Iris musulmanica Fomin	40, 44	tetra-	Arm-Atr	l	Salt marshes	D6.2

Species	2n	Ploidity[a]	Geoelement[b]	Belt[c]	Vegetation	Habitats[d]
Iris paradoxa Steven	20	di-	Arm-Atr	l-a	Semideserts, steppe, open forest	E1.2E, E1.33, E1.45, F5.34
Iris pseudocaucasica Grossh.	18	di-	Arm-Atr	l-m	Open forest, steppe	E1.2E, F5.34
Iris pumila L.	32	tetra-	Eur-A.Med	l-u	Semideserts, steppe	E1.2E, E1.33, E1.45
Iris reticulata M.Bieb.	20	di-	Cauc-Arm-Ir	l-a	Steppe, meadows, semideserts	E1.2E, E1.33, E1.45, E2.32
Ixioliriaceae						
Ixiolirion tataricum (Pall.) Schult. & Schult.fil.	48	hexa-	E.A.Med	l-sb	Semideserts, steppe, meadows	E1.2E, E1.33, E1.45, E2.32
Juglandaceae						
Juglans regia L.	32	di-	A.Med	l-m	Forest	G1.6H, G1.A1
Juncaceae						
Luzula spicata (L.) DC.	18	di-	Holarct	u-a	Meadows	E4.3A, E4.4
Luzula stenophylla Steud	24	di-	Cauc	sb-a	Wetlands	D4.1
Lamiaceae						
Ajuga orientalis L.	32	tetra-	E.A.Med	l-u	Steppe, meadows, forest edges	E1.2E, E2.32, E5.4
Betonica officinalis L.	16	di-	Palearct	l-m	Meadows, forest	E2.32, G1.6H, G1.A1
Betonica orientalis L.	16	di-	Arm-Ir	l-u	Meadows, steppe, forest edges	E1.2E, E2.32, E5.4
Dracocephalum botryoides Steven	12	di-	Cauc	a	Screes	H2.3
Dracocephalum multicaule Montbr. & Auch. ex Benth.	24	di-	Cauc-Arm-Ir	m-a	Rocks	H3.1

Lamium album L.	18	di-	Holarct	l-u	Steppe, forest, meadows	G1.6H, G1.A1, E1.2E, E2.32
Nepeta betonicifolia C.A.Mey.	36	tetra-	Cauc-Arm-Ir	m-u	Steppe	E1.2E
Nepeta mussinii Spreng.	18	di-	Arm-Ir	l-u	Semideserts, steppe	E1.2E, E1.33, E1.45
Nepeta nuda L.	18	di-	Palearct	l-u	Steppe, forest edges	E1.2E, E5.4
Nepeta sulphurea K.Koch	18	di-	E.A.Med	m-u	Steppe, meadows, forest edges	E1.2E, E2.32, E5.4
Nepeta supina Steven	18	di-	Cauc	sb-a	Rocks, screes	H3.1, H3.2, H2.3, H2.4
Phlomis orientalis Mill.	20	di-	Arm-Ir	l-m	Semideserts, steppe	E1.2E, E1.33, E1.45
Salvia armeniaca (Bordz.) Grossh.	18	di-	Arm-Atr	l-u	Semideserts, steppe	E1.2E, E1.33, E1.45
Salvia ceratophylla L.	22	di-	Arm-Ir	l-m	Semideserts, steppe	E1.2E, E1.33, E1.45
Salvia limbata C.A.Mey.	22	di-	Arm-Ir	l-m	Semideserts, steppe	E1.2E, E1.33, E1.45
Salvia syriaca L.	20	di-	E.A.Med	l-u	Semideserts, steppe	E1.2E, E1.33, E1.45
Salvia verticillata L.	32	di-	Eur-A.Med	l-u	Steppe, meadows	E1.2E, E2.32
Stachys iberica M.Bieb.	24	di-	E.Med	l-u	Steppe, meadows	E1.2E, E2.32
Stachys spectabilis Choisy ex DC.	24	di-	Arm-Ir	m-sb	Steppe, meadows	E1.2E, E2.32
Thymus rariflorus K.Koch	24	di-	Arm	l-u	Semideserts, steppe	E1.2E, E1.33, E1.45
Ziziphora raddei Juz.	18	di-	Arm	sb-a	Screes, rocks	H3.1, H3.2, H2.3, H2.4
Ziziphora tenuior L.	18	di-	A.Med	l-m	Semideserts, steppe	E1.2E, E1.33, E1.45

Species	2n	Ploidity[a]	Geoelement[b]	Belt[c]	Vegetation	Habitats[d]
Liliaceae						
Fritillaria armena Boiss.	24	di-	Arm	m-sb	Open forest, meadows	E2.32, F5.34
Fritillaria caucasica Adam	24	di-	Cauc-Arm-Atr	m-sb	Steppe, forest, meadows	G1.A1, E1.2E, E2.32
Fritillaria collina Adam	24	di-	Cauc	u-sb	Forest edges, meadows	E2.32, E5.4
Fritillaria kurdica Boiss. & Noe	24	di-	Arm-Ir	m-sb	Steppe, open forest	E1.2E, F5.34
Gagea alexeenkoana Miscz.	24	di-	Cauc	l-m	Steppe, semideserts	E1.2E, E1.33, E1.45
Gagea anisanthos K.Koch	72	high ploidy	Cauc	m-u	Steppe, meadows	E1.2E, E2.32
Gagea bulbifera (Pall.) Salisb.	24	tetra-	Palearct	l-m	Steppe	E1.2E
Gagea confusa A.Terracc.	36	tetra-	Arm-Ir	sb-a	Meadows	E2.32
Gagea joannis Grossh.	24	tetra-	Arm	u-a	Meadows	E4.3A, E4.4
Gagea tenuifolia (Boiss.) Fomin	24	tetra-	E.A.Med	l-u	Steppe, steppe shrubs	E1.2E, F2.33
Lilium armenum (Miscz. ex Grossh.) Manden.	24	tetra-	Arm	u-sb	Forest, meadows, tall grasses	G1.A1, E2.32, E5.5A
Lilium szovitsianum Fisch. & Ave-Lall.	24	tetra-	Arm	m-sb	Forest, meadows, tall grasses	G1.A1, E2.32, E5.5A
Tulipa biflora Pall.	24	tetra-	E.A.Med	l-sb	Semideserts, open forest, steppe	E1.2E, E1.33, E1.45, F5.34

Tulipa confusa Gabrielian	24	tetra-	Arm	m-sb	Forest, meadows	G1.A1, E2.32
Tulipa florenskyi Woronow	24, 24+2B	tetra-	Atr	l	Semideserts, open forest	E1.33, E1.45, F5.34
Tulipa julia K.Koch	24	tetra-	Arm	l-sb	Steppe, open forest, meadows	E1.2E, E2.32, F5.34
Tulipa sosnowskyi Achv. & Mirzoeva	24	tetra-	Arm	m-sb	Open forest, forest, rocks	F5.34, G1.A1, H3.1
Onagraceae						
Chamaenerion angustifolium (L.) Scop.	36	tetra-	Holarct	l-sb	Meadows, screes, wetlands	E2.32, H2.5, H2.6, D4.1
Epilobium alpinum L.	36	tetra-	Palearct	a	Wetlands	D4.1
Orchidaceae						
Dactylorhiza urvilleana (Steud.) H.Baumann & Kuenkele	80	octa-	Cauc-Arm-Atr	l-a	Forest, steppe shrubs, meadows, marshes	G1.A1, F2.33, E2.32, D2
Gymnadenia conopsea (L.) R.Br.	80	octa-	Palearct	m-a	Forest, meadows, steppe	G1.A1, E2.32, E1.2E
Listera ovata (L.) R.Br.	34	di-	Palearct	l-m	Forest	G1.6H, G1.A1
Papaveraceae						
Papaver fugax Poir.	14	di-	Cauc	l-u	Semideserts, steppe, ruderal habitats	E1.2E, E1.33, E1.45, I1.53
Papaver orientale L.	28	tetra-	Arm-Atr	m-sb	Steppe, meadows	E2.32, E1.2E
Plantaginaceae						
Plantago atrata Hoppe	12	di-	Eur-A.Med	sb-a	Meadows	E4.3A, E4.4

Species	2n	Ploidity[a]	Geoelement[b]	Belt[c]	Vegetation	Habitats[d]
Plantago lanceolata L.	12	di-	Palearct	l-sb	Semideserts, steppe, meadows	E1.2E, E1.33, E1.45, E2.32
Plantago maritima L.	12	di-	polychor	l-u	Semideserts, steppe	E1.2E, E1.33, E1.45
Plumbaginaceae						
Acantholimon takhtajanii Ogan.	32	tetra-	Arm	l-m	Semideserts, steppe, screes	E1.2E, E1.33, E1.45, H2.5
Poaceae						
Achnatherum bromoides (L.) P. Beauv.	24	di-	Med	l	Open forest, screes	F5.34, H2.5
Aegilops biuncialis Vis	28	tetra-	Med	l-m	Semidesert, steppe	E1.2E, E1.33, E1.45
Aegilops columnaris Zhuk.	28	tetra-	E.A.Med	l-m	Semidesert, steppe	E1.2E, E1.33, E1.45
Aegilops crassa Boiss.	42	hexa-	E.A.Med	l	Semideserts, rocks	E1.33, E1.45, H3.1
Aegilops cylindrica Host	14, 28	di-, tetra-	A.Med	l-u	Semideserts, steppe	E1.2E, E1.33, E1.45
Aegilops tauschii Coss.	14	di-	A.Med	l-u	Semideserts, steppe	E1.2E, E1.33, E1.45
Aegilops triaristata Willd.	28	tetra-	Med	l-m	Semideserts, steppe	E1.2E, E1.33, E1.45
Aegilops triuncialis L.	28	tetra-	A.Med	l-sb	Semideserts, steppe	E1.2E, E1.33, E1.45
Aegilops umbellulata Zhuk.	14	di-	E.A.Med	l	Semideserts	E1.33, E1.45
Aeluropus littoralis (Gouan) Parl.	20, 60	di-, hexa-	A.Med	l	Wetlands, salt marshes	D6.2, D4.1

Agropyron imbricatum (M.Bieb.) Roem. & Schult.	28, 42	di-, hexa-	Palearct	l-m	Meadows, steppe, semideserts	E1.2E, E1.33, E1.45, E2.32
Agrostis lazica Balansa	42	hexa-	Cauc	sb-a	Meadows	E4.3A, E4.4
Agrostis olympica (Boiss.) Bor	28	tetra-	Cauc	sb-a	Meadows	E4.3A, E4.4
Agrostis planifolia K.Koch	28	tetra-	Cauc	m-a	Meadows, screes	E4.3A, E4.4, H2.3
Alopecurus arundinaceus Poir.	42	hexa-	Palearct	l-m	Meadows	E2.32
Alopecurus brevifolius Grossh.	14	di-	Cauc	sb-a	Screes	H2.3
Alopecurus laguroides Balansa	14	di-	Cauc-Arm-Ir	a	Meadows	E4.3A, E4.4
Alopecurus myosuroides Huds.	14, 14+3B	di-	Eur-A.Med	l-m	Meadows, wetlands	H2.3, D4.1
Alopecurus tuscheticus Trautv.	42	hexa-	Cauc	sb-a	Screes	H2.3
Amblyopyrum muticum (Boiss.) Eig	14	di-	Arm	m	Steppe	E1.2E
Anisantha sterilis (L.) Nevski	14	di-	Eur-A.Med	l-u	Open forest, forest, ruderal habitats	F5.34, G1.A1, I1.53
Anisantha tectorum (L.) Nevski	14	di-	Eur-A.Med	l-sb	Semideserts, steppe, meadows	E1.2E, E1.33, E1.45, E2.32
Anthoxanthum odoratum L.	10, 10+4B	di-	Palearct	u-a	Meadows, steppe, rocks	E1.2E, E2.32, E4.3A, E4.4
Apera intermedia Hack.	14	di-	Arm-Ir	l-m	Semideserts, steppe, wetlands	E1.2E, E1.33, E1.45, D4.1

Species	2n	Ploidity[a]	Geoelement[b]	Belt[c]	Vegetation	Habitats[d]
Arrhenatherum elatius (L.) P. Beauv. ex J.Presl & C. Presl	28	tetra-	Palearct	m-a	Steppe, forest, meadows, tall grasses	E1.2E, G1.A1, E2.32, E5.5A
Arrhenatherum kotschyi Boiss.	14	di-	E.A.Med	m-sb	Screes	H2.6
Avena fatua L.	42	hexa-	Palearct	l-sb	Steppe, fields	E1.2E, I1.3
Bellardiochloa polychroa (Trautv.) Roshev.	14	di-	Arm-Ir	a	Meadows	E4.3A
Boissiera squarrosa (Banks & Sol.) Nevski	14	di-	E.A.Med	l-m	Semideserts, steppe	E1.2E, E1.33, E1.45
Bothriochloa ischaemum (L.) Keng	30	hexa-	A.Med	l-u	Steppe, medows, open forest	E1.2E, E2.32, F5.34
Briza elatior Sibth. & Sm.	14	di-	Cauc-Arm-Ir	m-sb	Forest, steppe, meadows	E1.2E, G1.A1, E2.32
Bromopsis indurata (Hausskn. & Bornm.) Holub	42	hexa-	Arm-Atr	l-u	Semideserts, steppe, screes, rocks	E1.2E, E1.33, E1.45, H2.5, H2.6, H3.1, H3.2
Bromopsis inermis (Leyss.) Holub	42	hexa-	Eur-A.Med	m-sb	Meadows, fields	E2.32, I1.3
Bromopsis variegata (M.Bieb.) Holub	14	di-	E.A.Med	u-a	Meadows, screes, rocks	E4.3A, E4.4, H2.3, H2.4, H3.1, H3.2
Bromus briziformis Fisch. & C.A.Mey.	14	di-	Cauc-Ir-Tur	l-m	Meadows, steppe, forest, open forest	E2.32, E1.2E, G1.A1, F5.34
Bromus commutatus Schrad.	14	di-	Eur-A.Med	l-m	Forest, open forest	G1.A1, F5.34
Bromus danthoniae Trin.	14	di-	Arm-Ir	l-u	Semideserts, steppe	E1.2E, E1.33, E1.45

Bromus japonicus Thunb.	14	di-	Palearct	l-sb	Forest, steppe, meadows, wetlands	G1.A1, G1.6H, E2.32, D4.1
Bromus pseudodanthoniae Drobow	28	tetra-	Arm-Ir	m-sb	Steppe, meadows	E1.2E, E2.32
Bromus scoparius L.	14, 28	di-, tetra-	Eur-A.Med	l-sb	Meadows, forest, wetlands	G1.A1, G1.6H, E2.32, D4.1
Bromus squarrosus L.	14	di-	E.A.Med	l-sb	Steppe, screes, rocks	E1.2E, H2.5, H2.6, H3.1, H3.2
Calamagrostis arundinacea (L.) Roth.	28	tetra-	Palearct	m-a	Meadows, forest, ruderal habitats	G1.A1, G1.6H, E2.32, I1.53
Catabrosa aquatica (L.) P.Beauv.	20, 30	tetra-, hexa-	Holarct	l-sb	Wetlands	D4.1
Catabrosella araratica (Lipsky) Tzvelev	42	hexa-	Arm	sb-a	Meadows, screes	E4.3A, E4.4, H2.3, H2.4
Catabrosella calvertii (Boiss.) Czerep.	10, 10+B, 10+2-3B	di-	Arm	l-a	Semideserts, steppe, meadows, forest	E1.2E, E1.33, E1.45, E2.32, G1.A1
Catabrosella fibrosa (Trautv.) Tzvelev	18	di-	Arm	a	Meadows, screes	E4.3A, H2.3
Chrysopogon gryllus (L.) Trin.	40	tetra-	A.Med	l	Semideserts, open forest	E1.33, E1.45, F5.34
Colpodium versicolor (Steven) Schmalh.	4	di-	Cauc-Arm-Ir	m-a	Meadows, screes	E2.32, E4.3A, E4.4, H2.3, H2.4
Cynodon dactylon (L.) Pers.	18	di-	polychor	l-m	Semideserts, ruderal habitats	E1.33, E1.45, I1.53
Cynosurus echinatus L.	14	di-	A.Med	l-sb	Forest, open forest, screes	G1.A1, F5.34, H2.5, H2.6

Species	2n	Ploidity[a]	Geoelement[b]	Belt[c]	Vegetation	Habitats[d]
Dactylis glomerata L.	14	di-	Eur-A.Med	l-a	Steppe, meadows, forest	E1.2E, E2.32, G1.A1, G1.6H
Deschampsia caespitosa (L.) P. Beauv.	26	di-	Holarct	m-a	Meadows, wetlands	E2.32, E4.3A, E4.4, D4.1
Echinaria capitata (L.) Desf.	18	di-	Eur-A.Med	l-m	Open forest, semideserts	F5.34, E1.33, E1.45
Echinochloa crusgalli (L.) P.Beauv.	36, 54	tetra-, hexa-	Holarct	l-u	Wetlands	C3.24, D4.1
Elytrigia elongatiformis (Drobow) Nevski	42	hexa-	E.A.Med	l-m	Forest, meadows	E2.32, G1.A1
Elytrigia intermedia (Host) Nevski	42	hexa-	Eur-A.Med	l	Steppe	E1.2E
Elytrigia obtusiflora (DC.) Tzvelev	42	hexa-	A.Med	l	Deserts, semideserts	E1.33, E1.45, H5.32
Elytrigia repens (L.) Nevski	42	hexa-	polychor	l-m	Meadows, forest, steppes, semideserts	E2.32, G1.A1, G1.6H, E1.2E, E1.33, E1.45
Elytrigia trichophora (Link) Nevski	28, 42	tetra-, hexa-	Eur-A.Med	l-m	Steppe, forest	E1.2E, G1.A1
Enneapogon persicus Boiss.	20	di-	A.Med	l	Semideserts, open forest, rocks	F5.34, E1.33, E1.45, H3.1
Eragrostis minor Host	40	tetra-	Eur-A.Med	l-u	Wetlands	D4.1
Eremopoa persica (Trin.) Roshev.	14	di-	A.Med	l-u	Semideserts, steppe, open forest	E1.2E, E1.33, E1.45, F5.34
Eremopoa songarica (Schrenk) Roshev.	28	tetra-	Eur-A.Med	l-a	Semideserts, steppe, meadows	E1.2E, E1.33, E1.45, E2.32

Eremopyrum bonaepartis (Spreng.) Nevski	14	di-	A.Med	l-m	Semideserts, steppe	E1.2E, E1.33, E1.45
Eremopyrum distans (K.Koch) Nevski	14	di-	A.Med	l-m	Semideserts, deserts, steppe	E1.2E, E1.33, E1.45, H5.32
Eremopyrum orientale (L.) Jaub. & Spach	28	tetra-	A.Med	l-m	Semideserts, deserts	E1.33, E1.45, H5.32
Eremopyrum triticeum (Gaertn.) Nevski	14	di-	Palearct	l-u	Semideserts	E1.33, E1.45
Festuca brunnescens (Tzvelev) Galushko	28	tetra-	Cauc-Arm-Ir	u-a	Meadows, steppe, forest, rocks	E1.2E, E2.32, G1.A1, H3.1
Festuca chalcophaea V.I.Krecz. & Bobrov	14	di-	Arm-Ir	sb-a	Meadows, steppe	E1.2E, E4.3A, E4.4
Festuca gigantea (L.) Vill.	42	hexa-	Eur-A.Med	l-u	Forest, wetlands	G1.A1, G1.6H, D4.1
Festuca ruprechtii (Boiss.) V.I.Krecz. & Bobrov	14, 42	di-, hexa-	Cauc	m-a	Meadows, screes, rocks	E2.32, H2.5, H2.6, H3.1, H3.2
Festuca sclerophylla Boiss. ex Bisch.	42	hexa-	Cauc-Arm-Ir	l-sb	Steppe, semideserts, open forest	E1.2E, E1.33, E1.45, F5.34
Festuca skvortsovii E.B.Alexeev	42	hexa-	Atr	l-u	Semideserts, open forest	E1.33, E1.45, F5.34
Festuca valesiaca Gaudin	14	di-	Palearct	l-sb	Steppe, open forest	E1.2E, F5.34
Festuca woronowii Hack.	14	di-	Cauc-Arm-Atr	m-a	Meadows, steppe, rocks, screes	E1.2E, E2.32, H3.1, H2.3

Species	2n	Ploidity[a]	Geoelement[b]	Belt[c]	Vegetation	Habitats[d]
Gaudinopsis macra (Steven ex M.Bieb.) Eig	14	di-	E.A.Med	l-m	Open forest, screes	F5.34, H2.5, H2.6
Glyceria nemoralis (Uechtr.) Uechtr. & Koern.	40	tetra-	Eur	m	Wetlands	D2, D4.1
Glyceria notata Chevall.	40	tetra-	Eur-A.Med	l-sb	Wetlands	D2, D4.1
Helictotrichon adzharicum (Albov) Grossh.	14	di-	Cauc	sb-a	Meadows	E4.3A, E4.4
Helictotrichon pubescens (Huds.) Besser	14	di-	Eur-A.Med .	m-a	Meadows, screes	E2.32, H2.3, H2.4
Hordeum bulbosum L.	28	tetra-	A.Med	l-u	Steppe, meadows, forest edges	E1.2E, E2.32, E5.4
Hordeum geniculatum All.	28	tetra-	Eur-A.Med	l-m	Semideserts	E1.33, E1.45
Hordeum glaucum Steud.	14	di-	A.Med	l-u	Ruderal habitats	I1.53
Hordeum hrasdanicum Gandilyan	42	hexa-	Arm	l-m	Ruderal habitats	I1.53
Hordeum marinum Huds.	14	di-	Eur-Med	l	Deserts, semideserts	H5.32, E1.33, E1.45
Hordeum murinum L.	28	tetra-	Eur-Med	l-m	Ruderal habitats	I1.53
Hordeum spontaneum K.Koch	14	di-	E.A.Med	l-m	Semideserts, ruderal habitats	E1.33, E1.45, I1.53

Hordeum violaceum Boiss. & Huet	14	di-	Arm-Ir	m-sb	Meadows, screes	E4.3A, E4.4, H2.3, H2.4
Koeleria albovii Domin	28, 42	tetra-, hexa	Cauc	m-a	Meadows	E2.32, E4.3A, E4.4
Koeleria macrantha (Ledeb.) Schult.	14	di-	Holarct	l-sb	Meadows, steppe, screes	E2.32, E1.2E, H2.5, H2.6
Lolium persicum Boiss. & Hohen.	14, 28	di-, tetra-	E.A.Med	l-u	Semideserts, steppe	E1.33, E1.45, E1.2E
Melica jacquemontii Decne.	18	di-	E.A.Med	l-u	Semideserts, steppe	E1.33, E1.45, E1.2E
Melica schischkinii Iljinsk.	18	di-	Arm	l-u	Semideserts, steppe	E1.33, E1.45, E1.2E
Melica taurica K.Koch	18	di-	Eur-Med	l-u	Steppe, screes, rocks	E1.2E, H2.6, H3.2
Melica transsilvanica Schur	18	di-	Palearct	l-m	Steppe, screes, rocks, forest edges	E1.2E, H2.6, H3.2, E5.4
Melica uniflora Retz	18	di-	Eur-Med	l-sb	Forest	G1.A1, G1.6H
Milium effusum L.	28	tetra-	Holarct	m-u	Forest, meadows	G1.A1, G1.6H, E2.32
Milium schmidtianum K.Koch	42	hexa-	E.Med	u-sb	Forest, meadows, wetlands	G1.A1, G1.6H, E2.32, D4.1
Milium vernale M.Bieb.	18	di-	Eur-A.Med	l-u	Forest, steppe, meadows	G1.A1, G1.6H, E2.32, E1.2E
Nardus stricta L.	24, 25, 26	aneu-	Holarct	u-a	Meadows	E4.3A, E4.4
Phalaroides arundinacea (L.) Rauschert	28	tetra-	Holarct	l-a	Wetlands	C3.24, D4.1
Phleum alpinum L.	14, 28	di-, tetra-	Holarct	m-a	Meadows	E2.32, E4.3A, E4.4
Phleum paniculatum Huds.	28	tetra-	Eur-A.Med	l-m	Open forest, forest	G1.A1, G1.6H, F5.34

Species	2n	Ploidity[a]	Geoelement[b]	Belt[c]	Vegetation	Habitats[d]
Phleum phleoides (L.) H.Karst.	28	tetra-	Eur-A.Med	l-a	Meadows, steppe, open forest	E2.32, E4.3A, E4.4, E1.2E, F5.34
Phleum pratense L.	42	hexa-	Palearct	l-sb	Meadows, steppe, forest	E2.32, E1.2E, G1.A1
Poa alpina L.	33-35	aneu-	Holarct	m-a	Meadows, wetlands, screes	E4.3A, E4.4, D4.1, H2.3, H2.4
Poa annua L.	28	tetra-	Holarct	l-a	Forest, steppe, meadows, wetlands	G1.A1, G1.6H, E1.2E, E2.32, E4.3A, E4.4, D4.1
Poa bulbosa L.	14, 28, 36, 42	di-,tetra,hexa-,aneu-	Palearct	l-a	Semideserts, steppe, meadows	E1.33, E1.45, E1.2E, E2.32, E4.3A, E4.4
Poa densa Troitsky	14	di-	Iran	n-sb	Steppe, semideserts, meadows	E1.33, E1.45, E1.2E, E2.32
Poa iberica Fisch. & C.A.Mey.	28	tetra-	Cauc	m-sb	Forest, open forest, meadows	G1.A1, F5.34, E2.32
Poa nemoralis L.	42	hexa-	Holarct	l-u	Forest, steppe shrubs	G1.A1, G1.6H, F2.33
Poa pratensis L.	36	aneu-	Palearct	m-sb	Meadows, screes	E2.32, H2.5, H2.6
Polypogon fugax Nees ex Steud.	42	hexa-	E.A.Med	l-m	Wetlands	C3.24, D4.1
Polypogon maritimus Willd.	28	tetra-	Palearct	l	Wetlands	C3.24, D4.1
Polypogon monspeliensis (L.) Desf.	28	tetra-	Palearct	l-m	Wetlands	C3.24, D4.1
Puccinellia sevangensis Grossh.	14, 42	di-, hexa-	Arm	l-m	Wetlands	C3.24, D4.1
Rhizocephalus orientalis Boiss.	14	di-	E.A.Med	l	Semideserts	E1.33, E1.45

Rostraria cristata (L.) Tzvelev	14	di-	A.Med	l	Semideserts	E1.33, E1.45
Sclerochloa dura (L.) P.Beauv.	14	di-	Eur-A.Med	l-u	Ruderal habitats	I1.53
Secale montanum Guss.	14	di-	Med	m-sb	Steppe, meadows	E1.2E, E2.32
Secale vavilovii Grossh.	14	di-	Arm-Ir	l-m	Steppe, screes, fields	E1.2E, H2.5, H2.6, I1.3
Setaria glauca (L.) P.Beauv.	18, 36	di-, tetra-	Holarct	l-u	Ruderal habitats	I1.53
Setaria italica (L.) P.Beauv.	18	di-	Eur-A.Med	l-m	Ruderal habitats	I1.53
Setaria verticillata (L.) P.Beauv.	18	di-	Eur-A.Med	l-m	Ruderal habitats	I1.53
Setaria viridis (L.) P.Beauv.	18	di-	polychor	l-sb	Ruderal habitats	I1.53
Stipa capillata L.	44	tetra-	Palearct	l-sb	Semideserts, steppe	E1.33, E1.45, E1.2E
Stipa gegarkunii P.A.Smirn.	44	tetra-	Arm (endemic)	u	Steppe	E1.2E
Stipa lessingiana Trin. & Rupr.	44	tetra-	Palearct	l-sb	Semideserts, steppe, open forest	E1.33, E1.45, E1.2E, F5.34
Taeniatherum crinitum (Schreb.) Nevski	14	di-	Med	l-u	Semideserts	E1.33, E1.45
Tragus racemosus (L.) All.	40	tetra-	Eur-A.Med	l-u	Open forest, ruderal habitats	F5.34, I1.53
Trisetum flavescens (L.) P.Beauv.	24, 28	tetra-	Eur-A.Med	m-a	Meadows, forest, steppe	E2.32, G1.A1, E1.2E

Species	2n	Ploidity[a]	Geoelement[b]	Belt[c]	Vegetation	Habitats[d]
Trisetum turcium Chrtek	28	tetra-	Cauc	m-a	Meadows, forest, wetlands	E2.32, G1.A1, D4.1
Triticum araraticum Jakubz.	28	tetra-	Arm-Ir	m	Steppe	E1.2E
Triticum boeoticum Boiss.	14	di-	E.A.Med	m	Steppe	E1.2E
Triticum urartu Tumanian ex Gandilyan	14	di-	Arm-Ir	m	Steppe	E1.2E
Vulpia hirtiglumis Boiss. & Hausskn.	28	tetra-	Iran	l	Deserts, semideserts	E1.33, E1.45, H5.32
Vulpia myuros (L.) C.C.Gmel.	42	hexa-	Eur-A.Med	l-m	Steppe, open forest	E1.2E, F5.34
Vulpia persica (Boiss. & Buhse) V.I.Krecz. & Bobrov	42	hexa-	E.A.Med	l-m	Semideserts, steppe	E1.33, E1.45, E1.2E
Zingeria biebersteiniana (Claus) P.A.Smirn.	4	di-	Euxin	u-sb	Meadows	E2.32
Zingeria kochii (Mez) Tzvelev	12	hexa-	Arm (endemic)	l-m	Wetlands	D4.1
Zingeria trichopoda (Boiss.) P.A.Smirn.	4, 8	di-, tetra-	Arm-Ir	m-u	Meadows, wetlands	E2.32, D4.1
Polygonaceae						
Calligonum polygonoides L.	54	hexa-	Arm-Ir	l	Deserts	H5.32
Oxyria digyna (L.) Hill	56	octa-	Palearct	a	Screes	H2.3, H2.4

Polygonum alpinum All.	20	di-	Palearct	u-a	Meadows	E4.3A, E4.4
Rumex alpinus L.	18	di-	Eur-Med	sb-a	Meadows	E4.3A, E4.4
Rumex angustifolius Campd.	40	tetra-	Eur-Med	l-m	Wetlands	D4.1
Rumex tuberosus L.	16	di-	A.Med	m-u	Meadows	E2.32
Primulaceae						
Androsace armeniaca Duby	20	di-	Arm-Atr	sb-a	Meadows, screes	E4.3A, E4.4, H2.3, H2.4
Androsace chamaejasme Wulff	20	di-	Holarct	a	Screes, rocks	H2.3, H2.4, H3.1, H3.2
Androsace maxima L.	60	hexa-	Eur-A.Med	l	Semideserts	E1.33, E1.45
Androsace raddeana Somm. & Levier	36	hexa-	Cauc	a	Meadows, screes	E4.3A, E4.4, H2.3, H2.4
Primula algida Adams	22	di-	Cauc-Arm-Ir	a	Meadows	E4.3A, E4.4
Primula macrocalyx Bunge	22	di-	E.A.Med	n-u	Steppe, forest, steppe shrubs, meadows	E1.2E, E2.32, F2.33, G1.6H, G1.A1
Ranunculaceae						
Adonis aestivalis L.	32	tetra-	Eur-A.Med	l-sb	Semideserts, steppe, fields	E1.33, E1.45, E1.2E, I1.3
Adonis parviflora Fisch. ex DC.	16	di-	E.A.Med	l-u	Semideserts, steppe, fields	E1.33, E1.45, E1.2E, I1.3
Anemone fasciculata L.	14	di-	Cauc	sb-a	Meadows	E2.32, E4.3A
Caltha polypetala Hochst.	32	tetra-	Cauc-Arm-Ir	m-a	Meadows, wetlands	E2.32, D4.1

Species	2n	Ploidity[a]	Geoelement[b]	Belt[c]	Vegetation	Habitats[d]
Ceratocephala falcata (L.) Pers.	12	di-	Eur-A.Med	l-sb	Semideserts, steppe	E1.2E, E2.32, F2.33
Consolida divaricata (Ledeb.) Schroeding.	16	di-	E.A.Med	l-m	Ruderal habitats	I1.53
Consolida glandulosa (Boiss. & Huet) Bornm.	16	di-	E.A.Med	l-m	Semideserts, steppe, fields	E1.33, E1.45, E1.2E, I1.3
Consolida orientalis (J. Gay) Schroeding.	16	di-	Med	l-u	Semidesert, steppe, open forest, fields	E1.33, E1.45, E1.2E, F5.34, I1.3
Delphinium foetidum Lomak.	16	di-	Atr	a	Screes	H2.3
Ficaria ficarioides (Bory & Chaub.) Halacsy	16	di-	E.Med	m-a	Meadows, steppe	E1.2E, E2.32
Pulsatilla albana (Steven) Bercht. & J.Presl	16	di-	Cauc	m-a	Meadows, steppe, steppe shrubs	E1.2E, E2.32, F2.33
Ranunculus aragazi Grossh.	16	di-	Arm	a	Meadows	E4.3A
Ranunculus caucasicus M.Bieb.	16	di-	Cauc	m-a	Meadows, forest	E2.32, G1.6H, G1.A1
Ranunculus grandiflorus L.	28	tetra-	Cauc	m-a	Forest, meadows, screes	G1.A1, E2.32, H2.5, H2.6
Ranunculus illyricus L.	32	tetra-	A.Med	l-sb	Steppe, meadows, open forest	E1.2E, E2.32, F5.34
Ranunculus oreophilus M.Bieb.	16	di-	Cauc	m-a	Meadows, forest	E2.32, G1.6H, G1.A1
Ranunculus oxyspermus Willd.	16	di-	A.Med	l-m	Semideserts, steppe	E1.33, E1.45, E1.2E

Ranunculus polyanthemos L.	16, 16+3B	di-	Palearct	l-m	Forest edges, meadows	E2.32, E5.4
Ranunculus szowitsianus Boiss.	16	di-	Arm-Ir	sb-a	Meadows	E4.3A, E4.4
Rosaceae						
Alchemilla caucasica Bus.	108	high ploidy	Cauc	a	Meadows	E4.3A, E4.4
Alchemilla epipsila Juz.	115	high ploidy	Cauc	u-a	Wetlands	D4.1
Amygdalus fenzliana (Fritsch) Lipsky	16	di-	Arm	l-m	Open forest, steppe	E1.2E, F5.34
Cotoneaster armenus Pojark.	68	hexa-	Arm	l-m	Open forest, steppe	E1.2E, F5.34
Cotoneaster integerrimus Medik.	68	hexa-	Eur	l-u	Steppe, open forest	E1.2E, F5.34
Crataegus armena Pojark.	51	tri-	Arm	m-sb	Open forest, forest edges	F5.34, E5.4
Crataegus atrosanguinea Pojark.	51	tri-	Arm-Ir	l	Open forest, forest edges	F5.34, E5.4
Crataegus pseudoheterophylla Pojark.	68	hexa-	Cauc	l-m	Open forest, forest edges	F5.34, E5.4
Potentilla gelida C.A.Mey.	14	di-	Cauc	sb-a	Meadows	E4.3A, E4.4
Potentilla porphyrantha Juz.	14	di-	Arm	sb-a	Rocks	H3.1
Potentilla raddeana (Th.Wolf) Juz.	42	hexa-	Arm	sb-a	Meadows	E4.3A, E4.4

Species	2n	Ploidity[a]	Geoelement[b]	Belt[c]	Vegetation	Habitats[d]
Potentilla recta L.	14	di-	Palearct	l-sb	Steppe, meadows, forest edges	E1.2E, E2.32, E5.4
Potentilla seidlitziana Bien.	21	tri-	Arm	a	Meadows	E4.3A, E4.4
Pyrus caucasica Fed.	34	di-	Cauc	l-m	Forest	G1.6H
Pyrus daralaghezii Mulk.	34	di-	Arm	m	Open forest	F5.34
Pyrus hyrcana Fed.	34	di-	Atr	m	Forest	G1.A1
Pyrus medvedevii Rubtsov	34	di-	Arm	m	Open forest	F5.34
Pyrus oxyprion Woronow	34	di-	Atr	m	Forest	G1.A1
Pyrus takhtajianii Fed.	34	di-	Arm (endemic)	m	Open forest	F5.34
Sibbaldia parviflora Willd.	14	di-	Cauc-Arm-Ir	a	Meadows	E4.3A, E4.4
Sibbaldia semiglabra C.A.Mey.	14	di-	Cauc	a	Meadows	E4.3A, E4.4
Sorbus aucuparia L.	34	di-	Eur	u-sb	Forest edges	E5.4
Sorbus hajastana Gabrielian	68	tetra-	Arm	u	Open forest	F5.34
Sorbus kusnetzovii Zinserl.	68	tetra-	Cauc	m-u	Open forest	F5.34
Sorbus persica Hedl.	68	tetra-	E.A.Med	m-sb	Open forest, forest edges	F5.34, E5.4
Sorbus roopiana Bordz.	68	tetra-	Arm-Ir	u-sb	Forest edges	E5.4
Sorbus subfusca (Ledeb.) Boiss.	34, 68	di-, tetra-	Cauc	u-sb	Forest edges	E5.4

Sorbus takhtajanii Gabrielian	68	tetra-	Arm	u-sb	Forest edges	E5.4
Sorbus tamamschjanae Gabrielian	51	tri-	Arm	u-sb	Forest edges	E5.4
Rubiaceae						
Callipeltis cucullaris (L.) DC.	22	di-	A.Med	l-m	Semideserts, open forests	E1.33, E1.45, F5.34
Crucianella gilanica Trin.	22	di-	Arm-Ir	l-sb	Steppe, semideserts	E1.2E, E1.33, E1.45
Sambucaceae						
Sambucus tigranii Troitzk.	38	di-	Arm	l-m	Open forest, steppe	E1.2E, F5.34
Saxifragaceae						
Saxifraga sibirica L.	18	di-	Palearct	m-a	Rocks	H3.1, H3.2
Scrophulariaceae						
Lagotis stolonifera (K.Koch) Maxim.	22	di-	Arm-Ir	l-sb	Wetlands	D4.1
Limosella aquatica L.	40	tetra-	Holarct	m-u	Wetlands	D4.1
Pedicularis armena Boiss. & Huet	16	di-	Cauc-Arm-Ir	a	Meadows	E4.3A
Pedicularis crassirostris Bunge	16	di-	Cauc	sb-a	Meadows	E4.3A, E4.4
Pedicularis sibthorpii Boiss.	16	di-	Arm-Ir	m-a	Meadows	E2.32, E4.3A, E4.4
Scrophularia chrysantha Jaub. & Spach	36	tetra-	Cauc-Arm	m-a	Rocks	H3.1

Species	2n	Ploidity[a]	Geoelement[b]	Belt[c]	Vegetation	Habitats[d]
Scrophularia olympica Boiss.	26	di-	Arm-Ir	m-a	Meadows	E2.32, E4.3A, E4.4
Veronica anagallis-aquatica L.	18	di-	E.A.Med	l-u	Wetlands	D4.1
Veronica armena Boiss. & Huet	16	di-	Arm	sb-a	Meadows, screes	E4.3A, E4.4, H2.3, H2.4
Veronica biloba Schreb.	28	di-	E.A.Med	m-a	Meadows, open forest, wetlands	E2.32, F5.34, D4.1
Veronica ceratocarpa C.A.Mey.	14	di-	Cauc-Arm-Atr	l-sb	Forest edges, meadows	E2.32, E5.4
Veronica gentianoides Vahl	32, 48	tetra-, hexa-	Cauc-Arm-Ir	m-a	Meadows	E2.32, E4.3A, E4.4
Veronica hederifolia L.	56	tetra-	Eur-A.Med	l-u	Forest edges, wetlands	E5.4, D4.1
Veronica microcarpa Boiss.	32	tetra-	Arm-Ir	l-m	Steppe, screes	E1.2E, H2.5, H2.6
Veronica multifida L.	32, 80	tetra-, octa-	Cauc-Arm-Ir	l-sb	Steppe, screes	E1.2E, H2.5, H2.6
Veronica orientalis Mill.	32, 48	tetra-, hexa-	Arm-Ir	l-a	Steppe, screes	E1.2E, H2.5, H2.6
Veronica pusilla Kotschy	14	di-	E.A.Med	m-a	Meadows, wetlands	E2.32, D4.1
Thymelaeaceae						
Daphne transcaucasica Pobed.	18	di-	Arm	m-u	Steppe, open forest	E1.2E, F5.34
Urticaceae						
Urtica dioica L.	48	hexa-	polychor	l-sb	Ruderal habitats	I1.53
Valerianaceae						

Valeriana officinalis L.	42	hexa-	Palearct	l-sb	Meadows, forest edges, steppe shrubs	E2.32, E5.4, F2.33
Valerianella sclerocarpa Fisch. & C.A.Mey.	32	tetra-	E.A.Med	l-m	Semideserts, steppe	E1.2E, E1.33, E1.45
Valerianella uncinata (M. Bieb.) Dufr.	16	di-	E.A.Med	m-u	Steppe, steppe shrubs, wetlands	E1.2E, F2.33, D4.1
Zygophyllaceae						
Zygophyllum atriplicoides Fisch. & C.A.Mey.	22	di-	Arm-Ir	l-m	Semideserts, deserts	E1.33, E1.45, H5.32

[a]Ploidity (cytoraces): *di* diploid, *tri* triploid, *tetra* tetraploid, *penta* pentaploid, *hexa* hexaploid, *octa* octaploid, *aneu* aneuploid, *mixo* mixoploid, *high ploidy* cytoraces for which $2n \geq 10$

[b]Geographical elements: *A.Med* Ancient Mediterranean, *Arm* Armenian, *Arm-Atr* Armeno-Atropatenian, *Arm-Ir* Armeno-Iranian, *Atr* Atropatenian, *Cauc* Caucasian, *Cauc-Arm* Caucaso-Armenian, *Cauc-Arm-Atr* Caucaso-Armeno-Atropatenian, *Cauc-Arm-Ir* Caucaso-Armeno-Iranian, *Cauc-E.A.Med* Caucaso-East Ancient Mediterranean, *Cauc-E.Med* Caucaso-East Mediterranean, *Cauc-Ir-Tur* Caucaso-Irano-Turanian, *E.A.Med* East Ancient Mediterranean, *E.Med* East Mediterranean, *E.Palearct* East Palearctic, *Eur* European, *Eur-A.Med* Euro-Ancient Mediterranean, *Eur-Cauc* Euro-Caucasian, *Eur-Med* Euro-Mediterranean, *Euxin* Euxinian, *Holarct* Holarctic, *Iran* Iranian, *Ir-Tur* Irano-Turanian, *Med* Mediterranean, *Palearct* Palearctic, *Pan-Pont* Pannono-Pontic, *Pan-Pont-Sarm* Pannono-Pontic-Sarmatian, *polychor* polychorous, *W.Palearct* West Palearctic

[c]Belt: *l* lower mountain belt (altitude < 1200–1300 m a.s.l.), *m* middle mountain belt (>1200<1800 m a.s.l.), *u* upper mountain belt (>1800<2100 m a.s.l.), *sb* subalpine belt (>2100<2700 m a.s.l.), *a* alpine belt (>2700 m a.s.l.)

[d]Habitats: C3.24 – Medium-tall non-graminoid waterside communities (*Acorus*), D2 – Valley mires, poor fens and transition mires, D4.1 – Rich fens, including eutrophic tall-herb fens and calcareous flushes and soaks, D6.2 – Inland saline or brackish species-poor helophyte beds normally without free-standing water, E1.2E – Irano-Anatolian steppes, E1.33 – East Mediterranean xeric grassland, E1.45 – Sub-Mediterranean wormwood steppes, E2.32 – Ponto-Caucasian hay meadows, E4.3A – Western Asian acidophilous alpine grassland, E4.4 – Ponto-Caucasian alpine grassland (limestone), E5.4 – Moist or wet tall-herb and fern fringes and meadows, E5.5A – Ponto-Caucasian tall-herb communities, F2.33 – Subalpine mixed brushes, F5.13 – Juniper matorral, F5.34 – Western Asian pseudomaquis, G1.6H – Caucasian beech forests, G1.A1 – Oak-ash-hornbeam woodland on eutrophic and mesotrophic soils, G3.93 – Grecian juniper – *Juniperus excelsa* – woods, H2.3 – Temperate-montane acid siliceous screes, H2.4 – Temperate-montane calcareous and ultra-basic screes, H2.5 – Acid siliceous screes of warm exposures, H2.6 – Calcareous and ultra-basic screes of warm exposures, H3.1 – Acid siliceous inland cliffs, H3.2 – Basic and ultra-basic inland cliffs, H5.32 – Stable sand with very sparse or no vegetation, I1.3 – Arable land with unmixed crops grown by low-intensity agricultural methods, I1.53 – Fallow un-inundated fields with annual and perennial weed communities, I2.22 – Subsistence garden areas

Annex 2: List of Bird Species Registered in Armenia

(occurring status and the occupied habitats are given for breeding species only).

	Latin names	English names	Breeding status	Deciduous forests	Juniper woodlands	Semideserts	Shrublands	Steppes	Meadows	Alpine meadows and carpets	Wetlands	Rocks & Screes	Riparian zone	Arable lands
	ANATIDAE													
1	*Cygnus olor*	Mute Swan	w, m – irregular											
2	*Cygnus columbianus*	Tundra Swan	w, m – irregular											
3	*Cygnus cygnus*	Whooper Swan	w, m – irregular											
4	*Anser albifrons*	Greater White-fronted Goose	w, m – irregular											
5	*Anser erythropus*	Lesser White-fronted Goose	yv – regular											
6	*Anser anser*	Greylag Goose	b – regular								X			
7	*Branta ruficollis*	Red-Breasted Goose	a											
8	*Tadorna ferruginea*	Ruddy Shelduck	yr – regular								X			
9	*Tadorna tadorna*	Common Shelduck	w, m – regular											
10	*Mareca penelope*	Eurasian Wigeon	w, m – irregular											
11	*Mareca strepera*	Gadwall	yr – regular								X			
12	*Anas crecca*	Common Teal	w, m – regular											
13	*Anas platyrhynchos*	Mallard	yr – regular								X		X	
14	*Anas acuta*	Northern Pintail	w, m – regular											
15	*Spatula querquedula*	Garganey	b – regular								X			

	Latin names	English names	Breeding status	Deciduous forests	Juniper woodlands	Semideserts	Shrublands	Steppes	Meadows	Alpine meadows and carpets	Wetlands	Rocks & Screes	Riparian zone	Arable lands
16	*Spatula clypeata*	Northern Shoveler	w, m – regular											
17	*Marmaronetta angustirostris*	Marbled Teal	b – regular								X			
18	*Netta rufina*	Red-crested Pochard	b – regular								X			
19	*Aythya ferina*	Common Pochard	b – regular								X			
20	*Aythya nyroca*	Ferruginous Pochard	b – regular								X			
21	*Aythya fuligula*	Tufted Duck	b – irregular								X			
22	*Aythya marila*	Great Scaup	a											
23	*Clangula hyemalis*	Long-tailed Duck	a											
24	*Melanitta fusca*	White-Winged Scoter	a***											
25	*Melanitta nigra**	Common Scoter*	un											
26	*Bucephala clangula*	Common Goldeneye	w, m – irregular											
27	*Mergellus albellus*	Smew	w, m – irregular											
28	*Mergus serrator*	Red-breasted Merganser	w, m – irregular											
29	*Mergus merganser*	Common Merganser	w, m – irregular											
30	*Oxyura leucocephala*	White-headed Duck	b – regular								X			
	PHASIANIDAE													
31	*Lyrurus mlokosiewiczi*	Caucasian Grouse	yr – regular						X	X				
32	*Tetraogallus caspius*	Caspian Snowcock	yr – regular							X				

33	*Alectoris chukar*	Chukar	yr – regular			X	X							
34	*Ammoperdix griseogularis*	See-see Partridge	b – regular			X								
35	*Francolinus francolinus*	Black Francolin	yr – regular										X	
36	*Perdix perdix*	Grey Partridge	yr – regular					X	X					
37	*Coturnix coturnix*	Common Quail	b – regular					X	X					X
38	*Phasianus colchicus*	Pheasant	yr – regular	X									X	
	GAVIIDAE													
39	*Gavia stellata*	Red-throated Loon	a											
40	*Gavia arctica*	Arctic Loon	a											
	PODICIPEDIDAE													
41	*Tachybaptus ruficollis*	Little Grebe	yr – regular								X			
42	*Podiceps cristatus*	Great Crested Grebe	yr – regular								X			
43	*Podiceps grisegena*	Red-necked Grebe	b – regular								X			
44	*Podiceps auritus*	Horned Grebe	a											
45	*Podiceps nigricollis*	Black-necked Grebe	yr – regular								X			
	PHALACROCORACIDAE													
46	*Phalacrocorax carbo*	Great Cormorant	yr – regular								X			
47	*Microcarbo pygmeus*	Pygmy Cormorant	yr – regular								X			
	PELECANIDAE													
48	*Pelecanus onocrotalus*	Great White Pelican	m – regular											
49	*Pelecanus crispus*	Dalmatian Pelican	b – regular								X			

	Latin names	English names	Breeding status	Deciduous forests	Juniper woodlands	Semideserts	Shrublands	Steppes	Meadows	Alpine meadows and carpets	Wetlands	Rocks & Screes	Riparian zone	Arable lands
	ARDEIDAE													
50	*Botaurus stellaris*	Great Bittern	b – unconfirmed**								X			
51	*Ixobrychus minutus*	Little Bittern	b – regular								X		X	
52	*Nycticorax nycticorax*	Black-crowned Night-heron	b – regular								X			
53	*Ardeola ralloides*	Squacco Heron	b – regular								X			
54	*Bubulcus ibis*	Cattle Egret	b – regular								X			
55	*Egretta garzetta*	Little Egret	b – regular								X		X	
56	*Ardea alba*	Great White Egret	w, m – regular											
57	*Ardea cinerea*	Grey Heron	yr – regular								X		X	
58	*Ardea purpurea*	Purple Heron	b – regular								X			
	CICONIIDAE													
59	*Ciconia nigra*	Black Stork	b – regular	X										
60	*Ciconia ciconia*	White Stork	yr – regular								X			
	THRESKIORNITHIDAE													
61	*Plegadis falcinellus*	Glossy Ibis	b – regular								X			
62	*Platalea leucorodia*	Eurasian Spoonbill	b – regular								X			
	PHOENICOPTERIDAE													
63	*Phoenicopterus roseus*	Greater Flamingo	m – irregular											
	ACCIPITRIDAE													
64	*Elanus caeruleus*	Black-winged Kite	a											
65	*Pernis apivorus*	European Honey-buzzard	b – regular	X										
66	*Pernis ptilorhynchus*	Oriental Honey-buzzard	a											

67	*Milvus migrans*	Black Kite	b – regular	X										
68	*Milvus milvus*	Red Kite	a											
69	*Haliaeetus leucoryphus**	Pallas's Fish Eagle*	un											
70	*Haliaeetus albicilla*	White-tailed Eagle	w, m – irregular***											
71	*Gypaetus barbatus*	Lammergeyer	yr – regular									X		
72	*Neophron percnopterus*	Egyptian Vulture	b – regular			X	X					X		
73	*Gyps fulvus*	Eurasian Griffon Vulture	yr – regular									X		
74	*Aegypius monachus*	Eurasian Black Vulture	yr – regular		X									
75	*Circaetus gallicus*	Short-toed Snake-eagle	b – regular	X	X									
76	*Circus aeruginosus*	Western Marsh-harrier	yr – regular								X		X	
77	*Circus cyaneus*	Northern Harrier	w – regular											
78	*Circus macrourus*	Pallid Harrier	m – regular											
79	*Circus pygargus*	Montagu's Harrier	b – regular						X					
80	*Accipiter gentilis*	Northern Goshawk	yr – regular	X										
81	*Accipiter nisus*	Eurasian Sparrowhawk	yr – regular	X										
82	*Accipiter badius*	Shikra	b – irregular										X	X
83	*Accipiter brevipes*	Levant Sparrowhawk	b – regular			X	X						X	X
84	*Buteo buteo*	Common Buzzard	b – regular	X										
85	*Buteo rufinus*	Long-legged Buzzard	yr – regular		X	X	X	X	X			X		

	Latin names	English names	Breeding status	Deciduous forests	Juniper woodlands	Semideserts	Shrublands	Steppes	Meadows	Alpine meadows and carpets	Wetlands	Rocks & Screes	Riparian zone	Arable lands
86	*Buteo lagopus*	Rough-legged Buzzard	w – irregular											
87	*Clanga pomarina*	Lesser Spotted Eagle	b – regular	X										
88	*Clanga clanga*	Greater Spotted Eagle	w, m – irregular											
89	*Aquila nipalensis*	Steppe Eagle	m – regular											
90	*Aquila heliaca*	Imperial Eagle	m – regular***											
91	*Aquila chrysaetos*	Golden Eagle	yr – regular		X							X		
92	*Aquila fasciata**	Bonelli's Eagle*	un											
93	*Hieraaetus pennatus*	Booted Eagle	b – regular	X										
	PANDIONIDAE													
94	*Pandion haliaetus*	Osprey	b – regular								X			
	FALCONIDAE													
95	*Falco naumanni*	Lesser Kestrel	b – regular					X						
96	*Falco tinnunculus*	Common Kestrel	yr – regular		X	X	X	X	X			X		X
97	*Falco vespertinus*	Red-footed Falcon	m – irregular											
98	*Falco columbarius*	Merlin	w, m – irregular											
99	*Falco subbuteo*	Eurasian Hobby	b – regular			X	X							X
100	*Falco eleonorae*	Eleonora's Falcon	a											
101	*Falco biarmicus*	Lanner Falcon	b – regular			X	X							
102	*Falco cherrug*	Saker Falcon	yr – regular					X						
103	*Falco peregrinus*	Peregrine Falcon	yr – regular	X	X	X	X					X		
	RALLIDAE													
104	*Rallus aquaticus*	Water Rail	yr – regular								X		X	
105	*Porzana porzana*	Spotted Crake	b – regular								X			

106	*Zapornia parva*	Little Crake	m – irregular											
107	*Zapornia pusilla*	Baillon's Crake	m – irregular											
108	*Crex crex*	Corn Crake	b – regular						X					
109	*Gallinula chloropus*	Common Moorhen	yr – regular								X		X	
110	*Porphyrio poliocephalus*	Grey-headed Swamphen	yr – regular								X		X	
111	*Fulica atra*	Common Coot	yr – regular								X		X	
	GRUIDAE													
112	*Grus grus*	Common Crane	b – regular								X			
113	*Anthropoides virgo*	Demoiselle Crane	m – regular											
	OTIDIDAE													
114	*Tetrax tetrax*	Little Bustard	m – irregular											
115	*Chlamydotis macqueenii*	Macqueen's bustard	ex***											
116	*Otis tarda*	Great Bustard	m – regular											
	HAEMATOPODIDAE													
117	*Haematopus ostralegus*	Eurasian Oystercatcher	m – regular											
	RECURVIROSTRIDAE													
118	*Himantopus himantopus*	Black-winged Stilt	b – regular								X			
119	*Recurvirostra avosetta*	Pied Avocet	b – regular								X			
	BURHINIDAE													
120	*Burhinus oedicnemus*	Eurasian thick-knee	b – regular			X					X			

	Latin names	English names	Breeding status	Deciduous forests	Juniper woodlands	Semideserts	Shrublands	Steppes	Meadows	Alpine meadows and carpets	Wetlands	Rocks & Screes	Riparian zone	Arable lands
	GLAREOLIDAE													
121	*Glareola normanni*	Black-winged Pratincole	m – regular											
122	*Glareola pratincola*	Collared Pratincole	b – regular								X			
	CHARADRIIDAE													
123	*Charadrius dubius*	Little Ringed Plover	b – regular								X			
124	*Charadrius hiaticula*	Common Ringed Plover	m – regular											
125	*Charadrius alexandrinus*	Kentish Plover	b – regular								X			
126	*Charadrius leschenaultii*	Greater Sand Plover	a											
127	*Charadrius asiaticus*	Caspian Plover	a											
128	*Charadrius morinellus*	Eurasian Dotterel	a											
129	*Pluvialis apricaria*	Eurasian Golden-plover	m – irregular											
130	*Pluvialis squatarola*	Grey Plover	m – irregular											
131	*Vanellus spinosus*	Spur-winged Lapwing	yr – regular								X			
132	*Vanellus gregarius*	Sociable Lapwing	m – irregular											
133	*Vanellus leucurus*	White-tailed Lapwing	b – regular								X			
134	*Vanellus vanellus*	Northern Lapwing	yr – regular								X			
	SCOLOPACIDAE													
135	*Calidris alba*	Sanderling	m – regular											

136	*Calidris minuta*	Little Stint	m – regular											
137	*Calidris temminckii*	Temminck's Stint	m – regular											
138	*Calidris ferruginea*	Curlew Sandpiper	m – regular											
139	*Calidris alpina*	Dunlin	m – regular											
140	*Calidris falcinellus*	Broad-billed Sandpiper	m – irregular											
141	*Calidris pugnax*	Ruff	m – regular											
142	*Lymnocryptes minimus*	Jack Snipe	m – irregular											
143	*Gallinago gallinago*	Common Snipe	m – regular											
144	*Gallinago media*	Greater Snipe	m – regular											
145	*Scolopax rusticola*	Eurasian Woodcock	yr – regular	X	X									
146	*Limosa limosa*	Black-tailed Godwit	m – regular											
147	*Limosa lapponica*	Bar-tailed Godwit	m – irregular											
148	*Numenius phaeopus*	Whimbrel	m – irregular											
149	*Numenius arquata*	Eurasian Curlew	m – regular											
150	*Tringa erythropus*	Spotted Redshank	m – irregular											
151	*Tringa totanus*	Common Redshank	yr – regular								X			
152	*Tringa stagnatilis*	Marsh Sandpiper	m – irregular											
153	*Tringa nebularia*	Common Greenshank	m – regular											
154	*Tringa ochropus*	Green Sandpiper	yr – regular								X		X	X
155	*Tringa glareola*	Wood Sandpiper	m – regular											
156	*Xenus cinerea*	Terek Sandpiper	m – irregular											
157	*Actitis hypoleucos*	Common Sandpiper	b – regular								X		X	X

	Latin names	English names	Breeding status	Deciduous forests	Juniper woodlands	Semideserts	Shrublands	Steppes	Meadows	Alpine meadows and carpets	Wetlands	Rocks & Screes	Riparian zone	Arable lands
158	*Arenaria interpres*	Ruddy Turstone	m – regular											
159	*Phalaropus lobatus*	Red-necked Phalarope	m – regular											
	STERCORARIIDAE													
160	*Stercorarius pomarinus*	Pomarine Jaeger	a											
161	*Stercorarius parasiticus*	Parasitic Jaeger	a											
162	*Stercorarius longicaudus*	Long-tailed Jaeger	a											
	LARIDAE													
163	*Ichthyaetus ichthyaetus*	Great Black-headed Gull	w – regular											
164	*Ichthyaetus melanocephalus*	Mediterranean Gull	m – irregular											
165	*Hydrocoloeus minutus*	Little Gull	w, m – regular											
166	*Chroicocephalus ridibundus*	Common Black-headed Gull	yr – regular								X			
167	*Chroicocephalus genei*	Slender-billed Gull	m – regular											
168	*Larus canus*	Mew Gull	m – irregular											
169	*Larus fuscus*	Lesser Black-backed Gull	a											
170	*Larus cachinnans*	Caspian Gull	a											
171	*Larus armenicus*	Armenian Gull	yr – regular								X			
172	*Sternula albifrons*	Little Tern	b – regular								X			

173	*Gelochelidon nilotica*	Gull-billed Tern	b – regular								X			
174	*Hydroprogne caspia*	Caspian Tern	m – irregular											
175	*Thalasseus sandvicensis*	Sandwich Tern	a											
176	*Sterna hirundo*	Common Tern	b – regular								X		X	X
177	*Sterna paradisaea*	Arctic Tern	a											
178	*Chlidonias hybridus*	Whiskered Tern	b – regular								X			
179	*Chlidonias niger*	Black Tern	m – regular***											
180	*Chlidonias leucopterus*	White-winged Tern	b – regular								X			
	PTEROCLIDIDAE													
181	*Pterocles orientalis*	Black-bellied Sandgrouse	b – regular			X								
182	*Pterocles alchata*	Pin-tailed Sandgrouse	a											
	COLUMBIDAE													
183	*Columba livia*	Rock Dove	yr – regular		X	X	X	X	X			X		
184	*Columba oenas*	Stock Dove	yr – regular			X	X							
185	*Columba palumbus*	Common Wood-pigeon	yr – regular	X	X								X	X
186	*Streptopelia decaocto*	Eurasian Collared-dove	yr – regular										X	X
187	*Streptopelia turtur*	European Turtle-dove	b – regular					X						
188	*Streptopelia senegalensis*	Laughing Dove	yr – regular										X	X
189	*Oena capensis*	Namaqua Dove	a											

	Latin names	English names	Breeding status	Deciduous forests	Juniper woodlands	Semideserts	Shrublands	Steppes	Meadows	Alpine meadows and carpets	Wetlands	Rocks & Screes	Riparian zone	Arable lands
	CUCULIDAE													
190	*Clamator glandarius*	Great Spotted Cuckoo	b – irregular			X	X							
191	*Cuculus canorus*	Common Cuckoo	b – regular	X	X	X	X	X	X		X	X	X	X
	TYTONIDAE													
192	*Tyto alba*	Common Barn-owl	b – unconfirmed**											
	STRIGIDAE													
193	*Otus scops*	Common Scops-owl	b – regular		X		X							X
194	*Bubo bubo*	Eurasian Eagle-owl	yr – regular		X	X	X	X	X			X		
195	*Athene noctua*	Little Owl	yr – regular		X							X		X
196	*Strix aluco*	Tawny Owl	yr – regular	X										
197	*Asio otus*	Long-eared Owl	yr – regular	X	X			X					X	X
198	*Asio flammeus*	Short-eared Owl	b – unconfirmed**											
199	*Aegolius funereus*	Boreal Owl	yr – regular	X										
	CAPRIMULGIDAE													
200	*Caprimulgus europaeus*	Eurasian Nightjar	b – regular		X	X	X	X	X			X	X	X
	APODIDAE													
201	*Apus melba*	Alpine Swift	b – regular									X		
202	*Apus apus*	Common Swift	b – regular		X	X	X	X	X	X	X	X	X	X
	ALCEDINIDAE													
203	*Alcedo atthis*	Common Kingfisher	b – regular								X		X	
204	*Ceryle rudis**	Pied kingfisher*	un											

	MEROPIDAE													
205	*Merops persicus*	Blue-cheeked Bee-eater	b – regular			X					X			
206	*Merops apiaster*	European Bee-eater	b – regular		X	X	X	X	X					
	CORACIIDAE													
207	*Coracias garrulus*	European Roller	b – regular		X	X	X							
	UPUPIDAE													
208	*Upupa epops*	Eurasian Hoopoe	b – regular		X	X	X	X	X			X	X	X
	PICIDAE													
209	*Jynx torquilla*	Eurasian Wryneck	b – regular	X										
210	*Picus viridis*	Eurasian Green Woodpecker	yr – regular	X										
211	*Dendrocoptes medius*	Middle Spotted Woodpecker	yr – regular	X										
212	*Dendrocopos major*	Greater Spotted Woodpecker	yr – regular	X										
213	*Dendrocopos syriacus*	Syrian Woodpecker	yr – regular		X								X	X
214	*Dendrocopos leucotos*	White-backed Woodpecker	b – unconfirmed**											
215	*Dryocopus martius*	Black Woodpecker	yr – regular	X										
216	*Dendrocopos minor*	Lesser Spotted Woodpecker	yr – regular	X										
	ALAUDIDAE													
217	*Melanocorypha calandra*	Calandra Lark	b – regular					X						

	Latin names	English names	Breeding status	Deciduous forests	Juniper woodlands	Semideserts	Shrublands	Steppes	Meadows	Alpine meadows and carpets	Wetlands	Rocks & Screes	Riparian zone	Arable lands
218	*Melanocorypha bimaculata*	Bimaculated Lark	b – regular			X								
219	*Melanocorypha yeltoniensis*	Black Lark	a											
220	*Calandrella brachydactyla*	Greater Short-toed Lark	b – regular			X	X							
221	*Alaudala rufescens*	Lesser Short-toed Lark	b – regular			X								
222	*Galerida cristata*	Crested Lark	yr – regular			X	X							X
223	*Lullula arborea*	Wood Lark	b – regular		X	X	X	X	X					X
224	*Alauda arvensis*	Eurasian Skylark	b – regular					X	X					X
225	*Alauda gulgula*	Oriental Skylark	a											
226	*Eremophila alpestris*	Horned Lark	yr – regular						X	X				
	HIRUNDINIDAE													
227	*Riparia riparia*	Sand Martin	b – regular								X		X	
228	*Ptyonoprogne rupestris*	Eurasian Crag Martin	b – regular	X	X	X	X	X				X		
229	*Hirundo rustica*	Barn Swallow	b – regular											X
230	*Cecropis daurica*	Red-ramped Swallow	a											
231	*Delichon urbica*	Northern House-martin	b – regular	X	X	X	X	X				X		X
	MOTACILLIDAE													
232	*Anthus richardi*	Richard's Pipit	a											
233	*Anthus campestris*	Tawny Pipit	b – regular			X	X	X						
234	*Anthus trivialis*	Tree Pipit	b – regular	X					X					
235	*Anthus pratensis*	Meadow Pipit	b – unconfirmed**											

236	*Anthus cervinus*	Red-throated Pipit	w, m – regular											
237	*Anthus spinoletta*	Water Pipit	b – regular							X				
238	*Motacilla flava*	Yellow Wagtail	b – regular								X			
239	*Motacilla citreola*	Citrine Wagtail	b – irregular								X			
240	*Motacilla cinerea*	Grey Wagtail	yr – regular										X	
241	*Motacilla alba*	White Wagtail	yr – regular								X		X	X
	BOMBYCILLIDAE													
242	*Bombycilla garrulus*	Bohemian Waxwing	w – irregular											
	CINCLIDAE													
243	*Cinclus cinclus*	White-throated Dipper	yr – regular										1	
	TROGLODYTIDAE													
244	*Troglodytes troglodytes*	Eurasian Wren	yr – regular	X										
	PRUNELLIDAE													
245	*Prunella modularis*	Dunnock	yr – regular	X										
246	*Prunella ocularis*	Radde's Accentor	b – regular					X						
247	*Prunella collaris*	Alpine Accentor	b – regular							X				
	TURDIDAE													
248	*Turdus torquatus*	Ring Ouzel	yr – regular		X				X	X				
249	*Turdus merula*	Eurasian Blackbird	yr – regular	X	X								X	X
250	*Turdus atrogularis*	Black-throated Thrush	w – irregular											
251	*Turdus philomelos*	Song Thrush	b – regular	X										
252	*Turdus iliacus*	Redwing	w – regular											

	Latin names	English names	Breeding status	Deciduous forests	Juniper woodlands	Semideserts	Shrublands	Steppes	Meadows	Alpine meadows and carpets	Wetlands	Rocks & Screes	Riparian zone	Arable lands
253	*Turdus viscivorus*	Mistle Thrush	yr – regular	X	X									
254	*Turdus obscurus*	Eyebrowed Thrush	a											
255	*Turdus pilaris*	Fieldfare	w – regular											
	MUSCICAPIDAE													
256	*Cercotrichas galactotes*	Rufous-tailed Scrub-robin	b – regular			X	X							
257	*Erithacus rubecula*	European Robin	yr – regular	X										
258	*Luscinia luscinia*	Thrush Nightingale	m – irregular											
259	*Luscinia megarhynchos*	Common Nightingale	b – regular	X	X								X	X
260	*Luscinia svecica*	Bluethroat	b – regular					X	X					
261	*Muscicapa striata*	Spotted Flycatcher	b – regular	X										
262	*Ficedula parva*	Red-breasted Flycatcher	b – regular	X										
263	*Ficedula semitorquata*	Semicollared Flycatcher	b – regular	X										
264	*Ficedula hypoleuca*	European Pied Flycatcher	m – irregular											
265	*Ficedula albicollis*	Collared Flycatcher	m – regular											
266	*Irania gutturalis*	White-throated Robin	b – regular			X	X							
267	*Phoenicurus ochruros*	Black Redstart	yr – regular		X	X	X	X	X	X		X		X
268	*Phoenicurus phoenicurus*	Common Redstart	b – regular	X										
269	*Phoenicurus erythrogaster**	White-winged Redstart*	un											

270	*Saxicola rubetra*	Whinchat	b – regular						X					
271	*Saxicola rubicola*	European Stonechat	b – regular					X	X		X			
272	*Saxicola maura*	Siberian Stonechat	yr – regular		X		X	X			X			
273	*Oenanthe isabellina*	Isabelline Wheatear	b – regular			X	X							
274	*Oenanthe oenanthe*	Northern Wheatear	b – regular					X	X	X				
275	*Oenanthe pleschanka*	Pied Wheatear	m – irregular											
276	*Oenanthe melanoleuca*	Eastern Black-eared Wheatear	b – regular		X	X	X							
277	*Oenanthe deserti*	Desert Wheatear	a											
278	*Oenanthe finschii*	Finsch's Wheatear	b – regular			X								
279	*Oenanthe chrysopygia*	Persian Wheatear	b – regular			X								
280	*Monticola saxatilis*	Rufous-tailed Rock-thrush	b – regular					X	X	X				
281	*Monticola solitarius*	Blue Rock-thrush	b – regular		X	X	X							
	SCOTOCERCIDAE													
282	*Cettia cetti*	Cetti's Warbler	yr – regular								X		X	X
	LOCUSTELLIDAE													
283	*Locustella naevia*	Grasshopper Warbler	b – regular						X					
284	*Locustella fluviatilis*	Eurasian River Warbler	m – irregular											
285	*Locustella luscinioides*	Savi's Warbler	b – regular								X			

	Latin names	English names	Breeding status	Deciduous forests	Juniper woodlands	Semideserts	Shrublands	Steppes	Meadows	Alpine meadows and carpets	Wetlands	Rocks & Screes	Riparian zone	Arable lands
	ACROCEPHALIDAE													
286	*Acrocephalus melanopogon*	Moustached Warbler	yr – regular								X			
287	*Acrocephalus schoenobaenus*	Sedge Warbler	b – regular								X			
288	*Acrocephalus scripaceus*	Eurasian Reed-warbler	b – regular								X		X	
289	*Acrocephalus palustris*	Marsh Warbler	b – regular						X		X		X	X
290	*Acrocephalus agricola*	Paddyfield Warbler	b – regular								X			
291	*Acrocephalus arundinaceus*	Great Reed-warbler	b – regular								X		X	
292	*Iduna caligata*	Booted Warbler	a											
293	*Iduna pallida*	Olivaceous Warbler	b – regular		X		X						X	X
294	*Hippolais languida*	Upcher's Warbler	b – regular			X	X							
295	*Hippolais olivetorum**	Olive-tree Warbler*	un											
296	*Hippolais icterina*	Icterine Warbler	b – regular	X										
	SYLVIIDAE													
297	*Sylvia atricapilla*	Blackcap	b – regular	X										
298	*Sylvia borin*	Garden Warbler	b – regular	X										
299	*Sylvia nisoria*	Barred Warbler	b – regular		X		X	X						
300	*Sylvia curruca*	Lesser Whitethroat	b – regular		X	X	X	X						X
301	*Sylvia crassirostris*	Eastern Orphean Warbler	b – regular			X	X							

302	*Sylvia communis*	Greater Whitethroat	b – regular	X				X	X	X				X
303	*Sylvia mystacea*	Menetries's Warbler	b – regular				X				X		X	X
	PHYLLOSCOPIDAE													
304	*Phylloscopus trochilus*	Willow Warbler	m – regular											
305	*Phylloscopus collybita*	Eurasian Chiffchaff	b – regular	X										
306	*Phylloscopus sindianus*	Mountain Chiffchaff	b – regular	X										
307	*Phylloscopus orientalis*	Eastern Bonelli's Warbler	a											
308	*Phylloscopus sibilatrix*	Wood Warbler	m – regular											
309	*Phylloscopus nitidus*	Green Warbler	b – regular	X										
310	*Regulus regulus*	Common Goldcrest	b – regular	X										
311	*Regulus ignicapillus**	Firecrest*	un											
	CISTICOLIDAE													
312	*Cisticola juncidis*	Zitting cisticola	a											
	PANURIDAE													
313	*Panurus biarmicus*	Bearded Reedling	yr – regular								X		X	
	AEGITHALIDAE													
314	*Aegithalos caudatus*	Long-tailed Tit	yr – regular	X										
	PARIDAE													
315	*Poecile lugubris*	Sombre Tit	yr – regular		X									
316	*Lophophanes cristatus**	Crested Tit*	un											

	Latin names	English names	Breeding status	Deciduous forests	Juniper woodlands	Semideserts	Shrublands	Steppes	Meadows	Alpine meadows and carpets	Wetlands	Rocks & Screes	Riparian zone	Arable lands
317	*Periparus ater*	Coal Tit	yr – regular	X										
318	*Cyanistes caeruleus*	Blue Tit	yr – regular	X										
319	*Parus major*	Great Tit	yr – regular	X	X								X	X
	SITTIDAE													
320	*Sitta europaea*	Eurasian Nuthatch	yr – regular	X										
321	*Sitta tephronota*	Eastern Rock-nuthatch	yr – regular			X	X							
322	*Sitta neumayer*	Western Rock-nuthatch	yr – regular		X	X	X	X				X		
	TICHODROMIDAE													
323	*Tichodroma muraria*	Wallcreeper	yr – regular							X				
	CERTHIIDAE													
324	*Certhia familiaris*	Eurasian Tree-creeper	yr – regular	X										
	REMIZIDAE													
325	*Remiz pendulinus*	Eurasian Penduline-tit	yr – regular								X		X	X
	ORIOLIDAE													
326	*Oriolus oriolus*	Eurasian Golden-oriole	b – regular				X						X	X
	LANIIDAE													
327	*Lanius phoenicuroides*	Red-tailed Shrike	a											
328	*Lanius collurio*	Red-backed Shrike	b – regular	X	X	X	X	X	X				X	X
329	*Lanius minor*	Lesser Grey Shrike	b – regular		X		X							X
330	*Lanius excubitor*	Great Grey Shrike	w – regular											

331	*Lanius senator*	Woodchat Shrike	b – regular			X	X							
332	*Lanius nubicus*	Masked Shrike	b – unconfirmed**											
	PYCNONOTIDAE													
333	Pycnonotus leucotis	White-eared Bulbul	a											
	CORVIDAE													
334	*Garrulus glandarius*	Eurasian Jay	yr – regular	X										
335	*Pica pica*	Black-billed Magpie	yr – regular	X	X	X	X	X					X	X
336	*Pyrrhocorax graculus*	Yellow-billed Chough	yr – regular							X				
337	*Pyrrhocorax pyrrhocorax*	Red-billed Chough	yr – regular		X	X	X	X	X			X		
338	*Corvus monedula*	Eurasian Jackdaw	yr – regular					X				X		
339	*Corvus frugilegus*	Rook	yr – regular											
340	*Corvus corone*	Carrion Crow	yr – regular										X	X
341	*Corvus corax*	Common Raven	yr – regular		X	X	X	X	X	X		X		
	STURNIDAE													
342	*Sturnus vulgaris*	Common Starling	yr – regular					X	X					
343	*Pastor roseus*	Rosy Starling	b – regular			X	X							
	PASSERIDAE													
344	*Passer domesticus*	House Sparrow	yr – regular											X
345	*Passer hispaniolensis*	Spanish Sparrow	yr – regular											X
346	*Passer montanus*	Eurasian Tree Sparrow	yr – regular			X	X	X						X
347	*Carpospiza brachydactyla*	Pale Rock-finch	b – regular			X	X							

	Latin names	English names	Breeding status	Deciduous forests	Juniper woodlands	Semideserts	Shrublands	Steppes	Meadows	Alpine meadows and carpets	Wetlands	Rocks & Screes	Riparian zone	Arable lands
348	*Petronia petronia*	Rock Sparrow	yr – regular		X	X	X	X	X			X		
349	*Montifringilla nivalis*	White-winged Snowfinch	yr – regular							X				
	FRINGILLIDAE													
350	*Fringilla coelebs*	Chaffinch	yr – regular	X										
351	*Fringilla montifringilla*	Brambling	w – regular											
352	*Serinus pusillus*	Red-fronted Serin	yr – regular		X									
353	*Serinus serinus*	European Serin	b – unconfirmed**											
354	*Chloris chloris*	European Greenfinch	yr – regular	X	X								X	X
355	*Carduelis carduelis*	European Goldfinch	yr – regular	X	X								X	X
356	*Spinus spinus*	Eurasian Siskin	b – unconfirmed**											
357	*Linaria cannabina*	Eurasian Linnet	yr – regular		X	X	X	X	X			X		
358	*Linaria flavirostris*	Twite	yr – regular							X				
359	*Loxia curvirostra*	Common Crossbill	yr – regular	X										
360	*Rhodopechys sanguinea*	Crimson-winged Finch	yr – regular					X		X		X		
361	*Bucanetes mongolicus*	Mongolian Finch	b – irregular			X								
362	*Bucanetes githaginus*	Trumpeter Finch	b – regular			X								
363	*Rhodospiza obsoleta*	Desert Finch	b – regular			X	X							
364	*Carpodacus erythrinus*	Common Rosefinch	b – regular	X					X					

365	*Pyrrhula pyrrhula*	Eurasian Bullfinch	yr – regular	X										
366	*Coccothraustes coccothraustes*	Hawfinch	yr – regular	X										
	CALCARIIDAE													
367	*Plectrophenax nivalis*	Snow Bunting	a											
	EMBERIZIDAE													
368	*Emberiza leucocephalos*	Pine Bunting	w – irregular											
369	*Emberiza citrinella*	Yellowhammer	w – regular											
370	*Emberiza cia*	Rock Bunting	yr – regular		X	X	X	X						
371	*Emberiza buchanani*	Grey-necked Bunting	b – regular			X								
372	*Emberiza hortulana*	Ortolan Bunting	b – regular					X	X					
373	*Emberiza schoeniclus*	Reed Bunting	yr – regular								X		X	
374	*Emberiza melanocephala*	Black-headed Bunting	b – regular		X	X	X	X	X				X	X
375	*Emberiza rustica*	Rustic Bunting	a											
376	*Emberiza calandra*	Corn Bunting	yr – regular		X	X	X	X	X				X	X
	Total numbers of breeding birds per habitat			61	50	60	56	46	38	17	78	25	50	48

The following symbols are used to describe the species' occurrence status: *b* breeding, *m* migrant, *w* wintering, *yr* year-round resident, *yv* year-round visitor, *a* accidental visitor, *ex* extinct in the country (the species has not been observed during last 50 years), *un* presence unconfirmed

*Species which have been mentioned by ornithologists or birdwatchers, but the record is not proven

**The current status in not confirmed, while the presence is proven. E.g. "b – unconfirmed" means that the species' occurrence was documented for Armenia, but its breeding status is not proven

***Former breeder

Annex 3: List of Butterfly Species Registered in Armenia

	Latin names	Status in Red Book of Armenia	Regional endemic		Deciduous forests	Juniper woodlands	Semideserts	Shrublands	Tragacanth Steppes	Grassy Steppes	Meadows	Alpine carpets	Wetlands	Rocks and Screes	Riparian zone	Arable lands
	Hesperiidae															
1	Erynnis tages	NE			X	X				X	X					
2	Erynnis marloyi	NE				X	X	X	X							
3	Carcharodus alceae	NE			X	X	X	X	X	X	X			X	X	X
4	Carcharodus lavatherae	NE				X			X							
5	Carcharodus flocciferus	NE			X											
6	Carcharodus orientalis	NE	X	X		X	X		X							
7	Carcharodus stauderi	NE	X	X		X			X							
8	Carterocephalus palaemon	NE									X					
9	Muschampia proto	NE	X	X		X	X		X							
10	Muschampia proteides	NE				X	X		X							
11	Muschampia poggei	NE	X	X			X	X								
12	Muschampia tessellum	NE								X	X					
13	Muschampia tersa	NE					X	X								
14	Spialia phlomidis	NE	X	X					X							
15	Spialia orbifer	NE				X	X	X	X							
16	Pyrgus malvae	NE			X						X					

17	Pyrgus melotis	NE				X			X	X	X				X	
18	Pyrgus sidae	NE			X	X					X					
19	Pyrgus cinarae	NE							X							
20	Pyrgus serratulae	NE			X	X				X	X					
21	Pyrgus armoricanus	NE			X					X	X					
22	Pyrgus alveus	NE							X		X					
23	Pyrgus jupei	NE	X	X								X				
24	Pyrgus cirsii	NE							X			X				
25	Eogenes alcides	NE													X	
26	Gegenes nostrodamus	NE													X	
27	Thymelicus lineola	NE								X	X					
28	Thymelicus sylvestris	NE			X	X			X	X	X					
29	Thymelicus hyrax	NE	X	X		X			X							
30	Thymelicus acteon	NE											X			
31	Ochlodes sylvanus	NE			X	X			X	X	X					
32	Hesperia comma	NE									X	X				
	Papilionidae															
33	Parnassius mnemosyne	VU			X						X					
34	Parnassius apollo	VU									X					
35	Iphiclides podalirius	NE			X	X									X	X
36	Papilio machaon	NE				X	X	X	X	X	X					
37	Papilio alexanor	VU				X			X							

	Latin names	Status in Red Book of Armenia	Regional endemic		Deciduous forests	Juniper woodlands	Semideserts	Shrublands	Tragacanth Steppes	Grassy Steppes	Meadows	Alpine carpets	Wetlands	Rocks and Screes	Riparian zone	Arable lands
	Pieridae															
38	Leptidea sinapis	NE			X										X	
39	Leptidea duponcheli	NE							X							
40	Anthocharis cardamines	NE			X	X					X				X	
41	Anthocharis gruneri	NE	X	X		X	X	X								
42	Anthocharis damone	NE	X	X					X							
43	Euchloe ausonia	NE				X	X	X	X							
44	Zegris eupheme	NE				X	X	X	X							
45	Aporia crataegi	NE			X	X									X	X
46	Pontia edusa	NE				X	X	X	X	X	X	X	X	X	X	X
47	Pontia chloridice	NE	X	X			X									
48	Pontia callidice	NE										X				
49	Pieris bryoniae	NE										X				
50	Pieris pseudorapae	NE			X	X	X	X	X	X	X	X	X	X	X	X
51	Pieris bowdeni	CR	X	X								X				
52	Pieris ergane	NE				X								X		
53	Pieris krueperi	NE	X	X										X		
54	Pieris rapae	NE					X	X								
55	Pieris brassicae	NE				X	X	X	X	X	X		X	X	X	X
56	Colias hyale	NE									X					
57	Colias alfacariensis	NE			X	X	X	X	X	X					X	X
58	Colias erate	NE					X									

59	Colias thisoa	NE	X	X								X				
60	Colias aurorina	VU	X	X		X			X							
61	Colias chlorocoma	VU	X	X					X							
62	Colias crocea	NE			X	X	X	X	X	X	X		X	X	X	X
63	Gonepteryx rhamni	NE			X											
64	Gonepteryx farinosa	NE				X	X	X	X							
	Lycaenidae															
65	Thecla betulae	NE			X											
66	Favonius quercus	NE			X											
67	Satyrium ledereri	NE	X	X			X	X								
68	Satyrium hyrcanica	NE	X	X			X	X								
69	Satyrium spini	NE				X	X	X	X							
70	Satyrium ilicis	NE			X											
71	Satyrium acaciae	NE			X	X			X							
72	Satyrium abdominalis	NE	X	X		X			X							
73	Callophrys rubi	NE				X										
74	Callophrys chalybeitincta	NE					X	X	X							
75	Callophrys paulae	NE	X	X					X							
76	Callophrys danchenkoi	NE	X	X		X			X							
77	Tomares romanovi	VU	X	X		X			X							
78	Tomares callimachus	NE				X	X	X	X							

	Latin names	Status in Red Book of Armenia	Regional endemic		Deciduous forests	Juniper woodlands	Semideserts	Shrublands	Tragacanth Steppes	Grassy Steppes	Meadows	Alpine carpets	Wetlands	Rocks and Screes	Riparian zone	Arable lands
79	Lycaena phlaeas	NE			X	X	X	X	X	X	X		X	X	X	X
80	Lycaena virgaurea	NE									X					
81	Lycaena tityrus	NE			X											
82	Lycaena candens	NE									X		X			
83	Lycaena dispar	CR											X			
84	Lycaena alciphron	NE							X	X	X					
85	Lycaena kurdistanica	NE	X	X		X	X	X	X	X						
86	Lycaena thersamon	NE								X	X					
87	Lycaena ochimus	NE	X	X			X	X	X							
88	Lycaena asabinus	NE	X	X			X									
89	Lycaena thetis	NE										X				
90	Athamanthia phoenicura	NE	X	X			X	X								
91	Lampides boeticus	NE			X	X			X						X	
92	Chilades trochylus	NE				X	X	X								
93	Tarucus balcanicus	NE					X	X								
94	Cupido minima	NE			X					X	X					
95	Cupido osiris	NE				X				X	X					
96	Celastrina argiolus	NE			X	X	X	X	X						X	X
97	Pseudophilotes vicrama	NE				X	X	X	X	X						

98	Pseudophilotes bavius	NE							X							
99	Turanana endymion	NE				X	X		X							
100	Glaucopsyche alexis	NE				X	X	X	X							
101	Iolana iolas	NE							X						X	
102	Maculinea rebeli	VU									X					
103	Maculinea arion	VU								X	X					
104	Maculinea nausithous	VU									X		X			
105	Plebeius argus	NE				X			X	X	X				X	X
106	Plebeius (idas) idas	NE							X							
107	Plebeius (idas) calliopis	NE										X				
108	Plebeius christophi	EN	X	X											X	
109	Plebejides sephirus	NE							X							
110	Plebejides zephyrinus	NE				X	X	X	X							
111	Eumedonia eumedon	NE									X					
112	Aricia agestis	NE			X	X	X	X	X	X	X			X	X	X
113	Aricia allous	NE									X					
114	Ultraaricia crassipuncta	NE	X	X							X					
115	Cyaniris bellis	NE			X					X	X					
116	Vacciniina alcedo	NE							X							

	Latin names	Status in Red Book of Armenia	Regional endemic		Deciduous forests	Juniper woodlands	Semideserts	Shrublands	Tragacanth Steppes	Grassy Steppes	Meadows	Alpine carpets	Wetlands	Rocks and Screes	Riparian zone	Arable lands
117	Albulina morgiana	NE				X			X							
118	Plebejidea loewii	NE				X	X	X	X							
119	Kretania eurypilus	NE							X							
120	Neolysandra coelestina	NE	X	X						X	X					
121	Neolysandra diana	EN	X	X							X					
122	Agriades pyrenaicus	NE										X				
123	Lysandra bellargus	NE			X	X	X	X	X	X					X	X
124	Lysandra corydonius	NE			X	X	X	X	X	X	X				X	X
125	Meleageria daphnis	NE			X	X			X	X	X					
126	Polyommatus (icarus) icarus	NE			X	X	X	X	X	X	X	X			X	X
127	Polyommatus (icarus) kashgharensis	NE					X	X								
128	Polyommatus amandus	NE							X	X	X					
129	Polyommatus dorylas	NE										X				
130	Polyommatus thersites	NE				X			X	X	X					
131	Polyommatus mirrha	NE	X	X								X				

132	Polyommatus (Agrodiaetus) ripartii	NE								X	X					
133	Polyommatus (Agrodiaetus) demavendi	NE	X	X	X	X	X	X	X	X						
134	Polyommatus (Agrodiaetus) eriwanensis	EN	X	X					X							
135	Polyommatus (Agrodiaetus) admetus	NE							X							
136	Polyommatus (Agrodiaetus) damon	NE							X	X	X					
137	Polyommatus (Agrodiaetus) cyaneus	NE	X	X					X							
138	Polyommatus (Agrodiaetus) firdussii	NE	X	X					X							
139	Polyommatus (Agrodiaetus) vanensis	NE	X	X					X							
140	Polyommatus (Agrodiaetus) surakovi	EN	X	X					X							
141	Polyommatus (Agrodiaetus) huberti	EN	X	X					X							
142	Polyommatus (Agrodiaetus) arasbarani	EN	X	X	X				X							
143	Polyommatus (Agrodiaetus) ninae	VU	X	X					X							

	Latin names	Status in Red Book of Armenia	Regional endemic		Deciduous forests	Juniper woodlands	Semideserts	Shrublands	Tragacanth Steppes	Grassy Steppes	Meadows	Alpine carpets	Wetlands	Rocks and Screes	Riparian zone	Arable lands
144	Polyommatus (Agrodiaetus) aserbeidschanus	NE	X	X								X				
145	Polyommatus (Agrodiaetus) altivagans	NE	X	X								X				
146	Polyommatus (Agrodiaetus) turcicus	VU	X	X					X		X					
147	Polyommatus (Agrodiaetus) iphigenia	EN									X					
148	Polyommatus (Agrodiaetus) damonides	CR	X	X		X	X									
	Nymphalidae															
149	Libythea celtis	NE				X									X	
150	Esperarge climene	NE			X											
151	Pararge aegeria	NE			X											
152	Lasiommata megera	NE				X	X	X	X	X	X	X		X	X	X
153	Lasiommata maera	NE				X			X					X		
154	Melanargia galathea	NE			X						X					
155	Melanargia russiae	NE								X	X					
156	Melanargia larissa	NE				X	X	X	X							
157	Coenonympha pamphilus	NE			X	X			X	X	X					

158	Coenonympha lyllus	NE	X	X		X	X	X								
159	Coenonympha symphita	NE	X	X							X					
160	Coenonympha saadi	NE	X	X		X	X	X								
161	Coenonympha glycerion	NE									X					
162	Coenonympha arcania	NE			X											
163	Coenonympha leander	NE									X					
164	Erebia aethiops	VU			X											
165	Erebia graucasica	NE	X	X								X				
166	Erebia medusa	NE								X	X					
167	Proterebia afra	NE				X	X	X	X							
168	Hyponephele lycaon	NE				X	X	X	X	X					X	
169	Hyponephele lycaonoides	NE	X	X							X					
170	Hyponephele lupina	NE				X									X	
171	Hyponephele naricoides	NE					X									
172	Maniola jurtina	NE			X	X	X	X	X	X	X				X	X
173	Hipparchia pellucida	NE			X	X										
174	Hipparchia syriaca	NE	X	X	X	X										
175	Hipparchia fatua	NE	X	X			X	X								
176	Hipparchia statilinus	NE							X	X						

	Latin names	Status in Red Book of Armenia	Regional endemic		Deciduous forests	Juniper woodlands	Semideserts	Shrublands	Tragacanth Steppes	Grassy Steppes	Meadows	Alpine carpets	Wetlands	Rocks and Screes	Riparian zone	Arable lands
177	Hipparchia parisatis	NE	X	X			X	X								
178	Brintesia circe	NE			X											
179	Arethusana arethusa	NE								X	X					
180	Satyrus dryas	NE			X											
181	Satyrus amasinus	NE	X	X		X	X	X	X							
182	Satyrus effendi	NE	X	X								X				
183	Pseudochazara pelopea	NE	X	X		X	X	X	X							
184	Pseudochazara schahrudensis	NE	X	X		X	X	X								
185	Pseudochazara beroe	NE	X	X					X			X				
186	Pseudochazara geyeri	NE	X	X					X	X						
187	Pseudochazara daghestana	NE	X	X						X	X					
188	Pseudochazara thelephassa	NE				X	X	X								
189	Chazara briseis	NE				X	X	X	X	X	X					
190	Chazara persephone	NE				X									X	
191	Chazara bischoffi	NE	X	X			X	X	X							
192	Thaleropis ionia	NE	X	X		X									X	
193	Limenitis reducta	NE			X	X									X	
194	Limenitis camilla	NE			X											
195	Neptis rivularis	NE			X	X									X	
196	Vanessa atalanta	NE				X	X	X	X	X	X				X	X

197	Vanessa cardui	NE				X	X	X	X	X	X				X	X
198	Inachis io	NE			X	X										
199	Polygonia c-album	NE			X	X									X	
200	Polygonia egea	NE				X	X	X						X	X	
201	Nymphalis xanthomelas	NE			X										X	
202	Nymphalis polychloros	NE			X										X	
203	Nymphalis antiopa	NE			X											
204	Aglais urticae	NE			X	X	X	X	X	X	X	X		X	X	X
205	Argynnis paphia	NE			X											
206	Argynnis pandora	NE			X	X									X	
207	Argynnis aglaja	NE			X						X					
208	Argynnis alexandra	NE									X					
209	Argynnis adippe	NE			X											
210	Argynnis niobe	NE				X	X	X								
211	Issoria lathonia	NE				X	X	X	X	X	X				X	X
212	Brenthis daphne	NE			X										X	
213	Brenthis hecate	NE								X	X					
214	Brenthis ino	VU									X					
215	Clossiana euphrosyne	NE			X											
216	Clossiana dia	NE			X											
217	Boloria caucasica	NE	X	X								X				
218	Euphydryas aurinia	NE				X			X	X	X				X	

	Latin names	Status in Red Book of Armenia	Regional endemic		Deciduous forests	Juniper woodlands	Semideserts	Shrublands	Tragacanth Steppes	Grassy Steppes	Meadows	Alpine carpets	Wetlands	Rocks and Screes	Riparian zone	Arable lands
219	Melitaea didyma	NE				X	X	X	X	X	X	X		X	X	X
220	Melitaea interrupta	NE							X	X						
221	Melitaea persea	NE	X	X		X	X	X	X							
222	Melitaea trivia	NE					X									
223	Melitaea cinxia	NE								X	X					
224	Melitaea arduinna	NE				X			X	X						
225	Melitaea phoebe	NE			X											
226	Melitaea ornata	NE					X									
227	Melitaea abbas	NE					X									
228	Melitaea vedica	EN	X	X			X									
229	Mellicta athalia	NE			X						X					
230	Mellicta caucasogenita	NE	X	X	X											
231	Mellicta aurelia	NE									X					